Y0-BDP-818

Multi-Arm Cooperating Robots

International Series on
MICROPROCESSOR-BASED AND INTELLIGENT SYSTEMS ENGINEERING

VOLUME 30

Editor

Professor S. G. Tzafestas, *National Technical University of Athens, Greece*

Multi-Arm Cooperating Robots

Robots

Dynamics and Control

by

M.D. ZIVANOVIC

Robotics Center,
Mihajlo Pupin Institute, Belgrade,
Serbia and Montenegro

and

M.K. VUKOBRATOVIC

Robotics Center,
Mihajlo Pupin Institute, Belgrade,
Serbia and Montenegro

 Springer

A C.I.P. Catalogue record for this book is available from the Library of Congress.

ISBN-10 1-4020-4268-X (HB)
ISBN-13 978-1-4020-4268-3 (HB)
ISBN-10 1-4020-4269-8 (e-book)
ISBN-13 978-1-4020-4269-0 (e-book)

Published by Springer,
P.O. Box 17, 3300 AA Dordrecht, The Netherlands.

www.springer.com

Printed on acid-free paper

TABLE OF CONTENTS

LIST OF FIGURES

PREFACE

Under the notion 'cooperative work', is understood, in a widest sense the realization of a coordinated action of several participants (cooperators) engaged in a given task. Cooperative work is performed by a cooperative system consisting of cooperators and work object.

Cooperative work incorporates the joint work of the cooperators, their coordinated action in task execution, contact with the environment, and mutual contact of the cooperators, either directly or indirectly via the work object.

In joint work, the action of individual participants in the cooperation cannot be independent in time and space from the work (action) of the other participants. It is assumed that the actions of the cooperation participants take place simultaneously and not consecutively. Thereby cooperation means that each participant in the joint work carries out its own work taking care of the state of the other cooperation participants. Namely, to every different state of an individual cooperator corresponds an equal number of different states of the other cooperation participants. It is assumed that each cooperator obtains, in some way or other, information about the state of the other participants. The object on which cooperative work is performed, along with all cooperation participants, represent to an individual participant a dynamic environment with which it interacts.

There are a lot of tasks that can be performed in cooperation. Most often they are related to manipulating bulky objects whose weights exceeds the working capacities of the individual participants in the cooperation. For example, assembly of mechanical blocks carried out by several participants is a common case in technological practice. A frequent task is passing an object from one participant or group of participants in the cooperation to another participant or group of participants. In cooperative work, the participants perform mutually coordinated actions, while ensuring different types of contacts or avoiding them.

If, however, the extremities of an animal are considered as participants in cooperative work (manipulation or locomotion), then such synchronized motion is a specific cooperative task. The same also holds for the work (cooperation) of the fingers of a hand holding an object.

Analogous to the cooperative work of an animal's extremities is the robotic manipulation performed by several robots or by the fingers of an artificial hand. While object grasping and transferring, as well as the work on it, are the tasks of manipulation cooperation, synchronized work of the lower extremities represents locomotion cooperation that enables motion of the locomotion platform (vehicle) in the form of a bipedal or, more frequently, multipedal gait. Therefore, cooperative work of artificial systems has its biological counterpart in locomotion-manipulation activities of living beings. It can be said that results of studying active locomotion-manipulation mechanisms and their cooperation counterparts with living beings can be generally used in the corresponding procedures of the synthesis of artificial gait and control systems in manipulation and locomotion robotics.

When cooperative manipulation is concerned, a fundamental research task is to find out the appropriate way to control the system of robots and object in the work space at any stage of cooperative work. This requires an exact understanding of the physical nature of the cooperative system and deriving the mathematical basis for its description. In the realization of this goal, two crucial problems are encountered. The first of them is the occurrence of kinematic uncertainty and the second one is the force uncertainty in the mathematical description of the physical nature of the cooperative system. These problems have been considered by a number of authors [1–5, 12–20, 42–46, 50–55], and they can be interpreted simply as the impossibility to uniquely determine contact forces, driving torques of the manipulation mechanism, as well as kinematic quantities of cooperating robots, starting from the required motion of the object of cooperative manipulation.

On the basis of their research in the domain of cooperative manipulation, the authors of this monograph have recently come up with several consistent solutions concerning cooperative system control. This was achieved by solving three separate tasks that are essential for solving the problem of cooperative manipulation as a whole. The first task is related to understanding the physical nature of cooperative manipulation and finding a way for a sufficiently exact characterization of the cooperative system statics, kinematics, and dynamics. After successfully completing this task, in the frame of a second task, the problem of coordinated motion of the cooperative system is solved. Finally, as a solution to the third task, the control laws of cooperative manipulation are synthesized.

The starting point in dealing with the above three tasks of cooperative manipulation was the assumption that the problem of force uncertainty in cooperative manipulation can be solved by introducing elastic properties into the cooperative system. This monograph is concerned with the case when elastic properties are introduced only in that part of the cooperative system in which force uncertainty arises. Coordinated motion and control in cooperative manipulation are solved as the problem of coordinated motion and control of a mobile elastic structure, taking into account the specific features of cooperative manipulation.

The contents of this monograph are organized into seven chapters.

Chapter 1 defines the notions and basic problems related to a cooperative system, cooperative manipulation and contact in cooperative manipulation. Also, co-ordinate systems used to describe the cooperative system characteristics are introduced.

In Chapter 2, some basic problems of cooperative manipulation are analyzed and a mathematical interpretation of the problem of kinematic uncertainty and force uncertainty is given.

Chapter 3 provides a concise systematization of previous solutions of the task of cooperative manipulation. It gives an analysis of the assumptions that are to be introduced in order to correctly solve the problem of cooperative manipulation under static conditions. It is shown that the problem cannot be solved without introducing the elastic properties of the loaded structure. Further, it is demonstrated that the cooperative system must be approximated by a mobile elastic structure. Also, it is shown how the problem of force uncertainty can be resolved by considering the deformation work of the elastic structure as a function of absolute coordinates. In other words, on the basis of such analysis, using a concrete simple example, a way is indicated for establishing a methodology of modeling dynamics of complex cooperative systems.

The difference between the way of considering statics and dynamics of the elastic structure of cooperative systems in the present book and in the available literature is in the following. In the literature [1–5], the authors start from the *a priori* implicit assumption that elastic displacements, needed to define the position of the elastic system in space, are not independent variables (state quantities), but they represent the displacements given in advance (like, for example, the known displacement of the support of an elastic structure when defining its statics [6, 7]). A consequence of such an *a priori* assumption is that the position of the unloaded state of elastic structure in the motion is known in advance, and the stiffness matrix of the elastic system is nonsingular. The elastic structure position in space can be defined by choosing any point, including a contact one. As a consequence, the manipulator internal coordinates that contact point belongs to, are given in advance, i.e. they are not state quantities. In deriving mathematical the model used in this work, it is assumed that all displacements of the elastic system (i.e. position of contact points and manipulated object mass center) are independent variables (state quantities), necessary and sufficient for describing elastic-system dynamics [8]. A consequence of such an assumption is that the stiffness matrix of the elastic part of the cooperative system is singular, i.e. it has to also contain the modes of motion of the elastic structure as of a rigid body.

Chapter 4 is concerned with the task of cooperative manipulation of a rigid object by an arbitrary number of rigid manipulators, a task that has been most often considered in the literature. The task was modified by introducing elastic interconnections between the object and manipulators. The problem of modeling cooperative manipulation is analyzed in detail. In order to make the cooperative system properties more comprehensible, assumptions are introduced by which the problem of modeling is significantly simplified. Namely, the cooperative system is divided into its rigid part (manipulators) and elastic part (object and elastic interconnections). Each part is modeled separately using Lagrange equations. The elastic system model is derived on the basis of the description of its deformation work as a function of internal forces defined in dependence of absolute coordinates (extended method of finite elements [9]). The cooperative system dynamics is modeled for the displacement with respect to the elastic system unloaded state. This means that the reference coordinate frame is attached to the unloaded state of the elastic system. The general motion of the cooperative system is described in terms of absolute (external) coordinates, and the mathematical forms of motion equations are generalized. Stationary and equilibrium states of the cooperative system are analyzed in detail. The results obtained by model testing for selected examples show the consistency of our approach to modeling cooperative manipulation.

The problem of the synthesis of cooperative system nominals is essentially made more complex by introducing the elastic properties of the cooperative system [10]. Solving this problem is the subject of Chapter 5, where the nominals are synthesized using the properties of cooperative manipulation, as well as the properties of macro and micro motions. The cooperative system motion, in which the object is firstly gripped and then transferred, whereby the manipulator's motion does not significantly disturb the gripping conditions (i.e. the geometric configuration realized at the end of the gripping phase is not significantly changed) is adopted as the system's coordinated motion. The coordinated motion of the cooperative system is synthesized in a two-stage procedure, in which contact loads of the elastic system are approximately determined. On the basis of the approximate values of contact forces or driving torques, adopted as nominals, procedures are proposed for the synthesis of the other nominal quantities of the overall cooperative system. The synthesis procedures are illustrated by a simple example.

The control of cooperative manipulation is analyzed in Chapter 6 for the model of cooperative manipulation dynamics with the problem of force uncertainty resolved. The analysis encompasses definitions and criteria of controllability and observability of linear systems from the point of view of mapping the domains of inputs, states, and outputs. It is shown that the conclusions about mapping of linear systems can be applied without any change onto the mapping of the domains of inputs, states and outputs of the nonlinear systems. This was the basis for deriving conclusions on the controllability and observability of cooperative systems. Results of this analysis are applied to perform mapping between two of any of the following sets: the set of internal coordinates, the set of external coordinates, the set of driving torques, the set of contact forces, and the set of elasticity forces. A systematization of the controlled outputs along with the typification of control tasks in cooperative manipulation is carried out. Two types of tasks are selected [11]. Control laws are proposed for the asymptotically stable tracking of the object nominal trajectory and nominal trajectories of contact points of the manipulators-followers, along with control laws for the asymptotically stable tracking of the object trajectory and nominal contact forces at the contact points of the followers. The analysis also encompasses the behavior of uncontrolled quantities. The choice of the control laws and behavior of the controlled cooperative system are illustrated with a simple example.

The concluding Chapter 7 provides a brief survey of the research results that have been achieved in studying cooperative manipulation, which is the subject of this monograph. The conclusions are grouped according to particular topics. Also, some possible directions of the future research are indicated.

A complete derivation of the elastic system dynamic model for its immobile and mobile states is given in Appendices A and B, respectively.

The authors are indebted to Professor Luka Bjelica for translating the manuscript and for editing and proofreading the complete text.

Milovan Živanović and Miomir Vukobratović
June 2005
Belgrade, Serbia and Montenegro

1 INTRODUCTION TO COOPERATIVE MANIPULATION

1.1 Cooperative Systems – Manipulation Systems

The term 'cooperative system' is generally understood as several coordinated participants simultaneously engaged in the execution of a given task.

In robotics, for example, the term cooperative system is understood as a manipulation system (Figure 1). The cooperation participants and the object may be either rigid or elastic. Rigid cooperators and objects are those that undergo deformation at an infinite load.

A cooperative system performs cooperative work. Cooperative work encompasses the joint work of the cooperators, their coordination on task execution, contact with the environment, and their direct contact or, due to the nature of the task, their indirect contact.

The joint work refers to the sum of works of all the individual cooperators, whereby the work of none of them can be independent in time and space from the work of the other participants. It is usually understood that the cooperative work is performed simultaneously. Cooperation means that each cooperator performs its own work, taking into account the state of the other participants in the cooperation. To each different state of one of the cooperators corresponds a different state of the other cooperator(s) and/or object. This assumes that each of the cooperators receives and possesses information about the state of the other cooperators and objects. To each of the cooperators, the object and other cooperators are, in principle, a dynamic environment with which it interacts, i.e. in contact. Apart from participating in cooperation, each cooperator is constantly in contact with the work space, which may impose, but not necessarily, some constraints on the motion of the object and/or cooperators.

The main objective of a cooperative system in robotics is to manipulate an object. Manipulation is performed with the aim of

- changing the space position of an object (transfer it from one place to another),

1

- tracking a given trajectory of the object at a given orientation along the trajectory and/or

- performing some work on a stationary or mobile object.

To explain the mode and stages of the work of a cooperative system, it is necessary to observe several stationary objects that should be jointly transferred by several manipulators, one by one, from one place to another along a predetermined trajectory, while not overturning or damaging them. At the initial moment, the manipulator grippers are at a distance from the object. The stages, the work content, and essential characteristics of the work of a cooperative system in the object manipulation are as follows:

1. planning of the approach,

2. approach to the object,

3. grasping,

4. gripping,

5. lifting,

6. transferring,

7. lowering,

8. releasing,

9. withdrawing.

At the stage of approach planning, free motion trajectories are chosen for each of the manipulators in order to avoid a collision during the motion prior to contacting the object. A different trajectory should be selected for each object.

By approaching, we mean the motion of each individual manipulator towards the object, which is terminated by touching the object, while no force is established between the manipulator gripper and object.

All the manipulators do not necessarily approach the object simultaneously. If the grasping started without the synchronous action of all the manipulators, this could lead to an uncontrolled displacement of the object. To prevent this, it is necessary to ensure termination of the grasping stage when all the manipulators have approached the object.

In the step of object gripping, the corresponding forces are established between the manipulator's tip and object, and these forces should be as such to cause no

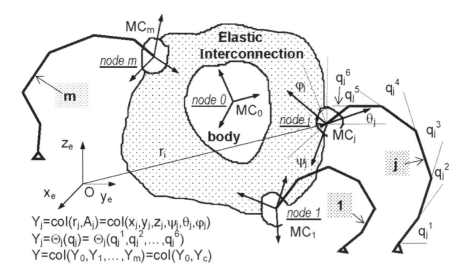

Figure 1. Cooperative manipulation system

object damaging. The gripping forces are internal forces of the system consisting of the manipulators and object (Figure 1). These forces arise as a consequence of elastic or plastic displacements (micro motion) of the structure of the object and manipulator. In the majority of cases, the inertial loads in gripping, due to the micro motion, are negligible in comparison with the loads produced by elastic or plastic displacements of the structures of the object and manipulator.

The steps of lifting, transferring, and lowering assume the motion in a macro space, whereby the object weight and all the forces produced by the motion of the object/manipulator are taken up by the manipulators. In these motions, the inertial loads are not negligibly small compared to the other loads.

The lifting step follows after the gripping. Prior to the steps of gripping and lifting, the object weight is distributed over the supports on the ground. Not later than at the beginning of the lifting stage, the object weight is distributed on as the load between the manipulators. In the lifting step, the object is raised from the support and the height of its position is gradually changed. The character of the motion of the cooperative system depends on the nature of the concrete task.

The transferring step consists of moving the object along a predetermined trajectory at a predetermined orientation. The manipulators move in such a way as to force the object to satisfy the preset motion requirements and/or produce the required gripping loads.

The object motion in manipulation is terminated by placing the object at a desired place. At the end of this step, the supports on the ground take over the object weight as the load, so that the manipulators retain only the load due to the deformations of their own structure and of the object.

In the lowering step, the load due to deformational displacements is reduced to zero, but without any additional motion of the already placed object.

Withdrawing assumes the motion of the manipulators to a safe distance from the placed object.

Then a new cycle is planned with the next object.

In this chapter we consider the cooperative system in the course of object gripping, lifting, and transferring. These stages of cooperative work assume that the manipulators are in contact with the object and thus with each other.

1.2   Contact in the Cooperative Manipulation

By contact in the cooperative manipulation is understood the mutual touching of the manipulators, touching of the manipulator and manipulated object, or some of these with obstacles. In this chapter we consider only the contact between the manipulator and object.

The site of interaction of the cooperators or of the cooperators and object is called contact (Figure 2). Contact represents the common boundary (interface) of the materials of the bodies being in contact.

A fundamental property of contact is its capacity to transfer information and loads between the contact participants. From the point of view of mechanics, the transfer of loads is the main interest. The transferred load is further conveyed to the structure (material) of the contact-making bodies, causing structural changes. Hence, the contact properties are to be defined separately in respect of the characteristics and behavior of the contiguous surfaces and separately in view of the characteristics of the structure adjacent to the interface (contact). In the general case, the structure adjacent to the contact is elastic. In the cooperative manipulation, a set of elastic environments of all contacts of the manipulators and object is called an elastic system.

1.3   The Nature of Contact

Physical contact consists of two contiguous surfaces and the space between them. The contacting surfaces belong to the contacting bodies, i.e. they are the envelopes of their structures. In a general case, the envelopes of the structures have not necessarily the same characteristics as the structures they envelop. When there exist one-to-one correspondence of points of the contacting surfaces (as if they were glued so that there is no void between them), the space between them is an empty space. In all other cases, there is a real space, either homogenous or inhomogeneous, between the surfaces in contact.

Contact properties are defined on the basis of the structural characteristics of the envelopes of the bodies in contact and on the basis of the mutual displacements taking place during the contact (Figure 2).

The envelopes of the contacting bodies may be either rigid or elastic. If the contact involves two bodies, it is possible to have four combinations of envelopes, and the contacts are named accordingly. A *rigid contact* is formed between two participants with rigid envelopes. An *elastic contact* is formed if the envelopes of both participants are elastic. If an elastic contacting surface of one contact participant is adjacent to a rigid contacting surface of the other participant, such contacts can also be treated as rigid.

During the contact, the contacting surfaces can be either translationally/rotationally fixed or mobile with respect to each other. Displacements in the contact are caused by a sliding or rotational macro motion of the contact participants. Depending on the type of the allowed motion, contacts can be either translational or rotational. If the contacting surfaces are mutually immobile during any general motion of the participants, we speak of a *stiff translational/rotational contact*. If, however, the contacting surfaces are mutually movable, then a *sliding translational/rotational contact* is in question.

One essential property of contact is that the loads between contacting bodies are transferred through it so that the contact is conceptually different from a kinematic pair. If the friction at the sliding contact is negligible, the load can be transferred only in those directions in which there is no relative displacement of contacting surfaces. This means that the requirement for load transfer imposes kinematic constraints on the motion of the bodies in contact. In a small vicinity of any point of contacting surfaces, there can exist maximally six constraints, three on translational and three on rotational motion, to ensure transfer of forces and moments. The number of motion constraints in a small environment of any point of contacting surfaces decreases by the number of different mutual motions of the contacting surfaces. Thus, for example, a ball contact imposes three constraints on translational and none on rotational displacements.

Transfer of loads in the contact is realized via the contacting surfaces of the contact participants. The load that is transferred at the contact is an acting load of the structure of each of the contact participants at their interface (Figure 2c). An essential characteristic of the contact is that all the loads appearing in it are internal loads of a system whose parts are the contact participants. Loads are transferred between the contact participants in the directions in which the contact imposes motion constraints. In the directions in which contact does not impose constraints, unpowered kinematic pairs (sliding and/or revolute) are formed, and the load can

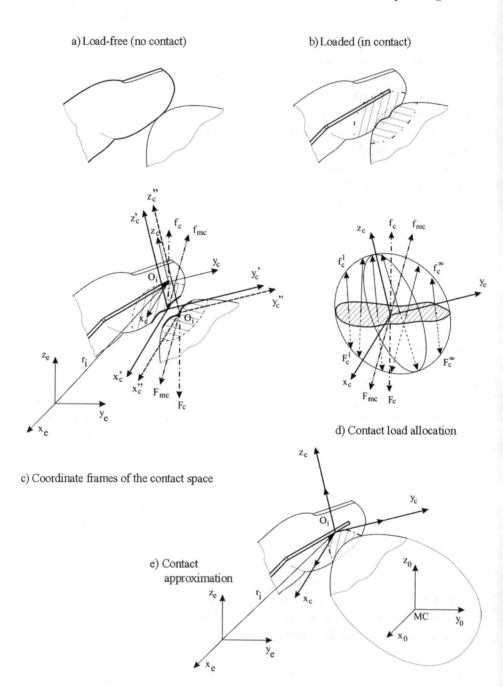

a) Load-free (no contact)

b) Loaded (in contact)

c) Coordinate frames of the contact space

d) Contact load allocation

e) Contact approximation

Figure 2. Contact

not be transferred. More precisely, in reality, in these directions appear the losses that are defined as friction, and they are usually neglected in the analysis.

A cooperative system may be represented by a kinematic chain having both powered and unpowered joints and/or by a kinematic chain having at least one link formed by all contact participants (e.g. the object and the cooperator's link in contact with it). This property of the cooperative system means that it can always have a smaller number of drives at joints than degrees of freedom (DOFs) (i.e. equations of motion).

Description of the contact must not be erroneous, as any error inevitably leads to erroneous conclusions about the mechanical characteristics of the cooperative system and automatically yields incorrect results on the basis of such a description.

The contact environment – elastic system. The three-dimensional space (structure) of a contact participant whose envelope is forming the contact, is the contact environment. The structure can be either rigid or elastic.

The contact load is an acting load of the structure of one of the contact participants at one of its interfaces. The boundary of the contact environment is chosen in accordance with the needs of the concrete task.

The motion and conditions in the contact environment are described by approximate models.

A rigid structure is approximated by a rigid body.

An elastic structure can be approximated by a continuous medium with an infinite number of infinitesimal material elements, with a finite series of elastically connected lumped masses, with a series of finite elements of different properties, etc. The points at which the elements are joined form the so-called *nodes* of the elastic structure. A series of nodes form a spatial grid. Nodes of the spatial grid can be either internal or external. Inertial properties can be ascribed to an elastic structure in the whole space or only at certain points, e.g. at all or only at some nodes, midway between them, at the gravity centers whose apexes are nodes, or at the gravity centers of the finite elements, etc.

In the cooperative manipulation, elastic properties can be assigned to the manipulators, to the object, or only to the environment of the manipulator-object contact (elastic interconnection). A set of selected approximations of the elastic structure of an individual manipulator and object in contact with the environment, is called an *elastic system*.

1.4 Introducing Coordinate Frames

A simple example of a cooperative system of the manipulation type is presented in Figure 3a. Three fingers – the thumb, index finger, and middle finger are gripping

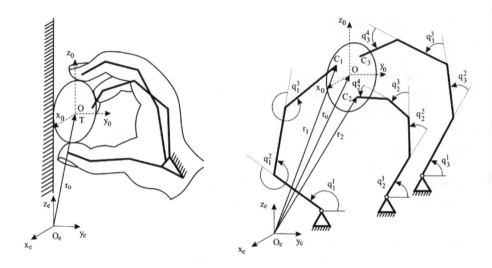

Figure 3. Cooperative work of the fingers on an immobile object

an object, making a rigid contact.

Properties of such a simple system are presented on the basis of the description of the kinematics, statics, and dynamics of the approximate cooperative system (Figure 3b). An approximate cooperative system (hereafter, cooperative system) will be the basis for all the analyses of the cooperative work. The analysis assumes such computations in which the calculated value assigned to a quantity in space can also be obtained (confirmed) by measurement. The quality of the adopted approximation determines the quality of the results of the analysis.

The cooperative system properties are described on the basis of the description of the kinematics, statics and dynamics of the approximate cooperative system. For that purpose, it is necessary to enumerate the cooperative system constituents and select coordinate frames in which this description will be made.

We say that the object bears the number '0' and that the manipulators have the numbers from 1 to m (in the example from Figure 3, $m = 3$). All the quantities related to the object have 0 as the last subscript, and all the quantities related to the manipulators have have as the last subscript an ordinal number, $i = 1, \ldots, m$ of the corresponding manipulator. The cooperative manipulator from which numbering begins is the leader.

The choice of coordinate frames depends on the selected approximation. Here we consider a cooperative system consisting of a rigid object and rigid manipulators with elastic interconnections between them. The motion of the rigid manipulators is described in the internal coordinates and the object motion in the external coordinates. The selected form of the approximate cooperative system allows us to easily obtain the relation to the known theoretical results of the robotics for the motion of rigid manipulators and the dynamics of rigid bodies and elastic systems. The introduction of elasticity only to the contact is of great technical significance. Such contact is convenient for the technical realization of some new and improvement of existing robotic systems by introducing the appropriate elastic inserts.

Replacement of a real cooperative system with an approximate one, as well as the enumeration and introduction of coordinate frames, will be discussed in the example illustrated in Figure 3. For a real cooperative system, we introduce the following assumptions: Let the palm be supported on the ground. Let all the palm links form an immobile link. Let all the finger links be rigid and let all the links of a finger lie in the same plane. In this plane only, let each link have one DOF with respect to the neighboring link. Under these assumptions, the natural cooperative system is approximated by a cooperative system formed by one three-DOF and two four-DOF manipulators connected with the object (Figure 3b). The properties of the joints and contact are defined separately for the concrete cases considered. We say has that the joints in this example are rigid and that the contact at the beginning is rigid. The cooperative system consists of four elements. The subscript '0' is assigned to the object, '1', '2', '3' to the fingers – manipulators, and 'e' to the support.

The external coordinate frame $Ox_e y_e z_e$ is immobile and fixed to the support (this is usually the ground) of the work space. This system of coordinates determines the position of every point on the object, but it does not allow the determination of the object's orientation. The object's position is defined with the aid of the position vector of one of its points, usually of the mass center (MC), given by the three coordinates $r_0 = col(r_0^{ex}, r_0^{ey}, r_0^{ez})$ with respect to the external coordinate frame and vector of its instantaneous orientation $\mathcal{A}_0 = col(\psi_0, \theta_0, \varphi_0)$ defined by three Euler angles of the coordinate frame attached to the object with respect to the external coordinate frame. This means that the object position in three-dimensional space is determined by the six-component vector

$$Y_0 = col(r_0, \mathcal{A}_0) = \begin{bmatrix} r_0 \\ \mathcal{A}_0 \end{bmatrix} = \begin{bmatrix} r_0^{ex} \\ r_0^{ey} \\ r_0^{ez} \\ \psi_0 \\ \theta_0 \\ \varphi_0 \end{bmatrix} \in R^6 \tag{1}$$

In an analogous way, we introduce the coordinates Y_{c1}, Y_{c2}, Y_{c3} for the position of the manipulator tips, i.e. the coordinate system fixed to the manipulator tip at the contact points C_1, C_2, C_3, whereby the subscripts stand for the ordinal number of the manipulator. The vectors $Y_{ci} = col(r_{ci}, \mathcal{A}_{ci} \in R^6, i = 0, 1, 2, 3$, represent the position vectors of the points in the six-dimensional coordinate frame, which we call the *natural coordinate frame of the object position.*

The motion equations are obtained on the basis of the quantities defined in the fixed inertial coordinate frame. If we neglect the motion of the natural coordinate frame with respect to the inertial coordinate frame, then the derivatives of position vector of any point in the system of external coordinates and the derivatives of the position vector of that point in the inertial coordinate frame will coincide. In that case, the system of external coordinates has the properties of the inertial coordinate frame, and its coordinates we call *absolute coordinates.* This allows us to derive the motion equations in the system of external coordinates in the same way as in the inertial coordinate frame.

Task space represents the work space in which the cooperative system moves. If the work space does not impose any constraints on the motion of any part of the cooperative system, then the work space coincides with the six-dimensional natural frame of the position coordinates. If the work space contains the obstacles imposing on the object motion d constraints (l_t for translation, $r_0 \in R^{3-l_t}$ and l_r for rotation, $\mathcal{A}_0 \in R^{3-l_r}$, $d = l_t + l_r$), then the free object motion takes place in the ($l = 6 - d$)-dimensional *free space.* For example, if we assume that during the gripping step the manipulator motion can take place only in the plane parallel to the coordinate plane $Oy_e z_e$ (Figure 3), then we have one constraint on translation ($x_e = $ const) and two constraints on rotation ($\psi_0 = $ const and $\varphi_0 = $ const). Free work space is then three-dimensional and the object can perform two-dimensional motion as a free motion. For the different cases considered, the task space is obtained by reducing the natural coordinate frame of the object.

The internal coordinate frame serves to describe the state of the manipulator. Internal coordinates represent the angles between individual links and their number is just equal to the number of DOFs of all the manipulator links. If all the manipulator joints are simple kinematic pairs (kinematic pairs of fifth class), then the number of internal coordinates is equal to the number of links.

In the example shown in Figure 3, for the known lengths of the particular links of the first manipulator l_1^j, $j = 1, 2, 3$, its tip position, as of a three-DOF manipulator, can be fully determined by the three angles: between the first link and the support q_1^1 and between the particular links q_1^2 and q_1^3. Analogously, we can introduce the internal coordinates of a four-DOF manipulator $q_2^1, q_2^2, q_2^3, q_2^4$ and $q_3^1, q_3^2, q_3^3, q_3^4$.

The general convention designation (symb)$_i^j$, $i = 1, \ldots, m$, $j = 1, \ldots, n_i$ is adopted, where (symb) stands for the symbol of the internal coordinate q or driving torque τ; m is the total number of manipulators, and n_i is the number of DOFs of the ith manipulator (in this example, $m = 3$, $n_1 = 3$, $n_2 = n_3 = 4$). In the general case, for m manipulators, of which every ith has m_i DOFs, the vectors $q_1 = \mathrm{col}(q_1^1, \ldots, q_1^{n_1}) \in R^{n_1}$ to $q_m = \mathrm{col}(q_m^1, \ldots, q_m^{n_m}) \in R^{n_m}$ form the space of the internal coordinates of the individual manipulators, and the vector

$$q = \mathrm{col}(q_1, \ldots, q_m) = \mathrm{col}(q_1^1, \ldots, q_1^{n_1}, q_2^1, \ldots, q_{m-1}^{n_{m-1}}, q_m^1, \ldots, q_m^{n_m}) \in R^n,$$

$$n = \sum_i^m n_i = n_1 + \ldots + n_m$$

forms the space of the internal coordinates of all the manipulators, i.e. *the space of internal coordinates of the cooperative system.*

In this example, the vectors of internal coordinates are

$$q_1 = \mathrm{col}(q_1^1, q_1^2, q_1^3) \in R^3, \quad q_2 = \mathrm{col}(q_2^1, q_2^2, q_2^3, q_2^4) \in R^4,$$

$$q_3 = \mathrm{col}(q_3^1, q_3^2, q_3^3, q_3^4) \in R^4 \quad \text{and}$$

$$q = \mathrm{col}(q_1, q_2, q_2) = (q_1^1, q_1^2, q_1^3, q_2^1, q_2^2, q_2^3, q_2^4, q_3^1, q_3^2, q_3^3, q_3^4) \in R^{11}$$

(see Figure 3).

Elastic system space is intended for the description of the elastic system motion. In the general case, the structure around each node may have a maximum six DOFs, i.e. the maximum allowed displacements of the elastic system. In concrete cases, depending on the given task, displacements can be allowed only in certain directions. Loads are introduced at the nodes depending on the concrete needs.

It is essential to point out that the number of allowed independent displacements of the elastic system nodes determines the total number of motion equations that describe its physical nature (statics and dynamics). For some particular cases, the local characteristics of the structure (e.g. composed of finite elements), nodes, and their allowed displacements, as well as mass distribution within the structure, may be a subject of choice.

An appropriate choice of elastic system suitable for technical application, consists of m finite elastic elements with the elastic properties defined in advance, placed at the external nodes (tips of the grippers) and with the object placed at the internal node.

We select an elastic system suitable for the presentation of the features of cooperative manipulation. The system has m external nodes and only one internal node. Six independent displacements of the elastic system are allowed at each node. All inertial properties of the elastic system are defined only by the nodes. The distribution of inertial characteristics may be different. It is assumed that the structure around each node has inertial properties possessed by rigid bodies. The selected presentation of the elastic system can be thought of as a system of $m + 1$ elastically connected rigid bodies. The suitability of the choice is revealed through the clear presentation of the consistent mathematical procedure of modeling statics and dynamics of the elastic system. If the inertial properties of the rigid bodies at external nodes are small compared to the inertial properties at the internal node, they can be neglected. Then, all the inertial properties are assigned to the internal node and to the rigid body placed there. The load is transferred through the external nodes between the gripper tip and elastic system as an external load of the elastic structure. For the internal node of the elastic system enter all the forces of the manipulated object.

When no loads are present at the elastic system nodes, displacements of the nodes are equal to zero. This state of the elastic system is called *state 0*.

Any load or displacement at some of the nodes causes displacement of the structure with respect to state 0, and thus determines the angles at the boundary surfaces of the contact-forming bodies, as well as the conditions at the elastic structure nodes through which the load is transferred onto the object. To describe the statics and dynamics of such an elastic system we need six quantities for each node, three for rotation and three for translation. These $6m + 6$ quantities define the space of an elastic system in cooperative manipulation.

In the theory of elasticity, the state of a loaded structure is described via the displacements of the loaded structure from the state 0 or from a pre-loaded state, known in advance, caused by the known load (e.g. by the elastic system weight). These displacements are defined in the local coordinate frame attached to the elastic part of the system and then, depending on the need, expressed in some global coordinate frame common to all the elements of the elastic system.

Cooperative manipulation takes place in the same space for all the cooperation participants. Space coordinates of the elastic system are adopted in the global coordinate frame, which is the same for all the parts of the elastic system.

The adopted elastic system is described in two coordinate frames.

The cooperative work done on the elastic system, whose unloaded state 0 is immobile, is described by the displacement coordinates denoted by the small letter y. Namely, in the unloaded state, at each node of the elastic system is placed a three-dimensional coordinate frame parallel to the coordinate frame of the external coordinates $Ox_e y_e z_e$ at each node. A fictitious rigid body having a certain initial orientation is placed. From these positions of rigid bodies, displacements are measured of the loaded state of the elastic system. Since connection of these rigid bodies is stiff, the displacements of the elastic systems are identical to the displacements of the rigid bodies. The displacement vector of the ith node is $y_i = \mathrm{col}(\Delta r_i, \Delta \mathcal{A}_i) \in R^6$. The displacement vector of all the nodes $y = \mathrm{col}(y_0, y_1, \ldots, y_m) \in R^{6m+6}$ represents the radius vector of the $(6m + 6)$-dimensional space of the elastic system, whose unloaded state 0 is fixed (immobile).

The work on the elastic system whose unloaded state performs general motion is described by means of coordinates of the nodes of loaded elastic system denoted by the capital Y. For the ith node, the vector $Y_i = \text{col}(r_i, \mathcal{A}_i) \in R^6$ describes the instantaneous position and orientation of the rigid body (in the sequel, the position and orientation will be termed attitude) placed at that node, i.e. the instantaneous attitude of the elastic system at that node with respect to the external coordinate system $Ox_e y_e z_e$. The position vector of all the nodes $Y = \text{col}(Y_0, Y_1, \ldots, Y_m) \in R^{6m+6}$ represents the radius vector of the $(6m+6)$-dimensional elastic system space whose unloaded state performs the general motion.

Contact space serves to describe the constraints imposed by the contact on the motion of the grippers.

The character and the number of quantities needed for the description of contact depends on the approximations introduced for particular classes of task, i.e., of the contact.

A precise description of motion constraints imposed by the contact assumes a precise description of the mutual motion of the contiguous surfaces of the contacting bodies, i.e. of the load transferred through the interface (Figure 2c,d). In this description, it is essential that those parts of the interface that are immobile with respect to each other, i.e. cannot move in some rotation/translation directions, have the same velocity in these directions, and any load at these points and in these directions represents the internal load of the overall contact structure. If we split the system along the mutually immobile parts of the interface, then in the split entities will act as loads of the same direction but in an opposite sense (Figure 3c,d). This means that the loads between the contact-forming bodies can be transferred only in the directions in which the contact imposes constraints on their motion.

Conditions at the contact are most simply and most correctly described by mutual displacements of the coordinate frames $C_i' x_c' y_c' z_c'$ and $C_i'' x_c'' y_c'' z_c''$, fixed to the elementary boundary surfaces of the contact-forming bodies (Figure 2c), with the axes $C_i' z_c'$ and $C_i'' z_c''$ in the normal direction. Let the origin of these coordinate frames be at the point $C_i = C_i' = C_i''$, whose radius vector in the coordinate frame $Ox_e y_e z_e$ is $r_{ci} = r_{ci}' = r_{ci}''$. Let the orientation of these coordinate frames with respect to the coordinate frame $Ox_e y_e z_e$ be $\mathcal{A}_{ci} = \mathcal{A}_{ci}' = \mathcal{A}_{ci}''$. It is supposed that in the initial moment of contact, these two coordinate frames are immobile and that they coincide. In these coordinate frames, we select arbitrary vectors ρ' and ρ'' which, at the initial moment, also coincide, $\rho' = \rho''$. If the contact is stiff, there is no relative translational displacement of the boundary points. This means that the coordinate frames $C_i' x_c' y_c' z_c'$ and $C_i'' x_c'' y_c'' z_c''$ coincide during the motion. For a stiff contact we can preset three conditions for translational $\dot{r}_{ci} = \dot{r}_{ci}' = \dot{r}_{ci}''$ and three conditions for rotational $\dot{\mathcal{A}}_{ci} = \dot{\mathcal{A}}_{ci}' = \dot{\mathcal{A}}_{ci}''$ relative motion of the coordinate frames $C_i' x_c' y_c' z_c'$ and $C_i'' x_c'' y_c'' z_c''$ at the point $C_i = C_i' = C_i''$. Hence, we say that the stiff contact imposes three constraints in respect of rotation and three constraints in respect of translation or, that the space of translation and rotation of the bodies in contact coincide at the point $C_i = C_i' = C_i''$.

If the contacting surface is rigid, then the radius vectors between any of its points during the motion are constant and can be expressed in any coordinate frame attached to the boundary surface. This allows us to express the properties of the contact boundary (usually surface) via the mutual motion of the coordinate frames attached at only one of its points C_i.

Therefore, to describe the constraints imposed by stiff and rigid contact, it is necessary to have six quantities that describe the space at the point C_i. If the coordinate system $C_i' x_c' y_c' z_c'$ is attached to a rigid gripper whose tip is at the point C_i, these quantities are the coordinates of position vector of the gripper tip in the natural coordinate frame connected to the ground, $Y_{ci} = \text{col}(r_{ci}, \mathcal{A}_{ci}) \in R^6$. In the cooperative manipulation involving n manipulators with rigid grippers, the space of the stiff and rigid contact of the cooperative system is formed by the subspaces of contact of all the manipulators. The space of the stiff and rigid contact of the cooperative system is defined by the following vector:

$$
\begin{aligned}
Y_c &= \text{col}(Y_{c1}, \ldots, Y_{cm}) = \text{col}(r_{c1}, \mathcal{A}_{c1}, \ldots, r_{cm}, \mathcal{A}_{cm}) \in R^{6m}, \\
Y_{ci} &= \text{col}(r_{ci}, \mathcal{A}_{ci}) \in R^6.
\end{aligned}
\tag{2}
$$

Sliding contact can be realized either as translational or rotational. If the contact is translational sliding, the coordinate frames $O x_c' y_c' z_c'$ and $O x_c'' y_c'' z_c''$ will move in parallel to each other and the radius vector of any point of these spaces can be expressed in its own coordinate frame as a function of the realized displacement expressed by only one increment vector $d\rho$, $\rho' = \rho'' + d\rho$ and $\rho'' = \rho' - d\rho$. In the case of sliding, the vector $d\rho$ has maximally two coordinates, the third coordinate being equal to zero. If the third coordinate is different from zero, the contact is broken. This means that the load in the translational sliding contact can be transferred at least in one and, at most, in two directions. In the case of sliding rotational contact, the radius vector ρ' of any point of the one coordinate frame can be obtained as an orthogonal transformation of the radius vector ρ'' of the other coordinate frame which, before transformation (prior to rotational sliding), coincided with the vector ρ'. In the directions in which orientation cannot be changed, load transfer is possible, while in the directions in which orientation can be changed, no load can be transferred.

As an example, we will consider the rigid contact at the point C_i of the rigid object and the ith rigid manipulator that is stiff in respect of translation and sliding in respect of rotation. The translation space is defined by the contact position vector given by the three coordinates $r_{ci} = \text{col}(r_{ci}^{ex}, r_{ci}^{ey}, r_{ci}^{ez}) \in R^3$ with respect to the external coordinate frame. The rotation space is defined by means of the vector of instantaneous orientation $\mathcal{A}_{ci} \in R^{l_{rci} \leq 3}$ determined by such a number of Euler angles of rotation of the coordinate frame attached to the contact with respect to the external coordinate frame that is equal to the number of constraints the contact imposes in respect of rotation. This means that the position of the ith contact in contact space is determined by the $(c_i = 3 + l_{rci} \leq 6)$-component vector

$$
Y_{ci} = \begin{pmatrix} r_{ci} \\ \mathcal{A}_{ci} \end{pmatrix} = \begin{pmatrix} r_{ci}^{ex} \\ r_{ci}^{ey} \\ r_{ci}^{ez} \\ \mathcal{A}_{ci} \end{pmatrix} \in R^{c_i}, \quad c_i \leq 6, \quad i = 1, \ldots, n . \tag{3}
$$

The vector $Y_c = \mathrm{col}(Y_{c1}, \ldots, Y_{cm}) \in R^c$, $c = \sum_i^m c_i$, forms the space of all contacts of the object and manipulators, i.e. the *cooperative system contact space*.

The description of contact space becomes much more complex if the contact is elastic or if the assumptions on contact properties are changed. With elastic contacts, the contiguous surface changes its form during the motion. Because of that, the vectors of the normals to the adjacent elementary surfaces of one of the contact-forming bodies are mutually displaced, and so are the coordinate frames attached to them. Thus, the conditions on contacting surfaces cannot be considered without loss in accuracy by taking into account only one contact point. Depending on the desired accuracy, the conditions at the elastic contact can be described not only by using an arbitrary finite number but also by using an infinite number of coordinates. However, elastic contacts are not the subject matter of this chapter.

It should be noticed that the motion constraints imposed by the contact due to rigid grippers are defined with respect to the mutual motion of the contiguous surfaces and not with respect to the properties of their environments. This allows the environment of the rigid grippers to be either rigid or elastic, irrespective of whether this environment is represented by elastic manipulators or the external environment of the gripper.

Cooperative system state space is determined by the necessary and sufficient number of independent quantities needed to describe its dynamics. In the analysis of the cooperative system's dynamics, the number of these quantities depends on the assumptions about the characteristics of the cooperative system constituents.

If we consider a cooperative system composed of m 6-DOF rigid manipulators handling a rigid object performing an unconstrained motion, the necessary and sufficient number of quantities needed to describe its motion will be $6m + 6$. The space state vector of a such cooperative system is

$$
Y = \mathrm{col}(Y_0, q_1, \ldots, q_m) \in R^{6m+6}. \tag{4}
$$

In the adopted approximation of the cooperative system, there appears the problem of the so-called force uncertainty (see Section 2.2).

If elastic bodies are inserted between the gripper tips and the object, to form an elastic system with $m + 1$ nodes, and if it is assumed that the grippers of non-redundant manipulators form a stiff and rigid contact with the elastic system at the contact points that coincide with the external nodes, then the state vector of such a cooperative system is identical to the state vector of the elastic system. The adopted state vector of the elastic system, i.e. of the cooperative system

$$y = \text{col}(y_0, y_{c1}, \ldots, y_{cm}) \in R^{6m+6} \tag{5}$$

will describe the gripping phase and the vector

$$Y = \text{col}(Y_0, Y_{c1}, \ldots, Y_{cm}) \in R^{6m+6} \tag{6}$$

will describe its general motion. As the manipulators are non-redundant, there is a unique functional dependence between the position of the manipulator gripper and internal coordinates, so that the vector (4) can also be adopted as the state vector of the cooperative system.

Such a choice of approximation of the cooperative system and its state quantities allows us to get a clear insight into the needs, differences, and consequences produced in the description of the cooperative system by introducing elastic properties in the part of the cooperative system consisting of rigid grippers and rigid object. The issue of recognizing the needs, differences, and consequences of the introduction of elastic properties is the main subject of this monograph.

If the assumption on the characteristics of contact and elastic system is changed, the number and character of state quantities of the cooperative system will be changed too.

1.5 General Convention on Symbols and Quantity Designations

In the description of the statics and dynamics, the load vector coordinates are the projections of this vector onto the axes of the coordinate frame used to describe the motion of that part of the cooperative system in which the given load is acting. Thus, the load vector coordinates (generalized forces) at the object MC, described in term of the natural coordinate frame coordinates will be

$$F_0 = \begin{pmatrix} \bar{F}_0 \\ M_0 \end{pmatrix} = \begin{pmatrix} F_0^{ex} \\ F_0^{ey} \\ F_0^{ez} \\ M_0^x \\ M_0^y \\ M_0^z \end{pmatrix} \in R^6 , \tag{7}$$

where F_0^{ex} [N], F_0^{ey} [N] and F_0^{ez} [N] are the force projections onto the axes of the fixed coordinate frame $Ox_e y_e z_e$, while M_0^x [Nm], M_0^y [Nm] and M_0^z [Nm] are the moment projections onto the axes of the coordinate frame $Ox_0 y_0 z_0$ fixed to the object. In an analogous way, we define the vector of contact force at the ith rigid contact $F_{ci} = \text{col}(F_{ci}^{ex}, F_{ci}^{ey}, F_{ci}^{ez}, M_{ci}^x, M_{ci}^y, M_{ci}^z) \in R^6$. For this contact, the vector of contact force of the cooperative system is formed by all contact forces of particular contacts, $F_c = \text{col}(F_{c1}, F_{c2}, \ldots, F_{cm}) \in R^{6m}$.

In order that the manipulator links could maintain their arbitrary position, move and perform work in a certain field of forces, active/resistance torques have to act at the joints. In the example shown in Figure 4 these torques are τ_i^j, $i = 1, 2, 3$, $j = 1, 2, 3$ for the first manipulator and $j = 1, 2, 3, 4$ for the second and third manipulators. If we assume that all the joints are powered and there are no losses at them, then the torques τ_i^j are driving torques of the fingers, i.e. manipulators. In the general case, for m manipulators, of which every ith one has n_i DOFs, the vectors from $\tau_1 = \text{col}(\tau_1^1, \ldots, \tau_1^{n_1}) \in R^{n_1}$ to $\tau_m = \text{col}(\tau_m^1, \ldots, \tau_m^{n_m}) \in R^{n_m}$ are driving torques of the individual manipulators, and

$$\tau = \text{col}(\tau_1, \ldots, \tau_m) = \text{col}(\tau_1^1, \ldots, \tau_1^{n_1}, \tau_2^1, \ldots, \tau_{m-1}^{n_{m-1}}, \tau_m^1, \ldots, \tau_m^{n_m}) \in R^n,$$

$$n = \sum_i^m n_i = n_1 + \ldots + n_m,$$

is the vector of driving torques of all the manipulators, i.e. the vector of driving torques of the cooperative system. In the example shown in Figure 4 the vectors of the driving torques are

$$\tau_1 = \text{col}(\tau_1^1, \tau_1^2, \tau_1^3) \in R^3,$$

$$\tau_\# = \text{col}(\tau_\#^1, \tau_\#^2, \tau_\#^3, \tau_\#^4) \in R^4, \quad \# = 2, 3$$

and

$$\tau = \text{col}(\tau_1, \tau_2, \tau_3)$$

$$= \text{col}(\tau_1^1, \tau_1^2, \tau_1^3, \tau_2^1, \tau_2^2, \tau_2^3, \tau_2^4, \tau_3^1, \tau_3^2, \tau_3^3, \tau_3^4) \in R^{11}.$$

By convention, the notation for all the quantities that are projected onto one and the same axis is determined with respect to the direction of the coordinate axis. For example, if the z-axis is vertical and oriented upwards, then the projections of all the vectors onto this axis are taken with this orientation as being positive.

For all linear displacements, linear velocities, linear accelerations and forces we assume the coordinates to be positive if their direction is in the sense of an increase of the coordinate onto which these quantities are projected.

All angular displacements, angular velocities, angular accelerations and moments are assumed to be positive if they tend to produce a positive rotational motion of the coordinate frame they are projected into.

1.6 Relation to Contact Tasks Involving One Manipulator

If the object from the example shown in Figure 3 is rigid and immobile, the cooperative work is reduced to the action of three independent manipulators. A still simpler case would be if, for example, the first and third manipulators were not active and were not in contact with the object. Then the problem of cooperative work would reduce to the problem of contact of the second manipulator and the environment. Obviously, the contact of one manipulator with environment is a particular case of cooperative work.

2 PROBLEMS IN COOPERATIVE WORK

The basic problems of cooperative work considered in the available literature are the problem of kinematic uncertainty and the problem of force uncertainty.

2.1 Kinematic Uncertainty

Kinematic uncertainties in cooperative manipulation arise as a consequence of the redundancy of manipulators and/or of contact characteristics.

2.1.1 Kinematic uncertainty due to manipulator redundancy

This instance of kinematic uncertainty in cooperative manipulation arises in the case of using redundant manipulators whose mobility index is higher than the number of DOFs of the manipulator gripper. This kinematic uncertainty is identical to the kinematic uncertainty of a redundant manipulator.

Let us explain this on the example of the second manipulator in Figure 3. Let the object and manipulator be rigid and let contact between the object and the terminal link of the manipulator be stiff. Let all four links move in one plane, $r_{c2}^{xe} = \text{const}$. The attitude of the contact on the object C_2 may be arbitrarily defined by defining the six-dimensional vector of the contact space $Y_{c2} = \text{col}(r_{c2}, \mathcal{A}_{c2}) \in R^6$. The contact space consists of the translation subspace and rotation subspace. Translation subspace is determined by the vector $r_{c2} = \text{col}(r_{c2}^{xe} = \text{const}, r_{c2}^{ye}, r_{c2}^{ze})$. Two coordinates of this vector can be arbitrarily chosen, i.e. we can arbitrarily choose the contact on the object in the plane $r_{c2}^{xe} = \text{const}$. The subspace of rotation (orientation) is determined by the rotation vector $\mathcal{A}_{c2} = \text{col}(\psi_{c2} = \text{const}, \theta_{c2}, \varphi_{c2} = \text{const})$, whereby the rotation about the axis x_e by the angle θ_{c2} can be arbitrary.

Let us determine the vector of manipulator tip position $Y_{c2}^f = \text{col}(r_{c2}^f, \mathcal{A}_{c2}^f) \in R^6$ as a function of the internal coordinates. The coordinates of the manipulator tip position are determined, via internal coordinates, by the following vector:

$$r_{c2}^f = \begin{pmatrix} r_{c2}^{xe} = \text{const} \\ y_{o2} + l_2^1 \cos q_2^1 + l_2^2 \cos(q_2^1 + q_2^2) + l_2^3 \cos(q_2^1 + q_2^2 + q_2^3) + l_2^4 \cos(q_2^1 + q_2^2 + q_2^3 + q_2^4) \\ z_{o2} + l_2^1 \sin q_2^1 + l_2^2 \sin(q_2^1 + q_2^2) + l_2^3 \sin(q_2^1 + q_2^2 + q_2^3) + l_2^4 \sin(q_2^1 + q_2^2 + q_2^3 + q_2^4) \end{pmatrix},$$

where y_{o2}, z_{o2} are the internal coordinates of the manipulator base O_2 and l_2^j, $j = 1, 2, 3, 4$ are the lengths of the manipulator links.

Orientation of the manipulator tip is determined by the vector

$$\mathcal{A}_{c2}^f = \begin{pmatrix} \psi_{c2} = \text{const} \\ q_2^1 + q_2^2 + q_2^3 + q_2^4 \\ \varphi_{c2} = \text{const} \end{pmatrix}.$$

If the contact is stiff, the position vectors of the contact point on the object r_{c2} and the vector of manipulator tip position r_{c2}^f are identical, $r_{c2} = r_{c2}^f$. For a stiff and rigid contact of the manipulator tip and object, the contact space vector $\mathcal{A}_{c2}$ and vector of the manipulator tip orientation $\mathcal{A}_{c2}^f$ are identical $\mathcal{A}_{c2} = \mathcal{A}_{c2}^f$. Hence, it follows that

$$Y_{c2} = \begin{pmatrix} r_{c2} \\ \mathcal{A}_{c2} \end{pmatrix} = \begin{pmatrix} r_{c2} \\ r_{c2}^{ze} \\ \psi_{c2} \\ \theta_{c2} \\ \varphi_{c2} \end{pmatrix}$$

$$= \begin{pmatrix} r_{c2}^{xe} = \text{const} \\ y_{o2} + l_2^1 \cos q_2^1 + l_2^2 \cos(q_2^1 + q_2^2) \\ \quad + l_2^3 \cos(q_2^1 + q_2^2 + q_2^3) + l_2^4 \cos(q_2^1 + q_2^2 + q_2^3 + q_2^4) \\ z_{o2} + l_2^1 \sin q_2^1 + l_2^2 \sin(q_2^1 + q_2^2) \\ \quad + l_2^3 \sin(q_2^1 + q_2^2 + q_2^3) + l_2^4 \sin(q_2^1 + q_2^2 + q_2^3 + q_2^4) \\ \psi_{c2} = \text{const} \\ q_2^1 + q_2^2 + q_2^3 + q_2^4 \\ \varphi_{c2} = \text{const} \end{pmatrix} \qquad (8)$$

$$= Y_{c2}^f,$$

where $r_{c2}^{xe} = \text{const}$, $\psi_{c2} = \text{const}$ and $\varphi_{c2} = \text{const}$.

For the case of a planar motion, upon eliminating constant coordinates, we obtain three coordinates of the contact space as a function of four internal coordinates of the manipulator

$$
\begin{pmatrix} r_{c2}^{ye} \\ r_{c2}^{ze} \\ \theta_{c2} \end{pmatrix} = \begin{pmatrix} y_{o2} + l_2^1 \cos q_2^1 + l_2^2 \cos(q_2^1 + q_2^2) \\ \qquad + l_2^3 \cos(q_2^1 + q_2^2 + q_2^3) + l_2^4 \cos(q_2^1 + q_2^2 + q_2^3 + q_2^4) \\ z_{o2} + l_2^1 \sin q_2^1 + l_2^2 \sin(q_2^1 + q_2^2) \\ \qquad + l_2^3 \sin(q_2^1 + q_2^2 + q_2^3) + l_2^4 \sin(q_2^1 + q_2^2 + q_2^3 + q_2^4) \\ q_2^1 + q_2^2 + q_2^3 + q_2^4 \end{pmatrix}
$$

$$
= \begin{pmatrix} r_{c2}^{ye}(q_2^1, q_2^2, q_2^3, q_2^4) \\ r_{c2}^{ze}(q_2^1, q_2^2, q_2^3, q_2^4) \\ \theta_{c2}(q_2^1, q_2^2, q_2^3, q_2^4) \end{pmatrix}. \tag{9}
$$

If the internal coordinates $q_2^1, q_2^2, q_2^3, q_2^4$ on the right-hand side of equality (8) or (9) are known, then the position of the contact point C_2 on the object is uniquely determined by the vector $(r_{c2}^{ye}, r_{c2}^{ze}, \theta_{c2})$, and is explicitly calculated from (9). If the contact point position is known (three quantities on the left-hand side of equality (9)) because of the existence of four unknown quantities and because of the periodicity of trigonometric functions, there is an infinite number of positions of the manipulator link that allow the manipulator tip to touch the object at a given point and with a given orientation of the terminal link. The uncertainty arising due to the periodicity of trigonometric functions is easily eliminated by the additional requirement that the joints constantly belong to a smooth function with an exactly determined second derivative (e.g. only to a concave or a convex function, Figure 4). In this case, we say that the manipulator is redundant and that kinematic uncertainty in the cooperative work is a consequence of the redundancy of the cooperation participants.

2.1.2 Kinematic uncertainty due to contact characteristics

Another situation arises when the contact does not impose kinematic conditions whose number is equal to the number of DOFs of the manipulator tip motion, irrespective of whether it is redundant or non-redundant (Figure 4). Let us explain this in the example of the first manipulator. If it is known that the manipulator joints constantly belong to a concave function, then the manipulator moving in the plane is also non-redundant (analogously to Equation (9), we obtain three equations with three unknowns). Let us suppose that the contact is stiff in respect of translation and sliding in respect of rotation. Then, the contact does not impose any constraints on the manipulator tip in respect of orientation, but only the requirement that the contact exists at a certain point. This is mathematically expressed by the requirement that the position vector r_{c1} at the given point C_1 on the object and the vector of manipulator tip position, r_{c1}^p, are identical, $r_{c1} = r_{c1}^p$ for an arbitrary tip

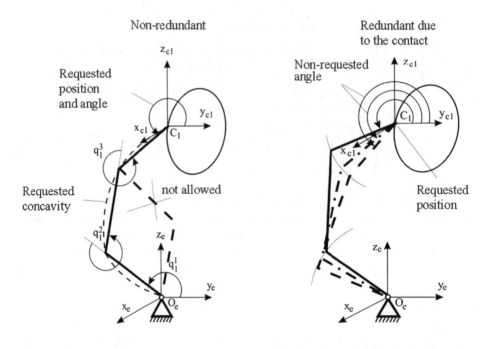

Figure 4. Kinematic uncertainty due to contact

orientation, $\forall \mathcal{A}_{c2}^p$. The contact imposes two constraints, in which two quantities depend of three quantities.

$$\begin{pmatrix} r_{c1}^{ye} \\ r_{c1}^{ze} \end{pmatrix} = \begin{bmatrix} r_{c1}^{pye}(q_1^1, q_1^2, q_1^3) \\ r_{c2}^{pze}(q_1^1, q_1^2, q_1^3) \end{bmatrix}. \tag{10}$$

As in the previous case of kinematic uncertainty, after stating the requirement concerning the object, there appears an infinite number of manipulator positions satisfying that requirement.

Kinematic uncertainty, however, is not essentially a problem of cooperative manipulation and will not be considered in this book.

2.2   Force Uncertainty

Let us consider the simplest example of the cooperative work of two manipulators with which we can explain the problem of force uncertainty. More correctly, this problem could be stated as the problem of distribution of the total load produced by the object in motion or at rest over the cooperation participants.

Let the two manipulators hold the object from the previous example and let

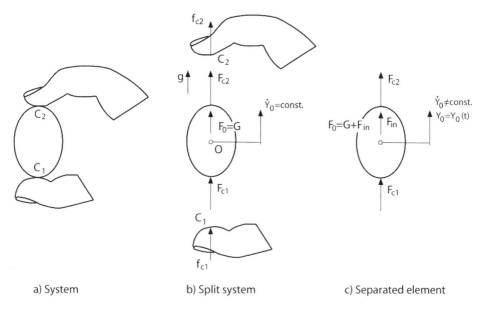

a) System   b) Split system   c) Separated element

Figure 5. Cooperative work of two manipulators on the object

them manipulate it without any constraint imposed on the object's motion (Figure 5).

Let the manipulators hold the object so that it is immobile. Let the contact points and object MC lie in the same vertical plane. Let the manipulators take up the object weight. Finally, let the manipulator tips be glued to the object, and let them transfer force only along the vertical.

With this example we will demonstrate the use of the general convention employed in this monograph. First, we adopt the reference coordinate frame and orientation of the position coordinate, e.g. of y, upward. The adopted orientation is positive and is marked by an arrow on the coordinate z. Projections of all vector quantities (g, f_{c1}, f_{c2}, F_{c1}, F_{c2}) on this direction is represented by the same directional. The application of the general notation convention in decomposing the system into subsystems and in the extraction of one of its elements, is illustrated in Figure 5b. The basic principle is that all the vector quantities are presented with the positive direction both at the points of their action in the overall system and on the singled-out element, irrespective of the fact that it may not be their real direction. The real direction is regulated by the values of the coordinates and additional (algebraic) conditions imposed by the system, i.e. contact.

In the object-manipulator contact, the realized forces may be of an arbitrary intensity and direction. Let us denote by capitals the contact forces operating as acting forces on the object, F_{c1} and F_{c2} and, by small letters, the contact forces operating as acting forces at the manipulator tips f_{c1} and f_{c2}. According to the general convention of notation used in the schematic, all the forces act in the same direction. Since the contact is preserved, the contact forces are internal forces of the cooperative system and, being the forces of action and reaction, they mutually annul, i.e. $F_{c1} = -f_{c1}$ and $F_{c2} = -f_{c2}$.

Let us consider the load of the disjointed system (Figure 5b). If the object is only in the gravitational field of force, the force balance can be expressed by the following vector equation:

$$F_0 + F_{c1} + F_{c2} = 0 \quad \Rightarrow \quad G = F_0 = -F_{c1} - F_{c2} = f_{c1} + f_{c2}, \qquad (11)$$

where $F_0 = G = \text{col}(0, 0, -mg)$ is the weight vector, m [kg] is the object mass, and g [m/s^2], gravitational acceleration. Since all forces are collinear, it is not necessary to write a moment equation. In the motion and/or cooperative work involving additional forces, nothing is essentially changed. In that case, the contact forces balance the result of the inertial, damping and all external forces acting on the object, $F_0 = G + F_{in} + F_s + \cdots$ (Figure 5c).

If the contact forces (right-hand side of Equation (11)) are known, the object weight G is uniquely determined. However, if the weight is known, there is an infinite number of ways of load distribution at the contacts, i.e. at the manipulator tips taking up the object's weight. This property of the cooperative system is known as '*the problem of force uncertainty*'.

If the object is rigid and if there is no danger of its breaking, the problem is easily solved by allowing the contact forces to be those that the manipulators can produce and so that condition (11) is satisfied. In order to have the problem of force distribution uniquely solvable, it is necessary to introduce the assumption on the elasticity of the system in that part where uncertainty appears. The relationship describing the elastic system properties is assigned to Equation (11). In a mathematical sense, the task becomes closed and all forces are uniquely determined. The solution of force uncertainty is given in Chapters 3 and 4.

2.3 Summary of Uncertainty Problems in Cooperative Work

In the robotics of cooperative systems, the problems of kinematic uncertainty and force uncertainty are treated as the impossibility of finding the kinematic quantities and forces on the manipulators when the kinematic quantities and forces for the manipulated object are known.

It should be noticed that the problem of kinematic uncertainty can be eliminated by an appropriate choice of manipulator characteristics and type of contact. Hence, this problem is not essentially the problem of cooperative work.

However, whatever choice of the type of contact and manipulator characteristics is made, the problem of force uncertainty will still exist. Hence, the problem of force uncertainty is a crucial problem of cooperative work, at least in the sense of how the problem has been defined.

It should be pointed out that neither the problem of kinematic uncertainty nor the problem of force uncertainty can exist if the kinematic quantities and forces exerted by the manipulator on the manipulated object are determined on the basis of the kinematic quantities and forces of the manipulators as cooperation participants.

2.4 The Problem of Control

We have discussed two problems that have been identified as crucial issues in cooperative manipulation. Here, several questions arise. First, whether these problems are a unique characteristic of cooperative work? Second, how these problems arise? Third, do these problems really exist?

In fact, the essential problem of cooperative work is not the kinematic uncertainty and force uncertainty but the control of the cooperative system. More precisely, the problem is how to synthesize the cooperative system control laws on the basis of existing knowledge.

It has already been mentioned that for the known right-hand sides of the relations (9), (10) and (11), their left-hand sides are uniquely determined. If the cooperative work is solved starting only from the information contained within the cooperation participants (internal coordinates and forces), then the problem of uncertainty in cooperative work does not exist, but the problem of the synthesis of cooperative system control does arise. The problem is the synthesis of control algorithms only on the basis of information from the sensors measuring the physical quantities that are also measured by the sensors of living beings.

The cooperative system and object move in the work space that is most simply described by means of a coordinate frame fixed to the support. The work space is also seen by the user of the cooperative system. By means of the coordinate system fixed to the support (inertial system) the user easily describes the requirements concerning the object motion in the work space. The dynamics and control laws for manipulators are usually described by means of internal coordinates. The problem of cooperative work thus stated lead unavoidably to the need of the existence of a mutually unique relation between the kinematic quantities and manipulator load and (required) information about the position and load of the body in the inertial system, i.e. it leads to the problem of force uncertainty.

Cooperative work always involves some sort of guidance. One part of the cooperative system imposes forced motion, on the other part, that is guidance. For example, the manipulators in Figure 3 can force the object to stand, to move, to get deformed, etc. If we bear in mind that a cooperative system always has the number of drives at joints smaller than the number of DOFs, the question is what are the quantities to be controlled and how one is to control them in order to guide the cooperative system? After finding the solution to the previous task, it is necessary to synthesize the logic and methodology of solving the problem of a cooperative system control.

The proposed control solutions are given on the basis of a dynamic model involving unresolved uncertainty problems, and are not consistent solutions of cooperative system control. A consistent solution of the law of cooperative system control is given in Chapter 6. That solution has been obtained on the basis of a model in which the problem of force uncertainty was solved (Chapter 3).

3 INTRODUCTION TO MATHEMATICAL MODELING OF COOPERATIVE SYSTEMS

In this chapter we present a consistent procedure for modeling a simple cooperative system consisting of two non-redundant manipulators handling a rigid object. We explain the origin of force uncertainty and present a method to solve this problem. It is shown that the problem of force uncertainty can be solved by introducing the assumption of elasticity of the cooperative system in its part where the force uncertainty arises. The problem of modeling, modeling procedure, and the model itself are illustrated by a simple example.

The basic problem in describing cooperative work is the determination of forces at the contact of the manipulator tip with the object and the determination of the object position on the basis of the known manipulator position and vice versa.

These problems can be defined as the problems of choice of the assumptions of the system's characteristics and behavior and the problems of a reliable mathematical description of the cooperative work based on these assumptions. In the majority of papers dealing with cooperative work, it is assumed that the manipulators and object are rigid. An unavoidable consequence of this assumption is the appearance of force uncertainty, which is manifested as the impossibility of establishing a unique relation between the force vector at the MC of the manipulated object and the force vector at the manipulator-object contact. Various approaches have been proposed to solve these problems, and a common feature of all of them is that contact forces are determined on the basis of the conditions proposed by the authors and not as a consequence of physical phenomena [12–18]. These conditions were given from the standpoint of the object requirements, manipulator requirements, or by a combination of both.

3.1 Some Known Solutions to Cooperative Manipulation Models

In [12] and [13], the vector of forces at contact points f_c was chosen as being completely independent of the manipulator dynamics, the criterion for choosing this vector being obtained on the basis of the requirements concerning the object. The solution adopted for contact force is the one that minimizes the square criterion

$$I_f = f_c^T W f_c, \qquad\qquad (*)$$

yielding the solution in the form

$$f_c = W^{-1} H^T (H W^{-1} H^T)^{-1} F_0, \qquad\qquad (**)$$

i.e.

$$f_{ci} = W_i^{-1} \left(\sum_{j=1}^{n} W_j^{-1} \right)^{-1} F_0, \quad i = 1, \dots, n,$$

where $W = \mathrm{diag}(w_1, \dots, w_n)$ is the weighting matrix; F_0 is the force vector at the object MC, and $H = (I\ I\ \dots\ I)$ is the block matrix of unit matrices, resulting from the relation $F_0 = \sum f_c = H f_c$.

In [14], a practical solution was given for the redistribution of the loads f_c onto the 'slave' manipulators as a function of the vector of internal forces f_I. In [13], the adopted distribution is the solution that minimizes the functional

$$\min\{\| f_c \|\}$$

under the condition of satisfying static friction conditions expressed by the inequality

$$e_{Ni} f_{ci} \geq \eta_i \| f_{ci} \|,$$

where

$$e_{Ni} = \mathrm{grad}\, S(p_i)/\|\mathrm{grad}\, S(p_i)\|$$

is the vector of the normal at the point p_i on the object surface, described by $S(x, y, z) = 0$, and η is the friction coefficient. In [15], the internal force requirements were selected so that they preserve the contact force within the friction cone.

The solution to the redundancy problem has also been sought as being independent of the object dynamics [16, 19]. In [19], the authors minimized the functional

$$\tau W \tau$$

solution of the minimization of contact forces. The work [16] considers the possible optimal splitting of the load between two industrial robots. As a possibility, a solution was proposed for the drives τ that satisfy the condition

$$OBJ > \|\tau\|^2,$$

where OBJ represents the criterion of minimal energy. For a uniform distribution of loads, it is proposed that contact forces are the same and also equal to one-half of the force at the object MC.

One possible approach is to consider both the object dynamics and manipulator dynamics. For the case of the absence of constraints on driving torques of the 'leader' and 'follower', by combining the right-hand sides of the behavior of the object and manipulator and minimizing the driving vector norm (which is equivalent to minimal energy), it was found in [16] that, because of extensive calculation, the solution for driving moments in the form $\tau^{l/f} = \tau^{l/f}(\tau, F_0, f_c)$ is almost inapplicable. As an alternative, the following distribution was considered

$$f_c^l = \alpha F_0, \quad f_c^f = (1 - \alpha) F_0, \quad 0 < \alpha < 1.$$

The minimization of the norm of contact forces

$$\min\{\|f_c\|\} = \min\{(\|f_c^l\|^2 + \|f_c^f\|^2)\}$$

yielded a solution as a function of the internal forces f_I

$$\alpha_r = \alpha_r(\tau, F_0, f_I).$$

In [20], driving torques were presented in the form

$$\tau = \tau' - J^T(I - G^+G)\epsilon,$$

where τ' represents the drive that ensures the motion along the trajectory; G is the transformation matrix of the expanded velocity vector of the contact points to the velocity vector of the object MC; G^+ is the generalized pseudo-inverse matrix (Moore–Penrose) of the matrix G, and ϵ is an arbitrary vector. The choice of the vector ϵ was made so as to allow the possibility of supervising internal forces, one possible choice being

$$\epsilon = \mathrm{sgn}(\tau')[J^T]_i^T,$$

where $[J^T]_i^T$ is the ith row of the transformation matrix J for transforming the velocity vector's internal coordinates into the expanded velocity vector of the contact points that yields a reduction of the manipulator load. For the case when the cooperative system mobility exceeds the dimensions of the operative space of the object, such a choice of vector ϵ was proposed that satisfies the condition $[P]$ of a certain sub-task described in the form

$$[P]\tau = \alpha.$$

The obtained solution

$$\epsilon = ([P] - J^T(I - G^+G))^+(\alpha - [P]\tau')$$

represents the generalization of the approach from [16].

From a formal point of view, until the system is not closed in a mathematical sense, the differential equations describing the cooperative manipulation behavior are to be supplemented by new equations.

3.2   A Method to Model Cooperative Manipulation

In the description of the cooperative system motion, there must always appear at least one relation that describes the equilibrium of the contact forces and forces at MC of the manipulated object. The form of this relation depends on the assumptions of the contact characteristics of the manipulators and object and of the structural properties of the environment.

If we assume that the manipulators and object are rigid and their contact is stiff and rigid, then only one vector relation, analoguous to (11), describes the equilibrium of m vectors of contact forces and one force vector that is acting at the object MC. If the contact force vectors are known, the force vector at the object MC is uniquely determined. If, however, the force vector at the object MC is known, the force vector at one contact point can be determined as a function of the known force vector at the object MC and $m - 1$ unknown force vectors at the other contact points.

The reason for the existence of only one relation for describing the equilibrium of the contact forces and forces at the object MC is that the description is based on the approximation of the cooperative system by rigid manipulators, rigid object, and rigid and stiff contact between them. A consequence of the existence of only one relation that describes the equilibrium of the contact forces and forces at the object MC is the impossibility of unique determination of contact forces as a function of only the forces acting at the object MC. In other words, the problem of the so-called 'force uncertainty' unavoidably arises.

Hence, the task is to consider some new assumptions that would ensure a unique solution of the cooperative system model, i.e. a unique distribution of forces at the contacts. The 'non-uniqueness' appears only in the description of the part of the system between the manipulator tips (grippers) and object. This suggests the

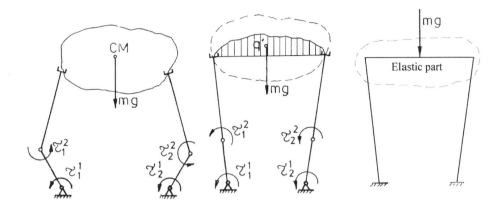

Figure 6. Reducing the cooperative system to a grid

conclusion that the approximation of this part of the cooperative system does not faithfully reflect its physical nature. Therefore, it is necessary to find some new approximation of the cooperative system between the manipulator tips (grippers) and object, from which will come our additional natural conditions that would ensure a unique mathematical description of the overall cooperative system model.

The mathematical model of a mechanical system should uniquely describe its kinematics, statics, and dynamics. A correct choice of the approximation of cooperative system is most simply made by analyzing the statics of the cooperative system, i.e. by analyzing the cooperative system's load in the state of rest. The appropriate choice of approximation of cooperative system leads to the solution of force uncertainty.

In the system at rest, the driving torques and forces at the contact of the tips of manipulators and object can be considered as a system of internal generalized forces, and gravitational forces as the system of external forces acting on the cooperative system. Then the cooperative system corresponds to a statically undetermined spatial grid made of the sticks fixed at one end to the support and at the other being in contact with the object (Figure 6). A detailed procedure for solving such a grid has been given in [6, 7, 21–24]. For the rest conditions, the results obtained in the mechanics of cooperative work should be in agreement with the results already obtained in other branches of mechanics (statics, dynamics, strength of materials, and structure theory).

Force uncertainty can be overcome by abandoning the assumption of the rigidity of the manipulators and object, or by retaining the same assumption but inserting elastic connections between the rigid manipulators and rigid object to satisfy the condition of deformation compatibility. According to the condition of defor-

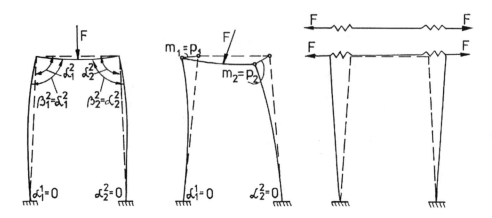

Figure 7. Approximation of the cooperative system by a grid

mation compatibility, each construction deforms so that no breaking of the connections between particular elements of the construction takes place. The number
of static uncertainties of the construction requires the same number of additional
explicit geometric conditions from which, for the known displacements, one can
determine unknown forces, or determine unknown displacements for the known
forces. The choice of geometric conditions depends on the concrete form of the
grid and character of the acting load (Figure 7). As a result, unique relations
between forces/moments and structural displacements at all of its points (cross-
sections) are obtained. In other words, none of the proposed criteria is adopted,
but the assumption on construction rigidity is abandoned, from which come some
additional geometric conditions.

There are several methods to solve the problems of static uncertainty. We will
consider the method of deformation work, implying from the principle of minimal
potential energy of the system. When considering the strength of materials, it is
assumed that the deformation is not accompanied by a change of the amount of
heat nor by acceleration of any particle of the material, i.e. the load changes are
very slow so that, due to the principle of energy conservation, the equation

$$\delta A_d + \delta Q = \delta T + \delta U \tag{12}$$

for any elastic system is reduced to the equality of the increments of deformation
work and potential energy

$$\delta A_d = \delta U, \tag{13}$$

where δU is the work increment due to external forces; δQ is the heat increment;
δT is the increment of kinetic energy, and δA_d is the increment of internal potential
energy (i.e. deformation work).

For small displacements of the elastic system, or for the displacements in the region of a linear relationship between the stress and dilatation, the deformation work is a homogeneous quadratic form of external forces and 'statically unknown' forces F_i (the law of superposition holds)

$$A_d = U = \frac{1}{2} \sum_{i=1}^{n} \sum_{j=1}^{n} \alpha_{ij} F_i F_j = F^T W^f F \tag{14}$$

or of the displacements u

$$A_d = U = \frac{1}{2} \sum_{i=1}^{n} \sum_{j=1}^{n} \beta_{ij} u_i u_j = u^T K u, \tag{15}$$

where W^f is the matrix of the so-called 'Maxwell's displacement influence numbers' α_{ij} (flexibility matrix), which represent the projection of the displacement of the acting point of the force F_i onto the direction of this force due to the unit force F_j; K is the stiffness matrix or the matrix of 'Maxwell's dual (reciprocal) coefficients' β_{ij}, representing the force that, by acting at the point j, produces a unit displacement at the point i, whereby the displacements at all other points equal zero, and u_i are the corresponding displacements (deflections).

According to the first and second Castigliano principles, the displacements and forces are determined as the derivative of deformation work with respect to forces and displacements

$$u_i = \sum_{k=1}^{n} \alpha_{ik} F_k = \frac{\partial A_d}{\partial F_i} = W_i^f F, \tag{16}$$

$$F_i = \sum_{k=1}^{n} \beta_{ik} u_k = \frac{\partial A_d}{\partial u_i} = K_i u, \tag{17}$$

where W_i^f and K_i are the ith rows of the matrices W^f and K, respectively.

Let us notice that the deformation work and deflections are inversely proportional to the elasticity module.

By comparing what was said above with the attempts to solve the problem of redundancy in cooperative work, it can be said that the criterion $(*)$ given in [12] is most similar to the expression for deformation work. However, it does not represent the deformation work itself, but an arbitrarily chosen criterion with a matrix of weighting elements and not of 'Maxwell's displacement influence numbers'. Even if that criterion would represent deformation work, it could be correctly applied only for static conditions of the cooperative system, and even then the forces at the contact of manipulator tips and object, as the grid internal forces could not be determined according to ($**$) [12], but according to (16).

On the basis of (16) and (17), we can derive several important conclusions.

Between displacements and forces, there exists a unique functional dependence. The relation is linear for small displacements and displacements that are in the area of the linear relation between the stress and dilatation. *If a prescribed force F_i is to be realized, it is necessary to realize the corresponding displacements (deflections), i.e. the grid position (of the cooperative system) with respect to the unloaded system, i.e. with respect to the position corresponding to the contact formation, for which the contact forces of the manipulator tips and object are equal to zero.* If the force increment is to be sought, it would be necessary to realize displacement increments with respect to the state for which displacements are considered. In other words, force control (at the contact too) is realized through position control, whereby potential force measurement allows us to find the grid position to which the measured force corresponds. From the point of view of technical realization, there appear the problem of precise control of displacements at the micrometer level, which are usually in the domain of the hysteresis of the position of regulation circuits. Such work is manifested as position oscillations in the domain of hysteresis and of the corresponding force oscillations. All this imposes the need for actuators of extremely high quality. In order to overcome this, it is convenient to have the terms with large displacements in the force expression (17), so that their influence is dominant in the force calculation, which is possible to realize provided the grid is made of a part that is very rigid and a part that is very elastic.

The influence coefficients α_{ij} are products of the dimensionless part (which is a function that comes out from the geometric configuration of grid nodes and system of forces) and the dimensional part (dimension [*position/force*]) that is inversely proportional to the elasticity module of the material and characteristics of the cross-section of the load gearing. In a similar way, we can also decompose the coefficients β_{ij}, whereby the dimensional part will be proportional to the elasticity module. Hence, it can be concluded that the values of forces and displacements can also be influenced by the appropriate choice of geometric configuration of the grid and a suitable choice of characteristics of the material and cross-section of the load gearing. In the case of a cooperative system, the sites of load action are given in advance, and the geometric arrangement of nodes is changeable. The choice of the geometric arrangement of nodes can be optimized so that, for example, force sensitivity at the contact of the manipulator tip and object to the internal coordinates is maximal, or that in no case does there arise the need for extremely small changes of internal coordinates.

During the cooperative system motion, the derivatives of coordinates are different from zero, so that there will appear forces dependent on these derivatives. This means that, unlike the static conditions, we cannot exclude from consideration the change of kinetic energy δT and dissipation (if it exists) as a function of velocity.

For a consistent description of the behavior of cooperative system motion, like for the one at rest, it is also necessary to form a correct set of assumptions on elastic properties of one part or entire cooperative system and characteristics of the contact of manipulator tips and object (stiff, hinged, spheric, point/surface, with/without friction). The adopted set of assumptions defines the geometric conditions for determining static/dynamic unknown quantities and, thus, the task of cooperative manipulation is classified.

Depending on the adopted assumptions, theoretical expressions of higher or lower complexity will be obtained for the kinetic energy T and deformation work $A_d = \Pi$ (potential energy). On the basis of them, the kinetic potential (Lagrange function) is formed, $L = T - \Pi$. It is important to notice that, in the cooperative system motion, all the conditions coming from the system's elastic properties must be simultaneously satisfied and all basic principles concerning the motion of a mechanical system must retain their validity. Because of that, the motion equations ought to be obtained by using some fundamental variational principle (e.g. Hamilton's integral principle or d'Alembert's differential principle) in the form of Lagrange, Newton, or Hamilton equations. In the resulting equations of motion, according to the Castigliano principles, elastic force is a derivative of deformation work with respect to displacement and, in each moment of motion, must be obtainable from the principle of minimal potential energy for the elastic system experiencing the action of the resulting external, inertial, and other elastic system loads existing at that moment.

In a number of works, the elastic properties of the manipulators and/or object have been considered without a clear and precise definition of the above physical properties, and without recognizing the need for introducing the elastic properties of the cooperative system or of manipulators in contact with the environment, but based only on profound research intuition. The models were formed for simple examples and for the cases of motion of an elastic system around the unloaded state of the elastic cooperative system [1–3, 25, 26]. In [4], an analysis was made of the cooperative system general motion but the resulting description of motion contained twice as many state quantities than was necessary.

In practical tasks, the problem of force uncertainty is solved in a simple way by considering both the manipulator and object as rigid bodies, whereas the connections of the object and manipulator are considered as an elastic body or a system of such bodies whose characteristics can be considered only in one direction and, if possible, without damping and with the link mass that is much smaller than the object mass. In that case, spatial inertial forces are reduced to the resulting inertial force of the rigid body with the acting point at the object MC, so that the object and the links can be considered as a grid under the action of a system of external forces (contact forces and object inertia force). All the volume integrals that appear in the description of kinetic and potential energy of elastic body are then transformed into definite integrals with the boundaries from the related straight line. Assuming the oscillation law as a sum of the products of orthogonal functions and generalized coordinates, kinetic and potential energies can be presented as homogeneous quadratic forms of these coordinates. For example, potential energy can be presented in the form (15), wherefrom differential equations of the behavior are obtained in the form of second-order Lagrange equations.

If, however, the mass of the link and its damping cannot be neglected, the corresponding forces produced should be projected onto the directions of the elastic forces, then superimposed with the elastic forces, to finally apply d'Alembert principle [1–3]. In other words, in practical technical applications, upon all these neglectings, the task is reduced to the use of the d'Alembert principle for solving dynamic tasks by static methods, constantly checking the equilibrium of active forces, forces of elastic interconnections (resistance of the supports/connections) and inertia forces during the motion.

In technical practice, for example, the airplane is considered as an elastic system under the action of external forces, the model being closest to the model of an object involved in cooperative work. In [27], the models were given for such a system. The simplest model is based on decomposing the plane into partial elastically interconnected masses, the so-called lumped-mass model. The other model, the so-called distributed-mass model, uses classical elasticity theory, whereby the plane is approximated by a set of elementary finite elements and takes a finite number of wave states (tones). The assumption on the elastic connection of the manipulators and object was adopted in modeling the one-dimensional horizontal translatory motion of the object [1–3]. However, the necessity of introducing such assumptions for resolving force uncertainty has not been explained and nor has there been presented a modeling methodology for such a simple example that could be potentially extended into the general case of cooperative system motion.

Therefore, if we want to abandon the assumption about rigid objects, rigid manipulators and their non-elastic contact, and retain the assumption about stiff contact, the problem of force uncertainty would be resolved, and kinematic uncertainty could appear only on the manipulators. The position vectors of contact points in the case of elastic contact become independent quantities so that we can speak of redundant manipulators, i.e. about the relationship between the number of DOFs of motion of the manipulators and the number of dimensions of free work space. If, however, kinematic uncertainty of the manipulators still appear, it is most suitable to overcome in a way similar to that nature has done with living beings.

If possible, one should choose manipulators whose number of DOFs is equal to the number of DOFs of the object's motion. Then there exists a unique kinematic relationship between the position and its derivatives for the tips of the manipulators and internal coordinates. If, however, this is not possible, then the kinematic chain should be decomposed into parts, the number of which is equal to the number of DOFs of the object motion. For each such part, it is necessary to define the work space in which a unique relationship will exist between the position of the decomposed part and points of the work space. In that space, it is necessary to choose, according to a certain criterion, the most favorable point or area for the tip of the decomposed part. Further on, the decomposed subspaces should be regarded as links of a fictitious manipulator, resolving kinematic relationship for the given subspaces. Then, within the given subspace, the relations are established between the subspace coordinates and internal coordinates of the concrete links. If the movements of the links are small, then their configuration should be set out so that they form a smooth curve (as with living beings). The analysis of possible ways of solving kinematic uncertainty of manipulators is not going to be considered in this chapter.

3.3   Illustration of the Correct Modeling Procedure

To illustrate the correct modeling procedure, we will consider an example that was dealt with in [1] and [4]. We will slightly modify the example by altering the axis along which the motion takes place. Namely, we will consider the object motion between the thumb and index finger as manipulators, along a vertical straight line, so that the object weight and manipulator reactions will lie on a straight line that is collinear to the gravitational acceleration. The alteration has been made to clearly define the characteristics under the conditions of rest of the cooperative system composed of two non-redundant manipulators and object. It is assumed that both the object and manipulators are rigid. Elastic interconnections will be placed between the manipulators and object, whereby the mass of these elastic interconnections is much smaller than that of the object, so it will be neglected. The damping properties are neglected, too. The adopted model is presented in Figure 8.

We shall first compose the model of the elastic system made of elastic interconnections and object. Applying displacement method [6, 7, 23, 24] we obtain

$$
\begin{aligned}
\textbf{(a)} \ \ y_1 \neq 0, \ \ y_2 = y_3 = 0 \ \ &\Rightarrow \ \ F_1 = c_p y_1 \\
\textstyle\sum F = 0 \ \ &\Rightarrow \ \ F_1 + F_2 = 0, \ \ F_3 = 0 \\
&\Rightarrow \ \ F_1 = -F_2, \ \ F_3 = 0,
\end{aligned}
$$

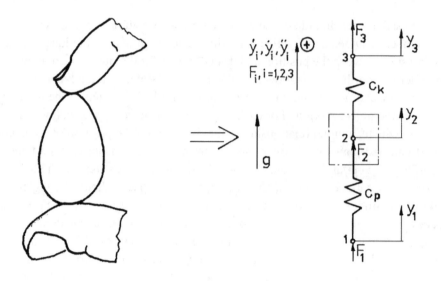

Figure 8. Linear elastic system

(b) $y_2 \neq 0$, $y_1 = y_3 = 0$ $\Rightarrow$ $F_2 = (c_p + c_k)y_2$
$\qquad\qquad\qquad\qquad\quad \Rightarrow$ $F_1 = -c_p y_2$ $\qquad\qquad$ (18)
$\qquad\qquad\qquad\qquad\quad \Rightarrow$ $F_3 = -c_k y_2$,

(c) $y_3 \neq 0$, $y_1 = y_2 = 0$ $\Rightarrow$ $F_3 = c_k y_3$
$\qquad\qquad\qquad\qquad\quad \Rightarrow$ $F_2 = -F_3 = -c_k y_3$
$\qquad\qquad\qquad\qquad\quad \Rightarrow$ $F_1 = 0$.

where c_p, c_k are the stiffnesses of the elastic interconnections, F_1, F_2, F_3 are the forces acting at the nodes 1, 2, and 3 of the elastic system, whose respective displacements are y_1, y_2, y_3. The displacements measured with respect to the geometric figure formed in the moment of establishing the manipulator-object contact, i.e. for the conditions under which forces at the manipulator-object contact are equal to zero (state 0).

If all the displacements take place simultaneously ($y_1 \neq 0$, $y_2 \neq 0$, $y_3 \neq 0$), then the superimposition yields the equations of force equilibrium for each node

$$F_1 = \quad c_p y_1 \qquad -c_p y_2 \qquad\qquad\qquad = -F_2 - F_3,$$

$$F_2 = \quad -c_p y_1 \quad +(c_p + c_k)y_2 \quad -c_k y_3 \quad = -F_1 - F_3,$$

$$F_3 = \qquad\qquad\qquad -c_k y_2 \qquad +c_k y_3 \quad = -F_1 - F_2, \qquad (19)$$

or, in matrix form,

$$
F_e = \begin{bmatrix} F_1 \\ F_2 \\ F_3 \end{bmatrix} = \begin{bmatrix} c_p & -c_p & 0 \\ -c_p & c_p + c_k & -c_k \\ 0 & -c_k & c_k \end{bmatrix} \cdot \begin{bmatrix} y_1 \\ y_2 \\ y_3 \end{bmatrix} = Ky,
$$

$$
K = K^T, \quad \text{rank } K = 2, \tag{20}
$$

from which the forces are easily expressed as a function of displacements.

As $\det K = 0$, rank $K = 2$, the system is kinematically unstable (mobile), so that the two displacements can be expressed as a function of the third one and of the acting forces. According to the theory of elastic systems, the matrix K contains the modes of rigid body motion. Expression (20) defines the relationship between the forces at m external nodes and one force at the internal node of the elastic system, where $m = 2$ is the number of manipulators. This expression defines the relationship between $2 \cdot (m + 1) \, (2 \cdot (2 + 1) = 6)$ quantities for the known constant stiffness matrix $K = \text{const}$, of which $m + 1 \, (2 + 1 = 3)$ are the quantities of displacement of nodes and $m + 1 \, (2 + 1 = 3)$ are the forces acting at the nodes of the elastic system. The origin of $m+1$ forces F_1, F_2 and F_3 is not essential (external forces transferred to the contact, gravitational, inertial, damping, etc, forces). It is essential to know that they, being the resultant acting forces, must act at the nodes of the elastic system to produce $m + 1$ displacements, y_1, y_2 and y_3.

It is necessary to make a distinction between the resultant forces acting on the elastic system and the external forces transferred to the contact, as of one of their components. Any resultant force acting on the elastic system at each node is balanced by the elastic force formed by the displacement of the elastic structure. This means that these forces are of the same intensity but acting in the opposite directions. Hence, the resultant forces acting on the elastic system will be called elastic forces. In the sections below, the elastic forces will be marked with a subscript 'e' (e.g. the elastic force at the ith node will be F_{ei}). The external forces transferred to the contact will be termed contact forces and they will bear the subscript 'c' (e.g. the force at the ith node will be F_{ci}).

Let A be part of the matrix K for which rank $A = 2$. Then (20) can be transformed to

$$
\begin{pmatrix} F_s \\ F_v \end{pmatrix} = \begin{pmatrix} A & b \\ c & d \end{pmatrix} \cdot \begin{pmatrix} y_s \\ y_v \end{pmatrix},
$$

$$
F_s = Ay_s + by_v \quad \Rightarrow \quad y_s = A^{-1}F_s - A^{-1}by_v \qquad = A^{-1}F_s|_{y_v=0},
$$

$$
F_v = cy_s + dy_v \quad \Rightarrow \quad F_v = \sum_{i=1}^{\text{rank } A} F_{si} + (d - cA^{-1}b)y_v = \sum_{i=1}^{\text{rank } A} F_{si}, \tag{21}
$$

where F_s is the vector of any two selected forces; y_s are the corresponding displacements at the points of action of the forces F_s; F_v and y_v are the remaining forces and displacement, respectively, while b, c, d are the corresponding vectors of the reduced matrix. In this example one can make three choices of forces: $F_s = F_c = \text{col}(F_1, F_3)$, $F_s = \text{col}(F_1, F_2)$, $F_s = \text{col}(F_2, F_3)$ and three corresponding displacements, $y_v = y_2$, $y_v = y_3$, and $y_v = y_1$, to define an elastic system in space. These properties of the elastic system represent the leadership principle in cooperative manipulation. The leader is that part of the cooperative system that defines the attitude of the elastic system in space. In this example, this can be either the object $(y_v = y_2)$ or one of the manipulators $(y_v = y_3$ or $y_v = y_1)$.

It should be noted that in order to determine the position of elastic system in space, it is necessary to provide $(m+1) - \text{rank } K \ (= 1)$ displacements of the elastic system nodes. This means that we can take as a known displacement of not only one node but of several nodes of the complex elastic system.

Let the displacement of the tip of the first (leader) manipulator be given, $y_v = y_1$. Displacement of the mass center of the solid object y_2 and the displacement of the tip of the second (slave) manipulator y_3 are determined as a function of the displacement y_1 and forces F_1 and F_2 according to (21). The force F_2 is equal to the sum of the weight and inertia of the object mass $F_2 = m(g + \ddot{y}_2)$ (provided $\ddot{y}_2$ is the absolute acceleration).

$$\begin{bmatrix} F_1 \\ F_2 \\ F_3 \end{bmatrix} = \begin{bmatrix} c_p & -c_p & 0 \\ -c_p & c_p + c_k & -c_k \\ 0 & -c_k & c_k \end{bmatrix} \cdot \begin{bmatrix} y_1 \\ y_2 \\ y_3 \end{bmatrix}, \tag{22}$$

$$\begin{pmatrix} y_2 \\ y_3 \end{pmatrix} = A^{-1} \begin{pmatrix} F_2 \\ F_3 \end{pmatrix} - A^{-1} b y_1$$

$$= \begin{pmatrix} c_p + c_k & -c_k \\ -c_k & c_k \end{pmatrix}^{-1} \begin{pmatrix} F_2 \\ F_3 \end{pmatrix} - \begin{pmatrix} c_p + c_k & -c_k \\ -c_k & c_k \end{pmatrix}^{-1} \begin{pmatrix} -c_p \\ 0 \end{pmatrix} y_1,$$

$$\begin{pmatrix} y_2 \\ y_3 \end{pmatrix} = \begin{pmatrix} \dfrac{1}{c_p} & \dfrac{1}{c_p} \\ \dfrac{1}{c_p} & \dfrac{1}{c_p} + \dfrac{1}{c_k} \end{pmatrix} \begin{pmatrix} F_2 \\ F_3 \end{pmatrix} + \begin{pmatrix} 1 \\ 1 \end{pmatrix} y_1$$

$$= A_f \begin{pmatrix} F_2 \\ F_3 \end{pmatrix} + \begin{pmatrix} 1 \\ 1 \end{pmatrix} y_1, \quad A_f = A^{-1}. \tag{23}$$

In the state of rest $F_2 = mg$, and the first manipulator (leader) does not announce the motion requirement (displacement $y_v = y_1 = y_1^0 = \text{const}$). The first manipulator tip represents support and the system is kinematically stable (fixed), so that the displacements are

$$\begin{pmatrix} y_2 \\ y_3 \end{pmatrix} = \begin{pmatrix} \dfrac{1}{c_p} & \dfrac{1}{c_p} \\ \dfrac{1}{c_p} & \dfrac{1}{c_p} + \dfrac{1}{c_k} \end{pmatrix} \begin{pmatrix} F_2 \\ F_3 \end{pmatrix} + \begin{pmatrix} 1 \\ 1 \end{pmatrix} y_1^0 = A_f \begin{pmatrix} F_2 \\ F_3 \end{pmatrix} + \begin{pmatrix} 1 \\ 1 \end{pmatrix} y_1^0.$$

From this simple example, we can derive the following conclusions.

To determine contact forces it is necessary to assume the existence of elastic interconnections. In order to calculate the forces, one needs to know the characteristics of the elastic connections and displacements of the elastic connections and object MC with respect to the state when the contact of the object and manipulators is just established. The values of displacements of nodes can be determined if the displacement of any node and the values of acting forces at the other nodes are known. The known values can be obtained either by measurement or by calculation.

Kinematically stable elastic systems are the subject matter of the theory of materials resistance, theory of constructions and theory of elasticity. The non-singular matrix A is adopted as a stiffness matrix for kinematically stable elastic systems. Its inverse matrix is the flexibility matrix $A_f = A^{-1}$ [6, 7, 23, 24]. A basic consequence of this assumption is that a kinematically stable elastic system is characterized by the existence of a unique relation between its displacements on the one hand and external and internal forces on the other.

An essential characteristic of the cooperative work is that the elastic system is not kinematically stable. This means that elastic displacements of all nodes in the elastic system motion change simultaneously. A consequence of kinematic instability is the singularity of the system's stiffness matrix K, which means that this matrix also contains modes of the rigid body motion.

Measurements of the displacements of contact points can be carried out in an indirect way, i.e. by the calculation based on the known internal coordinates of the manipulators. Displacement of the object MC is not suitable for measurement and it is better to do the calculations according to (23). This expression shows that if the number of DOFs of object motion is $l(= 1)$ and the number of nodes at which forces are acting is $m + 1 (= 2 + 1 = 3)$, then, on the basis of the known l displacements, one can calculate the displacements of the rest $(m + 1) \cdot l - l$ $(= 3 - 1 = 2)$ nodes. At that, it is also necessary to know the forces for which the displacements at the nodes are determined. Contact forces are easily measured with the aid of sensors built-in in the tips of the manipulators. If the measured quantity is given the superscript 'M', the measured contact force F_3 will be F_3^M. The force at the object MC cannot be measured but it can be calculated on the basis of the measured acceleration of the MC $(g + \ddot{y}_2))^M$ and object mass. The mass can be determined by measuring the displacement y_1^a of the node 1 for static conditions (a), for which $y_1 \neq 0$ and $y_2 = y_3 = 0$. Based on these conditions $m^M = -(c_p/g)y_1^a$, so that the force at the MC is determined by the expression

$$F_2^s = -\frac{c_p}{g}y_1^a(g + \ddot{y}_2)^M.$$

By introducing the known $(+)^P$, measured $(+)^M$, and calculated $(+)^S$ quantities into (23) one can calculate the displacements of $(m+1)\cdot l - l (= 2)$ nodes

$$\begin{pmatrix} y_2 \\ y_3 \end{pmatrix}^S = \begin{pmatrix} \dfrac{1}{c_p} & \dfrac{1}{c_p} \\ \dfrac{1}{c_p} & \dfrac{1}{c_p} + \dfrac{1}{c_k} \end{pmatrix}^P \begin{pmatrix} F_2^M \\ F_3^S \end{pmatrix} + \begin{pmatrix} 1 \\ 1 \end{pmatrix} y_1^P,$$

i.e. on the basis of knowing the acceleration at the object MC, contact force at one node and displacement at the other, we can determine the displacements at the other two nodes. By introducing the calculated displacements into (22), we can determine the force at the first node, F_1, which, (like with node 2), because of the absence of mass and connection damping, is equal to contact force. It is easy to check whether the replacement of the last expression in (22) yields the identity $(F = F \in R^{m+1(=3)})$, i.e. whether the procedure is correct. Analogous procedure could also be employed for more complex examples under the assumption of motion realization as in the considered example.

A case interesting for technical practice appears when the displacements of manipulators tips and forces at them are known (or measured), as well as the distance between the tips of the manipulators. Let us suppose that, according to the static conditions (a) from (18), the object mass has been determined and that the rigidity of the manipulator-object connection was known in advance. For the determination of all $2(m+1)$ quantities in the expression (20) (i.e. (21)), for determining $m+1$ connections, it is necessary to know (measure) $m+1(= 3)$ quantities. The simplest case for calculation is when we know $m+1(= 3)$ displacements of the elastic system nodes. If the displacement of the object MC is not measured, and since $\det K = 0$, then $m+1-l(= 2)$ measured quantities should be displacements/forces and l-measured quantities should be forces/displacements at the contact of the manipulators and object. Let us suppose that $l = 1$ contact forces (F_v) and $(m+1)\cdot l - l = 2$ displacements of contact points (y_s) have been measured. Then the number of displacements that were not measured corresponds to the number of DOFs of the object motion. The selection of l-measured contact forces should be carried out so that, after introducing the measured quantities into (20) or (21), one obtains a non-homogeneous system of equations from which it is possible to uniquely obtain the remaining l-unmeasured displacements, based on the overall displacements and remaining $(m + 1)l - l$ unmeasured forces. This can be done provided the rank of the matrix of the obtained non-homogeneous system is equal to l. For example, let y_1^M, y_3^M and F_1^M be measured and let y_2, F_2 and F_3 be determined. From (22), (i.e. from (19)) we have

$$c_p y_2 = c_p y_1^M + 0 \cdot y_3^M - F_1^M \quad \Rightarrow \quad y_2^S = y_1^M - \frac{F_1^M}{c_p}.$$

After determining the l non-measured displacements, along with the $(m + 1)l - l$-measured ones, the overall displacements of all the nodes are known. By introducing them into (22), we also have all the forces at both the contact points and object's MC. On the basis of the determined forces at the MC and object mass we can determine acceleration at the object MC.

In the above analysis, we assumed that all the motions are generated around an immobile figure, known in advance, formed at contact points at the moment of contact formation, which has been considered in [1]. If the manipulators move together with the object, the problem becomes more complex. In order to explain these phenomena, a complete mathematical model has been derived for the object motion along a vertical straight line to which belong all the contact points, as well as the object's MC. Also, it is supposed that breaking of connections cannot take place for any value of displacements of contact points 1 and 3 (tips of the manipulators are glued to the object). Thus, the analysis of the friction forces and characteristics of different types of contact is avoided.

In the example considered, the displacements y_1, y_2 and y_2 are independent quantities (generalized coordinates), so that Equation (20) can be derived using the Lagrange equations.

Kinetic energy is given by the expression

$$T = \frac{1}{2}m\dot{Y}_2^2 = \frac{1}{2}m(\dot{Y}_{20} + \dot{y}_2)^2 = \frac{1}{2}m(\dot{Y}_{10} + \dot{y}_2)^2, \tag{24}$$

where $\dot{Y}_2$ is the absolute velocity of the object MC; $\dot{Y}_{20}$, $\dot{Y}_{10}$ are the velocities of the points at which the object MC and contact point 1 would be found if the elastic system moved as a rigid body, or as a system with the displacements realized in the moment of contact formation ($f_{ci} = 0$, $i = 1, 2$), and $\dot{y}_2$ is the MC velocity with respect to the state of contact formation (Figure 9). If the masses of elastic interconnections were not neglected, the total kinetic energy would be equal to the sum of kinetic energies of elastic interconnections and of the manipulated object, given by the expression (24).

When $f_{ci} = 0$, $i = 1, 2$, no change will occur in the geometric figure formed by the contact points 1, 2 and 3, and determined by the node coordinates y_{10}, y_{20}

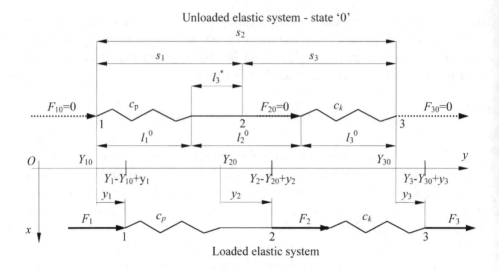

Figure 9. Approximating a linear elastic system

and y_{30} ($s_i = $ const, $i = 1, 2, 3$), so that, in accordance with the designations in Figure 9, we have

$$\begin{aligned}
Y_{20} &= Y_{10} + l_1^0 + l_2^* & &= Y_{10} + s_1, \\
Y_{30} &= Y_{10} + l_1^0 + l_2^0 + l_3^0 & &= Y_{10} + s_2, \\
Y_{30} &= Y_{20} + l_2^0 + l_3^0 - l_2^* & &= Y_{20} + s_3,
\end{aligned}$$

i.e.

$$\begin{bmatrix} -1 & 1 & 0 \\ -1 & 0 & 1 \\ 0 & -1 & 1 \end{bmatrix} \begin{bmatrix} Y_{10} \\ Y_{20} \\ Y_{30} \end{bmatrix} = \begin{bmatrix} s_1 \\ s_2 \\ s_3 \end{bmatrix} \Rightarrow \dot{Y}_{10} = \dot{Y}_{20} = \dot{Y}_{30}, \quad \ddot{Y}_{10} = \ddot{Y}_{20} = \ddot{Y}_{30}.$$

$$(25)$$

The coordinates and derivatives of coordinates of the position vectors of the acting points of the active forces F_1, weight mg, and F_3 are

$$Y_i = Y_{i0} + y_i \;\; \Rightarrow \;\; \dot{Y}_i = \dot{Y}_{i0} + \dot{y}_i \;\; \Rightarrow \;\; \ddot{Y}_i = \ddot{Y}_{i0} + \ddot{y}_i, \quad i = 1, 2, 3, \qquad (26)$$

from which one obtains the relative displacements of the elastic system nodes Δy_{12} and Δy_{23}:

$$\begin{aligned}
y_1 - y_2 &= \Delta y_{12} = Y_1 - Y_2 + Y_{20} - Y_{10} = Y_1 - Y_2 + s_1 \\
&= \left(1 - \frac{Y_{10} - Y_{20}}{||Y_1 - Y_2||} \right) (Y_1 - Y_2),
\end{aligned}$$

$$y_2 - y_3 \;=\; \Delta y_{23} = Y_2 - Y_3 + Y_{30} - Y_{20} = Y_2 - Y_3 + s_3$$

$$= \left(1 - \frac{Y_{20} - Y_{30}}{||Y_2 - Y_3||}\right)(Y_2 - Y_3), \tag{27}$$

because $(Y_i - Y_j)/(||Y_i - Y_j||)$, $(i, j) = (1, 2), (2, 3)$ is the unit vector along which the motion is performed.

It is assumed that the potential energy of elastic system Π is the energy of its deformation $\Pi = A_d$. Deformation energy is the accumulated energy of the elastic structure, i.e. of the elastic system (springs). This energy is defined as the overall work of the internal forces F_{ij} on their entire path, i.e. by the displacements y_{ij}. It is important to emphasize that these displacements are measured from the state of the elastic system in which displacements are equal to zero (state 0 of the elastic system). The deformation energy is defined by the expression

$$\Pi \;=\; \frac{1}{2}F_{12}\Delta y_{12} + \frac{1}{2}F_{23}\Delta y_{23} = \frac{1}{2}c_p\Delta y_{12}^2 + \frac{1}{2}c_k\Delta y_{23}^2$$

$$= \frac{1}{2}c_p(y_1 - y_2)^2 + \frac{1}{2}c_k(y_2 - y_3)^2$$

$$= \frac{1}{2}c_p(Y_1 - Y_2 + s_1)^2 + \frac{1}{2}c_k(Y_2 - Y_3 + s_3)^2$$

$$= \frac{1}{2}\pi_{12}(Y_1 - Y_2)^2 + \frac{1}{2}\pi_{23}(Y_2 - Y_3)^2, \tag{28}$$

where the spring forces (internal forces of the elastic system) are given by the expressions

$$F_{12} \;=\; c_p\Delta y_{12} = c_p(y_1 - y_2) = c_p(Y_1 - Y_2 + s_1)$$

$$= c_p\left(1 - \frac{Y_{10} - Y_{20}}{||Y_1 - Y_2||}\right)(Y_1 - Y_2),$$

$$F_{23} \;=\; c_k\Delta y_{23} = c_k(y_2 - y_3) = c_k(Y_2 - Y_3 + s_3)$$

$$= c_k\left(1 - \frac{Y_{20} - Y_{30}}{||Y_2 - Y_3||}\right)(Y_2 - Y_3), \tag{29}$$

whereas the generalized stiffnesses are defined by

$$\pi_{12} = \pi_{21} = c_p\left(1 + \frac{s_1}{||Y_1 - Y_2||}\right)^2 = c_p\left(1 - \frac{Y_{10} - Y_{20}}{||Y_1 - Y_2||}\right)^2,$$

$$\pi_{23} = \pi_{23} = c_k\left(1 + \frac{s_3}{||Y_2 - Y_3||}\right)^2 = c_k\left(1 - \frac{Y_{20} - Y_{30}}{||Y_2 - Y_3||}\right)^2. \tag{30}$$

It is important to notice that generalized stiffnesses are products of two factors. The first is the stiffness of the elastic structure (springs c_p and c_k) that can be measured or determined by one of the methods based on the considering the properties of the elastic system and its local coordinate frame with respect to the fixed unloaded state. The other factor defines the relation by which the stiffness from the local coordinate frame is transposed into the absolute coordinate frame using the information about the instantaneous absolute coordinates $Y_1 - Y_2$, $Y_2 - Y_3$ and information about the known state of the unloaded elastic system $Y_{10} - Y_{20}$, $Y_{20} - Y_{30}$.

In a geometrical sense, potential (deformation) energy represents the sum of the areas of the right-angle triangles. The number of triangles is equal to the number of internal forces, i.e. relative displacements of the elastic system nodes. In each triangle, the cathetuses make one internal force and the corresponding relative displacement of the elastic system node in the direction of action of that force.

In matrix form, according with (27) and (28), the potential (deformation) energy is

$$\Pi = \frac{1}{2}\begin{bmatrix} y_1 - y_2 \\ y_2 - y_3 \end{bmatrix}^T \begin{bmatrix} c_p & 0 \\ 0 & c_k \end{bmatrix} \begin{bmatrix} y_1 - y_2 \\ y_2 - y_3 \end{bmatrix} = \frac{1}{2}\epsilon_y^T \pi_\epsilon \epsilon_y,$$

$$\epsilon_y = \mathrm{col}(y_{12}, y_{23}), \quad \pi_\epsilon = \mathrm{diag}(c_p, c_k),$$

$$\Pi = \frac{1}{2}\begin{bmatrix} Y_1 - Y_2 \\ Y_2 - Y_3 \end{bmatrix}^T \begin{bmatrix} \pi_{12} & 0 \\ 0 & \pi_{23} \end{bmatrix} \begin{bmatrix} Y_1 - Y_2 \\ Y_2 - Y_3 \end{bmatrix} = \frac{1}{2}\epsilon^T \pi_\epsilon \epsilon,$$

$$\epsilon = \epsilon(Y) = \mathrm{col}(y_{12}(Y), y_{23}(Y)),$$

(31)

$$\Pi = \frac{1}{2}\begin{bmatrix} y_1 \\ y_2 \\ y_3 \end{bmatrix}^T \begin{bmatrix} c_p & -c_p & 0 \\ -c_p & c_p + c_k & -c_k \\ 0 & -c_k & c_k \end{bmatrix} \cdot \begin{bmatrix} y_1 \\ y_2 \\ y_3 \end{bmatrix} = \frac{1}{2}y^T K y,$$

$$y = \mathrm{col}(y_1, y_2, y_3),$$

(32)

$$\Pi = \frac{1}{2}\begin{bmatrix} Y_1 \\ Y_2 \\ Y_3 \end{bmatrix}^T \begin{bmatrix} \pi_{12} & -\pi_{12} & 0 \\ -\pi_{12} & \pi_{12} + \pi_{23} & -\pi_{23} \\ 0 & -\pi_{23} & \pi_{23} \end{bmatrix} \cdot \begin{bmatrix} Y_1 \\ Y_2 \\ Y_3 \end{bmatrix} = \frac{1}{2}Y^T \pi(Y) Y,$$

$$Y = \mathrm{col}(Y_1, Y_2, Y_3).$$

(33)

According to the Castigliano principle (17), elastic forces at the elastic system nodes given in terms of displacements y, are

$$F = \frac{\partial \Pi}{\partial y} = Ky = \begin{bmatrix} c_p & -c_p & 0 \\ -c_p & c_p + c_k & -c_k \\ 0 & -c_k & c_k \end{bmatrix} \cdot \begin{bmatrix} y_1 \\ y_2 \\ y_3 \end{bmatrix}$$

$$= \begin{bmatrix} c_p y_1 - c_p y_2 \\ -c_p y_1 + (c_p + c_k) y_2 - c_k y_3 \\ -c_k y_2 + c_k y_3 \end{bmatrix} = \begin{bmatrix} F_1 \\ F_2 \\ F_3 \end{bmatrix}. \tag{34}$$

The same forces are obtainable by using the absolute coordinates. By applying the Castigliano principle in expression (28), the potential energy expressed with the aid of absolute coordinates will be

$$F = \frac{\partial \Pi}{\partial Y} = \begin{bmatrix} c_p(Y_1 - Y_2 + s_1) \\ -c_p Y_1 + (c_p + c_k) Y_2 - c_k Y_3 - c_p s_1 + c_k s_3 \\ c_k(Y_3 - Y_2 - s_3) \end{bmatrix}$$

$$= \begin{bmatrix} c_p y_1 - c_p y_2 \\ -c_p y_1 + (c_p + c_k) y_2 - c_k y_3 \\ -c_k y_2 + c_k y_3 \end{bmatrix} = \begin{bmatrix} F_1 \\ F_2 \\ F_3 \end{bmatrix}. \tag{35}$$

If the potential energy is expressed in matrix form (33), then the elastic forces at the nodes are defined by the following expression:

$$F = \frac{\partial \Pi}{\partial Y} = \frac{1}{2} \frac{\partial (Y^T \bar{\pi}(Y) Y)}{\partial Y} + \pi(Y) Y, \tag{36}$$

where $\partial(Y^T \bar{\pi} Y)/\partial Y$ is the vector of the derivative of the quadratic form (scalar) $Y^T \pi Y$ with respect to the vector Y, whereby the macron denotes that the partial derivative is taken over the matrix π. According to (36), the elastic force F_1 at the first node will be determined by the expression

$$F_1 = \frac{\partial \Pi}{\partial Y_1} = \frac{1}{2} Y^T \frac{\partial \bar{\pi}(Y)}{\partial Y_1} Y + \pi_{1st\_row}(Y) Y,$$

where $\pi_{1st\_row}(Y)$ denotes the first row of the matrix $\pi(Y)$.

The generalized forces are defined by the expressions

$$Q_1 = F_1 \frac{\partial Y_1}{\partial Y_1} = F_1 \frac{\partial Y_1}{\partial y_1}\bigg|_{Y_1 = \lambda_1 Y_{10} = \lambda_2 y_1} = F_1,$$

$$Q_2 = (-mg) \frac{\partial Y_2}{\partial Y_2} = (-mg) \frac{\partial Y_2}{\partial y_2}\bigg|_{Y_2 = \lambda_1 Y_{20} = \lambda_2 y_2} = -mg,$$

$$Q_3 = F_3 \frac{\partial Y_3}{\partial Y_3} = F_3 \frac{\partial Y_3}{\partial y_3}\bigg|_{Y_3 = \lambda_1 Y_{30} = \lambda_2 y_3} = F_3. \tag{37}$$

The kinetic potential is

$$L = T - \Pi = \frac{1}{2}m(\dot{Y}_{10} + \dot{y}_2)^2 - \frac{1}{2}c_p(y_1 - y_2)^2 - \frac{1}{2}c_k(y_2 - y_3)^2 \tag{38}$$

$$= \frac{1}{2}m(\dot{Y}_{10} + \dot{y}_2)^2 - \frac{1}{2}c_p(Y_1 - Y_2 + s_1)^2 - \frac{1}{2}c_k(Y_2 - Y_3 + s_3)^2,$$

so that the derivatives of the kinetic potential are

$$\frac{\partial L}{\partial \dot{Y}_1} = \frac{\partial L}{\partial \dot{y}_1} = 0 \qquad\qquad \Rightarrow \quad \frac{d}{dt}\frac{\partial L}{\partial \dot{Y}_1} = 0,$$

$$\frac{\partial L}{\partial \dot{Y}_2} = \frac{\partial L}{\partial \dot{y}_2} = m\dot{Y}_2 = m(\dot{Y}_{10} + \dot{y}_2) \Rightarrow \frac{d}{dt}\frac{\partial L}{\partial \dot{Y}_2} = m\ddot{Y}_2 = m\ddot{Y}_{10} + m\ddot{y}_2,$$

$$\frac{\partial L}{\partial \dot{Y}_3} = \frac{\partial L}{\partial \dot{y}_3} = 0 \qquad\qquad \Rightarrow \quad \frac{d}{dt}\frac{\partial L}{\partial \dot{Y}_3} = 0,$$

$$\frac{\partial L}{\partial Y_1} = -c_p(Y_1 - Y_2 + s_1) = -c_p(y_1 - y_2) = \frac{\partial L}{\partial y_1},$$

$$\frac{\partial L}{\partial Y_2} = c_p(Y_1 - Y_2 + s_1) - c_k(Y_2 - Y_3 + s_3)$$

$$= c_p(y_1 - y_2) - c_k(y_2 - y_3) = \frac{\partial L}{\partial y_2},$$

$$\frac{\partial L}{\partial Y_3} = c_k(Y_2 - Y_3 + s_3) = c_k(y_2 - y_3) = \frac{\partial L}{\partial y_3}.$$

Such simple relations are obtained only because the fact that only translatory motion is considered, taking the example in which position vectors have only one coordinate. This allows us to decompose in a simple way the motions that would correspond to the motion of elastic system as a rigid body and the motion at deformation. In the case of pure translation, we have

$$\frac{d}{dt}\frac{\partial L}{\partial \dot{Y}_2} = \frac{d}{dt}\frac{\partial L}{\partial \dot{y}_2} = m\ddot{Y}_{10} + m\ddot{y}_2 = f_{20}(\ddot{Y}_{10}) + f_2(\ddot{y}_2)|_{\ddot{y}_{20}=\ddot{Y}_{10}},$$

$$\frac{\partial L}{\partial Y_i} = \frac{\partial L}{\partial y_i}, \quad \frac{\partial L}{\partial Y_i} = \bar{f}_{i0}(Y_0) + \bar{f}_i(Y), \quad Y_0 = col(Y_{10}, Y_{30}, Y_{30}), \ i = 1, 2, 3.$$

$$\tag{39}$$

In the case of a rotational motion, both the kinetic and potential energies are non-linear functions of the absolute coordinates. Hence, the decomposition of the motion can be carried out without essential loss in accuracy in the dynamics description. The reason lies in the fact that

$$\frac{d}{dt}\frac{\partial L}{\partial \dot{Y}_i} = \frac{d}{dt}\frac{\partial L}{\partial(\dot{Y}_{i0} + \dot{y}_i)} \neq f_{i0}(\ddot{Y}_{i0}) + f_i(\ddot{y}_i),$$

$$\frac{\partial L}{\partial Y_i} = \frac{\partial L}{\partial(Y_{i0} + y_i)} \neq \bar{f}_{i0}(Y_0) + \bar{f}_i(y), \quad i = 1, 2, 3, \tag{40}$$

so that the question arises as to the correctness of the results obtained in [4].

As damping properties are neglected, their dissipation energy D is equal to zero, $D = 0$.

After introducing the obtained expressions into the Lagrange equations

$$\frac{d}{dt}\frac{\partial T}{\partial \dot{Y}_i} - \frac{\partial T}{\partial Y_i} - \frac{\partial \mathcal{D}}{\partial \dot{Y}_i} + \frac{\partial \Pi}{\partial Y_i} = Q_i, \quad i = 1, 2, 3,$$

$$\frac{d}{dt}\frac{\partial L}{\partial \dot{Y}_i} - \frac{\partial L}{\partial Y_i} = Q_i, \quad L = T - \Pi, \quad D = 0, \quad i = 1, 2, 3, \tag{41}$$

we obtain a model of an elastic system in the coordinates that characterize deformation y_1, y_2, y_3 and coordinates characterizing the motion of elastic system described as a rigid body Y_{10} described by the expressions

$$\begin{array}{rclcl} c_p y_1 & -c_p y_2 & & = & F_1, \\ m\ddot{y}_2 \quad -c_p y_1 & +(c_p + c_k)y_2 & -c_k y_3 & = - & m(g + \ddot{Y}_{10}), \\ & -c_k y_2 & +c_k y_3 & = & F_3, \end{array} \tag{42}$$

or, in absolute coordinates,

$$\ddot{Y}_2 + \frac{c_p + c_k}{m}Y_2 = \frac{c_p}{m}Y_1 + \frac{c_k}{m}Y_3 + \frac{c_p}{m}s_1 - \frac{c_k}{m}s_3 - g,$$

$$F_1 = c_p(Y_1 - Y_2 + s_1),$$

$$F_2 = -m\ddot{Y}_2 - mg = -c_p Y_1 + (c_p + c_k)Y_2 - c_k Y_3 - c_p s_1 + c_k s_3,$$

$$F_3 = c_k(Y_3 - Y_2 - s_3). \tag{43}$$

By its form, model (42) is identical to expression (20) for the description of an elastic system under static conditions, whereby in this case the force at the MC is defined as

$$F_2 = -m(g + \ddot{Y}_{10} + \ddot{y}_2) = -mg - m\ddot{Y}_2, \tag{44}$$

i.e. the dependence $F = \partial \Pi / \partial y = Ky$ has been fully preserved. Thus, the principle of the minimum of deformation (potential) energy (13) is preserved at any moment, which means that the quasi-static conditions of elastic system have been preserved at any moment of the motion.

Equations (25), (26) and (42) or (43) determine in full the dynamic model of the elastic system composed of elastic interconnections and object. The drives for the manipulators are driving torques at joints, so that the output quantities of the manipulators are positions of contact points 1 and 3. Hence, the input quantities to the model of the elastic system are instantaneous absolute positions of the contact points with the manipulators Y_1 and Y_3. If the masses of elastic interconnections are neglected, the state quantities of the elastic system are identical to the state quantities of the manipulated object. In that case, the state quantities are the position and velocity of the object MC Y_2 and $\dot{Y}_2$. The elastic forces are at the same time the contact forces $F_1 = -f_{c1}$ and $F_3 = -f_{c2}$ (f_{c1} and f_{c2} are the forces at the tips of the manipulators) and can be adopted as output quantities of the elastic system. However, problems appear if the elastic system model is presented in the form (42). The number of state quantities (positions and velocities) is exactly twice the number of DOFs of the object motion. That number of state quantities is necessary and sufficient for the description of the overall object dynamics. In (42), it is convenient to select y_2 and $\dot{y}_2$ as state quantities, but then the acceleration $\ddot{Y}_{10} = \ddot{Y}_{20}$ remains undetermined. As the number of state quantities cannot exceed two, the quantities related to the motion of unloaded elastic system as a rigid body (here, the acceleration is $\ddot{Y}_{10} = \ddot{Y}_{20}$) have to be taken as known or measured, as was done in [4].

Let us assume that the manipulators are rigid and non-redundant and let their contact with the manipulated object be rigid and stiff. Let the mathematical model of manipulators be given by $H_i(q_i)\ddot{q}_i + h_i(q_i, \dot{q}_i) = \tau_i + J_i^T f_{ci}$ and let the mathematical form of kinematic relationship between the internal and external coordinates be $Y_i = \Theta_i(q_i) \in R^{6 \times 1}$, $i = 1, \ldots, m$ (the complete mathematical model of the manipulators is given in Section 4.8 and kinematic relations in Section 4.9).

A correct model of the cooperative manipulation, without any uncertainty, is determined by the elastic system model (43), model of manipulators, and kinematic relationships between the internal and external coordinates, with the remark that $f_{c1} = -F_1$ and $f_{c2} = -F_3$. A block diagram of this model is given in Figure 10.

From the block diagram it is evident that, for solving the cooperative system dynamics, it is necessary to know:

- model parameters (e.g. mass of the manipulated object m, stiffnesses $c_p, c_k, g, \ldots$),

- distances s_1 and s_3 between the nodes 1–2 and 2–3 of the elastic system in

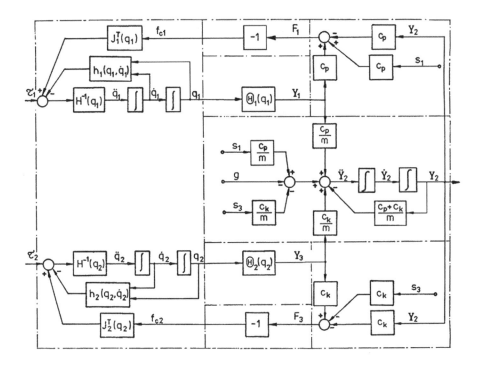

Figure 10: Block diagram of the model of a cooperative system without force uncertainty

  its unloaded state, in which all displacements are zero, and

- input quantities represented by the driving torques τ_1 and τ_2.

  Therefore, the input to the cooperative system model is only the driving torques, as in the reality, and all other quantities are uniquely determined without any uncertainty.

3.4   Simulation of the Motion of a Linear Cooperative System

In order to demonstrate the correctness of the modeling process, we simulated the 'linear cooperative system' dealt with in [1] and [4]. The model was expanded by introducing dissipative properties of the elastic interconnections.
  The dissipation function was taken in the form

$$\mathcal{D} = -\frac{1}{2}d_p(\dot{y}_1 - \dot{y}_2)^2 - \frac{1}{2}d_k(\dot{y}_2 - \dot{y}_3)^2 = -\frac{1}{2}d_p(\dot{Y}_1 - \dot{Y}_2)^2 - \frac{1}{2}d_k(Y_2 - Y_3)^2.$$

To describe the motion in the gripping phase, it is convenient to use the model of the elastic system described with the aid of coordinates with respect to the deviation y from the unloaded state (if it is fixed, then $\ddot{Y}_{10} = 0$). The elastic system model in the coordinates with respect to the deviation from the unloaded state 0 is

$$\ddot{y}_2 + \frac{(d_p + d_k)}{m}\dot{y}_2 + \frac{(c_p + c_k)}{m}y_2 = \frac{d_p}{m}\dot{y}_1 + \frac{d_k}{m}\dot{y}_3 + \frac{c_p}{m}y_1 + \frac{c_k}{m}y_3 - g - \ddot{Y}_{10},$$

$$
\begin{aligned}
F_{e1} &= c_p y_1 - c_p y_2, \\
F_{e2} &= -c_p y_1 + (c_p + c_k)y_2 - c_k y_3 \\
&= -m(\ddot{Y}_{10} + \ddot{y}_2) - mg + d_p \dot{y}_1 - (d_p + d_k)\dot{y}_2 + d_k \dot{y}_3, \\
F_{e3} &= -c_k y_2 + c_k y_3, \\
F_{c1} &= d_p \dot{y}_1 - d_p \dot{y}_2 + c_p y_1 - c_p y_2, \\
F_{c2} &= -d_k \dot{y}_2 + d_k \dot{y}_3 - c_k y_2 + c_k y_3,
\end{aligned}
$$

where d_p and d_k are the damping coefficients of connections, $F_{ei}, i = 1, 2, 3$ are the elasticity forces generated at the nodes, and $F_{cj}, j = 1, 2$ are the contact forces.

To describe the general motion of the elastic system, one should use the model presented in the absolute coordinates Y, given by the relations

$$\ddot{Y}_2 + \frac{(d_p + d_k)}{m}\dot{Y}_2 + \frac{(c_p + c_k)}{m}Y_2 = \frac{d_p}{m}\dot{Y}_1 + \frac{d_k}{m}\dot{Y}_3 + \frac{c_p}{m}Y_1 + \frac{c_k}{m}Y_3 - g + \frac{c_p}{m}s_1 - \frac{c_k}{m}s_3,$$

$$
\begin{aligned}
F_{e1} &= c_p Y_1 - c_p Y_2 + c_p s_1, \\
F_{e2} &= -c_p Y_1 + (c_p + c_k)Y_2 - c_k Y_3 - c_p s_1 + c_k s_3 \\
&= -m\ddot{Y}_2 - mg + d_p \dot{Y}_1 - (d_p + d_k)\dot{Y}_2 + d_k \dot{Y}_3, \\
F_{e3} &= -c_k Y_2 + c_k Y_3 - c_k s_3, \\
F_{c1} &= d_p \dot{Y}_1 - d_p \dot{Y}_2 + c_p Y_1 - c_p Y_2 + c_p s_1, \\
F_{c2} &= -d_k \dot{Y}_2 + d_k \dot{Y}_3 - c_k Y_2 + c_k Y_3 - c_k s_3.
\end{aligned}
$$

Models of the one-DOF linear manipulators are taken in the form

$$
\begin{aligned}
m_1 \ddot{q}_1 + m_1 g &= \tau_1 + f_{c1}, & f_{c1} &= -F_{c1}, \\
m_2 \ddot{q}_2 + m_2 g &= \tau_2 + f_{c2}, & f_{c2} &= -F_{c2}.
\end{aligned}
$$

Kinematic relations between the external and internal coordinates are given by the following expressions:

$$q_1 = Y_1 = Y_{10} + y_1, \qquad q_2 = Y_3 = Y_{30} + y_3,$$
$$\dot{q}_1 = \dot{Y}_1 = \dot{Y}_{10} + \dot{y}_1 = \dot{y}_1|_{Y_{10}=\text{const}}, \qquad \dot{q}_2 = \dot{Y}_3 = \dot{Y}_{30} + \dot{y}_3 = \dot{y}_3|_{Y_{30}=\text{const}},$$
$$\ddot{q}_1 = \ddot{Y}_1 = \ddot{Y}_{10} + \ddot{y}_1 = \ddot{y}_1|_{Y_{10}=\text{const}}, \qquad \ddot{q}_2 = \ddot{Y}_3 = \ddot{Y}_{30} + \ddot{y}_3 = \ddot{y}_3|_{Y_{30}=\text{const}}.$$

By coupling the kinematic relations and models of elastic system dynamics and manipulators, one obtains the model of cooperative manipulation. For the general motion, the model of cooperative manipulation expressed via absolute coordinates is

$$m_1 \ddot{Y}_1 + d_p \dot{Y}_1 - d_p \dot{Y}_2 + c_p Y_1 - c_p Y_2 + m_1 g + c_p s_1 = \tau_1,$$
$$m_2 \ddot{Y}_3 - d_k \dot{Y}_2 + d_k \dot{Y}_3 - c_k Y_2 + c_k Y_3 + m_2 g - c_k s_3 = \tau_2,$$
$$m \ddot{Y}_2 - d_p \dot{Y}_1 + (d_p + d_k) \dot{Y}_2 - d_k \dot{Y}_3 - c_p Y_1 + (c_p + c_k) Y_2 - c_k Y_3 + mg - c_p s_1 + c_k s_3 = 0,$$
$$d_p \dot{Y}_1 - d_p \dot{Y}_2 + c_p Y_1 - c_p Y_2 + c_p s_1 = F_{c1},$$
$$-d_k \dot{Y}_2 + d_k \dot{Y}_3 - c_k Y_2 + c_k Y_3 - c_k s_3 = F_{c2}.$$
$$(45)$$

The compact form of the model

$$
\begin{aligned}
m_1 \ddot{Y}_1 + m_1 g + F_{c1} &= \tau_1 \\
m_2 \ddot{Y}_3 + m_2 g + F_{c2} &= \tau_2 \\
m \ddot{Y}_2 + mg - F_{c1} - F_{c2} &= 0
\end{aligned}
\quad \Rightarrow \quad
\begin{aligned}
m_1 \ddot{Y}_1 + m_1 g &= \tau_1 + f_{c1}, \\
m_2 \ddot{Y}_3 + m_2 g &= \tau_2 + f_{c2}, \\
m \ddot{Y}_2 + mg &= -f_{c1} - f_{c2},
\end{aligned}
$$
$$(46)$$

shows that the mathematical form of the cooperative system (all rigid) model has been preserved. The introduced elasticity property gives the meaning to contact forces as a function of the current (relative) position of manipulator tips and object.

Numerical values of the parameters of elastic system (Figure 8) are $s_1 = s_2 = 0.05$ [m], $m = 25$ [kg], $c_p = 20 \times 10^3$ [N/m], $c_k = 10 \times 10^3$ [N/m], $d_p = 500$ [N/(m/s)] and $d_k = 1000$ [N/(m/s)]. Numerical values of the manipulator model parameters are $m_1 = 12.5$ [kg] and $m_2 = 12.5$ [kg].

The initial position of the cooperative system prior to the gripping process is determined by the nodes coordinates $Y_{10} = 0.150$ [m], $Y_{20} = 0.200$ [m] and $Y_{30} = 0.250$ [m].

Results obtained by simulating a linear cooperative system are presented in Figure 11. The selected driving torques perform gripping, lifting, and further oscillatory motions of the object. Since the cooperative system is not stabilized, the absolute positions of contact points diverge, retaining though the necessary mutual distances.

In all the diagrams, the independent variable (on the abscissa) is the simulation time in seconds. The dependent variables are the inputs and simulation results.

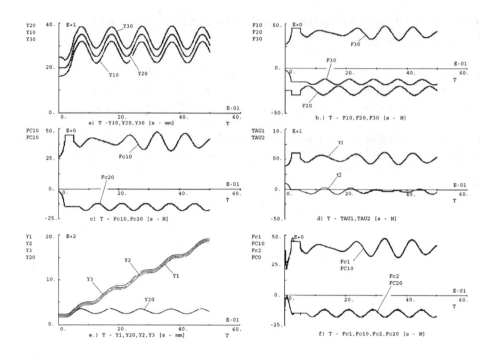

Figure 11. Results of simulation of a 'linear' elastic system

The explanations at the bottom of each diagram give first the independent variable (T) and then the dependent variable and its dimension. The letter denotes physical quantity used in simulation, while the numeral gives the ordinal number of the physical quantity vector. The symbols for the MC position and force of the manipulated object are X_0, Y_0 and F_{10}, whereas Y_i, F_i, F_{ci} and τ_i, $i = 1, 2$ are the displacements of contact points, elastic forces, contact forces and manipulator drives, respectively. Symbols for the first and second derivatives are obtained by adding the letters 'S' and 'SS' to the basic symbol of the quantity. Thus, for example, the symbols for the first and second derivatives of Y are Y_{1S} and Y_{1SS}, respectively.

3.5 Summary of the Problem of Mathematical Modeling

Based on the introductory consideration concerning the consistent mathematical procedure for modeling a simple cooperative system it is possible to derive the following general conclusions that could serve as landmarks in the process of modeling complex cooperative systems:

- The problem of force uncertainty is to be solved by introducing the assumption on elasticity of that part of the cooperative system in which that uncertainty appears.

- It is convenient to model an elastic system separately in order to ensure an easier and more correct description of its (quasi)statics and dynamics.

- In modeling an elastic system, it is necessary to first solve the static conditions on the basis of the minimum of potential (deformation) energy ($\delta A_d = \delta U$, (13)).

  As a result of this step, we get:

    - the relation $F = Ky$ between the elastic forces F and stiffness characteristics K and displacement of the elastic system with respect to its unloaded state y,

    - the number of state quantities of elastic system n_y equal to the dimension of the vector $y \in R^{n_y}$,

    - singular stiffness matrix K (det $K = 0$, rank $K < n_y$),

    - kinematically unstable (mobile) elastic system,

    - arbitrary choice $n_y -$rank K of displacements of the leader for the given elastic system in space.

- The relation $F = Ky$ is to be transposed into the dependence of elastic force on the absolute coordinates $F = K(Y)Y$ and deformation energy determined as a function of the absolute coordinates Y, the energy needed to perform the general motion of the elastic system.

- The kinetic and deformation energies and generalized forces should be determined as a function of absolute coordinates Y and Lagrange formalism is to be applied to generate the equation of motion of the elastic system.

- A model of the cooperative system dynamics is to be formed by coupling the model of elastic system motion with the models of manipulators and relations describing the contact conditions.

4 MATHEMATICAL MODELS OF COOPERATIVE SYSTEMS

4.1 Introductory Remarks

In Chapter 3 we stated that the problem of force uncertainty in cooperative manipulation can be eliminated and that a consistent description of cooperative manipulation is achieved by considering the cooperative system as an elastic system. Thus the problem arises of introducing the elastic properties and modeling dynamics of the complex cooperative system as an elastic system.

A model of a cooperative system dynamics can be formed on the basis of equations coming from integral principles (e.g. Hamilton principle) or equations coming from differential principles (e.g. D'Alembert principle, principle of virtual displacements). Thus, one obtains Lagrange's, Hamilton's, or Newton's equations, which are mutually equivalent [28].

In order to apply differential principles, it is necessary to know all the forces/moments. When applying the D'Alembert principle, the equations of motion are obtained by forming equations of dynamic equilibrium of forces and moments, by supplementing active forces and connection forces with all the forces that appear during the motion (inertial forces, forces of damping, and elasticity). The application of the principle of virtual displacements requires knowing the forces/moments as a function of kinematic quantities for the chosen DOFs as arguments. Equations of dynamic equilibrium are obtained by equating to zero the total work of forces/moments on the virtual displacements.

To apply integral principles, it is necessary to know Lagrange's function (kinetic and potential energies), dependent on the generalized coordinates and their derivatives, dissipation function, and generalized forces. The advantage of this principle is that Lagrange's function, its components, and dissipation function are scalar quantities (work, energy) and their values for the overall system are obtained simply by adding particular values of these functions given for the system components. It is only essential that these functions are given in the same system of inertial (or absolute, Section 1.4) coordinates.

In this chapter, cooperative manipulation dynamics will be modeled on the basis of Lagrange's equations derived by applying Hamilton's principle. The number of generalized coordinates and the number of equations obtained are exactly equal to the number of DOFs of the system's motion. To obtain these equations it is necessary to determine first the Lagrange function which is equal to the difference between the kinetic and potential energy, and then dissipation function and generalized forces. The choice of the form of these equations depends on the degree of accuracy of description of the physical nature one wants to achieve by the mathematical model of the cooperative system. The accuracy will depend on the introduced assumptions on the characteristics of the cooperative system, first of all of its elastic properties.

As already mentioned, the problem of force uncertainty does not exist if at least one part of the cooperative system is elastic. This means that a cooperative system must be elastic. An elastic cooperative system can be considered as composed of

- elastic components (manipulators and object),

- elastic manipulators and rigid object,

- rigid manipulators and elastic (deformable) object, and

- rigid manipulators, rigid object and elastic interconnections at the contacts.

For technical application, manipulators should be rigid enough. Hence, it is assumed that the manipulators in cooperative manipulation are rigid.

Manipulated objects may have very diverse elastic properties, which depends of both the nature of the material and object shape (structure). In modeling, one should allow for the possibility of elastic properties of the object. However, the same manipulators should be capable of manipulating both rigid and elastic objects. If only elastic properties of the object are assumed, then the choice of control law has to be based on the detailed knowledge of the elastic properties of each manipulated object. This means that control tasks should be solved separately for each manipulated object, which is not acceptable in practice. Hence, it is adopted that the connections between the object and manipulators are elastic, whereas the object is either rigid or has elastic properties at least in the neighborhood of the contact points, so that such an object, together with elastic interconnections, can be considered as an ideal elastic whole (body).

In this way, the cooperative system is decomposed into an elastic part and a rigid part. Rigid part consists of the subsystems of interconnected rigid bodies made of manipulator links. The elastic part of the cooperative system (in the sequel, elastic system) is represented by elastic interconnections at the contact and rigid manipulated object or by the elastic part of the manipulated object in the neighborhood of the contact points, along with the rest (rigid) part of the object.

In the preceding chapter, we described the correct procedure of modeling, assuming that the cooperative system can be decomposed first into its constitutive parts. Then, each of the parts is modeled separately. The overall model of cooperative system is obtained by joining the separate models with the aid of relations that describe the conditions at contact of the decomposed parts (equality of forces, positions, velocities, and accelerations).

Models of manipulators, especially of non-redundant ones, have been dealt with in many papers and can be without alterations taken over for the model of that part of the cooperative system.

A specific feature of a cooperative system is the existence of the manipulator-object contacts and necessity of introducing elasticity.

Mathematical description of contact is trivial if the contact is rigid and stiff and very complex when it is elastic and sliding. This means that the problem of contact is not essentially a problem of mathematical modeling of cooperative systems, as it can be overcome by an appropriate choice of contact characteristics.

The introduction of elasticity cannot be avoided if one wants to get a consistent solution to the problem of force uncertainty, i.e. if exact distribution of the loads at the manipulator tips and object in the course of cooperative system motion is sought. This means that a key problem in mathematical modeling of cooperative system dynamics is the modeling of the dynamics of an elastic system.

Properties of elastic systems can be described in different ways and thus the resulting description of their motion will have a higher or lower degree of accuracy. An exact description of cooperative system motion will be obtained by the exact description of the properties and motion of elastic parts of the system, which is a subject dealt with in the theory of elasticity and theory of oscillations of continual bodies [6, 7, 23, 24, 29, 30]. According to the theory of elasticity, the object elastic properties are judged on the basis of comparison of the kinematic characteristics of an elementary volume before and after the deformation, whereby the conditions are found for the load that can produce the given deformed state. To make conclusions about the overall body, it is necessary to consider all the volume elements, which means integrating the entire volume. The smaller the volume elements, the more exact the description and the more extensive calculations are needed.

The motion of an elementary volume of cooperative system can be mathematically described as the displacement of a spatial vector consisting of one translation, one rotation, and one deformation. In other words, if position vectors are associated with the unloaded and loaded states of an elementary volume, then it is possible to obtain one vector by transforming the coordinates of the other vector. For small displacements of an elementary volume, the transformation is linear and non-linear for the overall volume. The skew-symmetric part of the transformation describes rotation, while the symmetric one describes deformation. If the non-deformed state vector is known and if this state is fixed, the study of static/dynamic displacements of the elastic system is reduced to studying only the symmetric part of transformation, which is the subject matter of all the disciplines dealing with displacements around a known unloaded state in space (strength of materials, aeroelasticity, dynamics of constructions, oscillation theory). Then the position of any point of the loaded elastic system can be described in Cartesian coordinates defined by the displacements with respect to the unloaded state, or by the coordinates of the initial unloaded state which, for the loaded state, represent curvilinear coordinates. The coordinates of any point of the initial unloaded state can be expressed as a function of position coordinates of that point in the loaded state, and then the Cartesian coordinates of the loaded state are curvilinear coordinates of the unloaded state. In other words, if the fixed frame $Oxyz$ and the frame $O_{i0}x_{i0}y_{i0}z_{i0}$ describing the undeformed state are assumed to be Cartesian, then the absolute coordinates of any point of the strained elastic system are curvilinear coordinates of that point in the unloaded state. And contrary, if the absolute coordinates of any point of the loaded elastic system are adopted to be Cartesian, then the coordinates of that point in the unloaded state represent curvilinear coordinates of the absolute coordinates of that point. The problem of determining stress and deformation for a known shape, dimensions, material characteristics and system of external forces for a loaded ideally elastic system assumes finding 15 functions, viz., six stress components, six deformation components and three displacement components at each point of the body. The exact method of solving is reduced to solving Lamé equations that contain only displacements, or Beltrami–Mitchell equations defining the given stress state, or the Saint Venant semi-inverse method. In engineering practice, wide applications have found approximate methods, which can be divided in two groups. The first group includes the methods based on the approximate solving of the system of differential equations for an approximate model with a finite number of DOFs. To these methods belong the finite-difference method, iteration method, and finite-element method, in the frame of the latter being the developed force method, displacement method, and method of direct stiffness as a generalization of the displacement method. The other group encompasses variational methods, of which most well known are the Rayleigh–Ritz and Galerkin methods. The Rayleigh–Ritz method is based on the approximate representation of the task extremal by a class of a finite number of suitably selected functions that satisfy the contour conditions and for which unknown coefficients should be determined, which eventually reduces to solving an algebraic system of equations. The application of any of these methods requires extensive calculations.

A description of the dynamics of elastic systems is primarily needed to study their oscillatory motion around a selected position, with the aim of finding the form of oscillations and characteristic frequencies of that elastic body. In engineering practice, elastic bodies are usually constructions of mechanical and civil engineering. The dynamics of continual elastic bodies is described by integro-differential equations or by Lagrange's equations, whereby each elementary part of the body possesses its own DOF, so that an elastic body has an infinite number of DOFs. This description allows us to recognize all the forms of oscillations (modal forms) and all the characteristic frequencies of the elastic body. Upon decomposing the body into smaller elastically interconnected parts, the number of DOFs of the elastic body becomes limited. Then the number of characteristic frequencies that one can observe and the number of corresponding oscillation forms are exactly equal to the number of adopted DOFs. The degree of qualitative and quantitative agreement of the results with reality will depend on the choice of the decomposed structure as a substitute for the continual structure and inertial properties assigned to the components of that structure. For example, if the decomposed mass of the grid is replaced with the concentrated masses at the grid nodes, we will obtain a sufficiently good approximation of at least first (basic) modal form.

The cooperative manipulation proceeds relatively slowly. As the bandwidths of the actuators of the control system are in the range from parts to about 15 Hz, it means that the manipulation drives cannot be of high frequency. A question arises as to what frequency the model should faithfully describe the dynamics of the elastic system handled by the manipulators. If, in the domain of the bandwidth of manipulator actuators, there are no characteristic oscillations (modes) of the elastic system, or if there are no other modes (except for the first one) close to it, then it suffices to describe the elastic system motion via a high-quality presentation of the first modal form. A consequence of a such conclusion is that the cooperative system dynamics is sufficiently well characterized when the elastic system is considered as a system with a finite number of DOFs of motion.

Another problem is the choice of the DOFs of motion and the arrangement of the elastic system inertial properties that are assigned to the adopted DOFs.

An elastic system can be approximated by a discontinual structure in different ways.

A basic goal of this chapter is not to attain high accuracy of the model but to find a consistent method of solving the problem of force uncertainty in cooperative work, so that a minimal number of DOFs will be sought. The simplest way to find this number of DOFs is to replace the elastic system with a space grid whose external nodes coincide with the contact points, whereas the only internal node coincides with the MC of the manipulated object. The links between the nodes of the grid thus formed are simply approximated by different forms of elementary beams. The choice of link characteristics is given through the selection of characteristics of the corresponding matrices of rigidity and elasticity in the domain of the linear stress-deformation relationship, which represents the subject of special studies. Here we describe an elastic system by using the results of the direct stiffness method and the displacement method from the group of finite-element methods.

In a multi-robot work involving m manipulators, there are m contact points, so that there also exist the same number of external nodes of the grid thus formed. Hence, the total number of external and internal nodes is $m + 1$. If all the mass of external elements is placed between the grid nodes, then the DOFs of the connecting elements between nodes will be lost and only DOFs of the motion of the nodes will remain.

A minimal number of DOFs of such a grid will correspond to the point masses at the grid nodes and will be $3(m + 1)$. If it is adopted that the grid nodes coincide with the MCs of the rigid bodies that, by their inertial properties, replace the inertial properties of the continual elastic system, then the number of grid DOFs will be $6(m + 1)$. The appropriateness of placing rigid bodies at the grid nodes is related to the simple introduction of external loads consisting of three force components and three moment components, simplicity of the application of finite-element method, and forming a model with easily measurable quantities. In accordance with the above, *it is adopted that the elastic system can be approximated by a spatial grid at the nodes of which act external loads. It is assumed that each node has six DOFs and that the nodes coincide with the MCs of the solid rigid objects as representatives of the inertial properties of the elastic interconnections at the contact and manipulated object, i.e. the total number of DOFs of the elastic system is $6m + 6$* (Figure 12).

Let us consider a choice of coordinate frames suitable to describe the dynamics in cooperative work and elastic system dynamics. In its basic approach, the existing theory of elasticity considers the motion of an elastic system with respect to the unloaded state. *In order to directly apply the results of the theory of elasticity, in the case where the cooperative work involves an object whose unloaded state is immobile, the cooperative manipulation dynamics will be described in the system of coordinates representing displacements of the loaded state with respect to the unloaded state.* The choice of this coordinate frame for the general case of an elastic system motion would mean that the coordinate system should be attached to the unloaded (undeformed) state of the elastic system, to seek the position of that state during the motion, and with respect to it, determine the loaded (deformed) state. The advantage of such a choice of coordinates stems from the developed procedure of the theory of elasticity for describing elastic systems and the possibility of decoupling the motion into a transmission motion of the rigid unloaded system and relative motion due to deformation of the loaded cooperative system. A drawback would be the fact that the introduction of this coordinate frame requires finding the

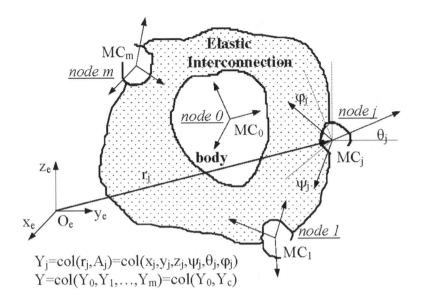

Figure 12. Elastic system

position of the unloaded state 0 during the motion.

In the general case, the manipulated object performs a general motion. The manipulator position is usually given in the internal coordinates. For a known position of a manipulator, i.e. for its known internal coordinates, the position of the manipulator tip is uniquely determined in the external (absolute) coordinate frame. As kinetic energy is described in terms of absolute velocities, it is necessary to know the absolute coordinates of the inertial parts of the system, i.e. of both the manipulator and the elastic system. Hence, the cooperative system dynamics can be conveniently described in absolute coordinates. If the manipulators in cooperative manipulation are non-redundant, the choice of internal coordinates to describe the motion is equivalent to the choice of absolute coordinates. Contact forces arise as a consequence of the instantaneous state of the mobile elastic system, i.e. of the instantaneous state of the absolute coordinates of the elastic system, which is also an advantage of selecting absolute coordinates.

However, there arises the problem of describing the elastic system's motion due to deformation, as the above choice of coordinates assumes that the coordinates of elastic system's loaded state, and not those of the unloaded state, are taken as basic coordinates. Knowing the position of the absolute coordinates of the loaded state, the problem is how to determine the accumulated deformation energy and energy of dissipation that are needed in forming Lagrange equations. In other words, the problem is how the choice of absolute coordinates connected to the deformed state can ensure the description of the deformation of the elastic system and establish connections with the existing theory of elasticity given in the Cartesian system of coordinates attached to the unloaded state.

The objective of introducing assumptions on the cooperative system characteristics is to propose a simple modeling procedure based on specified properties of the system, which would potentially enable avoiding the exact solving of the system of differential equations used to describe deformation of the elastic system.

A specific feature of cooperative work is that it is to be performed in steps. The manipulated object has to be approached to perform its grasping and gripping. Positions of the cooperative system in these work stages have to be stored, so that they can be used as known. Of special importance are all the pieces of information about the non-deformed state of the elastic system. These data are related to the instant of establishing contact between the tips of the manipulators and the object. On the basis of information about the state of internal coordinates of the manipulators it is possible to uniquely determine the positions of all contact points at the moment of contact formation, i.e. positions of the nodes of the unloaded space grid, their mutual distances and orientation, are uniquely determined. For the known masses of the links and manipulated object in the instant of contact formation, it is possible to determine the static position of the elastic spatial grid, for which there is no system of forces acting at it and where all node displacements are equal to zero (in the sequel, unloaded state 0 or only state 0). That position corresponds to the non-deformed space grid, free from the action of any load. The non-deformable grid's motion in space takes place purely geometrically, as the motion of a non-deformable geometric figure. In the real motion, any deformed state of the elastic system whose coordinates during the motion are known, is obtained by deforming that figure, whose exact position during the motion is not known. Although the position of the undeformed state 0 during the motion is not known, information exists about the relative position of its nodes (see Figure 9 and expressions (26) and (27)). Along with the known absolute coordinates of the loaded state, there are enough pieces of information (finite number of coordinates (28), (31) to (33)) for the approximate formation of the expressions for the energies of deformation and dissipation during the deformation as a function of the absolute coordinates.

Therefore, by knowing the absolute coordinates of the unloaded state, along with information about the relative distances and orientation of contact points in the instant when contact between the manipulators and the object is established, the problem of describing the elastic system is simplified, hence *the absolute coordinates are adopted to describe the general motion of the elastic system.*

4.2 Setting Up the Problem of Mathematical Modeling of a Complex Cooperative System

Consider the cooperative work of m rigid manipulators with six DOFs that are manipulating the object whose general motion in three-dimensional space takes place without any constraint. It is assumed that either the manipulator-object connections are elastic and the object is rigid or that the object is elastic (Figure 1). For both cases, it is assumed that each connection or the part of the manipulated object in the vicinity of the contact point can be represented by a rigid body and its elastic environment. Contact forces, gravitation forces, and the forces of damping and elasticity are acting at the MC of the rigid body.

A separate body with an elastic environment can be considered as a system of $m + 1$ elastically interconnected rigid bodies (Figure 12). Thus, the system formed has $m + 1$ nodes. Let the elastic system nodes and rigid bodies MCs coincide, and let six DOFs of motion be allowed to each body. Displacement of the elastic system at the nodes is identical to the displacement of the rigid body whose MC is connected to that node. Gravitational and contact forces are considered as external forces acting at the object MC, i.e. they represent the acting forces of the elastic structure at its nodes (Figure 13).

The following task is encountered: assuming that the mutual positions of the manipulator tips at the moment of formation of the manipulator-object contact are known, derive a mathematical model of the cooperative system motion describing the steps of gripping, lifting, general motion, lowering, and releasing.

To model the general motion of a cooperative system consisting of m rigid manipulators with six DOFs, rigid object, and elastic interconnections between the manipulator tips and the object, it is necessary to do the following:

- Carry out modeling of the dynamics of the cooperative work for which unloaded state 0 of the elastic system is immobile during the cooperative work.

- Knowing the mutual position of the manipulator tips at the moment of forming the manipulator-object contact, carry out mathematical modeling of the cooperative manipulation dynamics for the general case of motion of the manipulated object as a function of the absolute coordinates of the contact points and the MC of the manipulated object.

- Synthesize a mathematical model of the cooperative system from the previously derived mathematical models for elastic system, manipulators, and relations that describe kinematic connections and load at the connections of these subsystems.

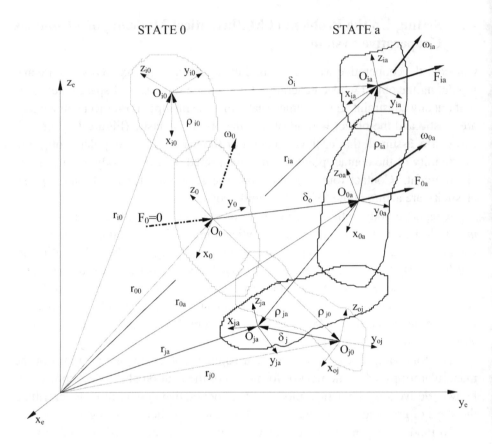

Figure 13. Displacements of the elastic system nodes – the notation system

The mathematical model has to be formed in such a way that the same model sufficiently describes the cooperative system under static and dynamic conditions. Neglecting mass, damping, and elastic properties in the model of the general motion of the elastic system, model the general motion of the manipulated object as a rigid body. Zero values of all elastic forces must be obtainable whenever the loaded state coincides with the unloaded state 0.

4.3   Theoretical Bases of the Modeling of an Elastic System

The motion of the cooperative system can be considered as the motion of a system of bodies in a complex field of forces composed of the fields of gravitational and elastic forces. The gravitational field is of a potential character, whereas the potentiality of the elastic field holds only for the linear stress-deformation relationship in the close neighborhood of the unloaded state 0. To form the equations of motion in this field, it is necessary to know the accumulated potential energy in the system of bodies. If the potential gravitational forces (they constantly act in the direction of the Oz axis of the absolute coordinate frame) in Lagrange's equations are associated with non-potential forces and are considered as a system of unknown external forces, then it remains only to determine the potential energy of the elastic forces (deformation energy).

The purpose of the introduced assumptions and proposed modeling procedure is to avoid solving a system of equations that describes the deformation of the elastic system and, using approximate methods, derive the model of the cooperative system only on the basis of known absolute coordinates of the MCs and their derivatives, along with gripping points at the initial moment, i.e. the contact points (tips of the manipulators) and the MC of the manipulated object.

The idea of modeling in the system of absolute coordinates is based on the following. As it is assumed that all the mass is concentrated at the elastic system nodes, inertial and external forces (represented by gravitational and contact forces) act at these nodes. The links between particular nodes are massless, so that the dissipation forces of the elastic system are also associated with the forces at other nodes. As we do not deal with manipulation in a resistive environment, there are no surface resistance forces. Hence, the forces acting at each node can be replaced with one resulting force. These resulting forces act at the nodes of the elastic system. To each deformed state corresponds only one system of node forces. The instantaneous deformed state can be obtained by static deformation of the unloaded state 0 involving the same system of forces, which enables one to calculate deformation energy by using static procedures. The work of the external forces is equal to the work of the internal forces, i.e. to the deformation energy. Components of the balancing elastic forces are equal to the derivatives of deformation work with respect to the corresponding coordinate and are equal to the components of the resulting node forces. The resulting forces are decomposed along the axis of the absolute coordinate frame, so that deformation energy also has to be expressed in the same coordinate frame as a global frame for the elastic system.

Deformation energy is a function of the properties of the concrete shape and elastic system material, and can be determined using exact or approximate methods. When adopting assumptions needed to form the mathematical model, cooperative manipulation should be considered as a system with a finite number of DOFs and, hence, the deformation energy (i.e. the stiffness matrix) should be determined by some approximate methods.

The basic notion of describing the deformation energy via the absolute coordinates will be illustrated using the finite-element method.

The theoretical basis of the finite-element method is the principle of minimum energy for varying displacements (the principle of virtual displacements), whereby the increments of the works of the external and internal forces are the same [6, 7, 23, 24]. Generally, the method consists of decomposing the structure into characteristic elementary finite elements, separately forming stiffness equations for each of the finite elements in the local coordinate frame, and forming equations of global stiffness of the overall structure in the joint (global) coordinate frame for all the elements, whereby it is necessary to take into account the conditions of interconnection of the finite elements into a whole (the conditions of force equilibrium and compatibility of displacements).

The procedures of forming equations of individual stiffnesses have been described in detail in [6, 7, 23, 24]. The equation of individual stiffness of the ith finite element is of the form

$$F^{ei} = K^{ei} \Delta^{ei}, \qquad K^{ei} = (K^{ei})^T, \tag{47}$$

where F^{ei} is the vector of node forces acting on a finite element; K^{ei} is the square matrix of individual stiffness of the finite element, and Δ^{ei} is the vector of node displacements of the finite element that defines the number of DOFs of the finite-element motion in the direction of the node force F^{ei}. The number of DOFs of the finite-element motion depends on the choice of the type of load or displacement that is to be taken into consideration. For a given choice of displacements Δ^{ei}, the matrix K^{ei} for one finite element is determined only once. If the stiffness matrix also contains the motion modes of the finite element as a rigid body, then it is singular. If this equation is given in the local coordinate frame of the finite element attached to the element position in the elastic structure, an orthogonal transformation has to be applied to transpose it into the global coordinate frame.

By uniting all the equations of the finite elements, one obtains the following system of equations

$$F^e = K^e \Delta^e, \qquad K^e = (K^e)^T, \tag{48}$$

with the disassembled stiffness matrix $K^e = \text{diag}(K^{e1}, K^{e2}, K^{e3}, \ldots)$, and expanded vectors of the force $F^e = \text{column}(F^{e1}, F^{e2}, F^{e3}, \ldots)$ and displacement $\Delta^e = \text{column}(\Delta^{e1}, \Delta^{e2}, \Delta^{e3}, \ldots)$. If the equations of stiffness of each finite element are given in a global coordinate frame, the conditions of structure assembly are reduced to equating the forces and displacements of the finite elements at the common mode and eliminating redundant rows and columns from the disassembled stiffness matrix (method of direct stiffness [7]). On the contrary, one seeks the matrix a of the global kinematic conditions of the connection of node displacements (continuity) of the elastic structure Δ in the common (global) coordinate frame of the node displacements of finite elements Δ^e which, for the statically determined systems ($a = a_0$), is represented by the algebraic relation (displacement method)

$$\Delta^e = a_0 \Delta. \tag{49}$$

This relation defines how the finite elements are assembled in the structure and it is easily obtained for the statically determined systems. Elements of the matrix a_0 are obtained by considering the geometry of the relation between the node displacements of finite elements Δ^e and individual unit displacements in the direction of each displacement Δ_i as known, whereby all the other displacements Δ_j, $j \neq i$ are zero. If the system is statically indeterminate, then from the viewpoint of kinematics, the system is indeterminate too. Then, it is not possible to define the above relation on the basis of kinematic observations but is necessary to also take into account the equilibrium conditions from which kinematically indeterminate quantities are associated with the displacements Δ. If the kinematically uncertain quantities are denoted by Δ_k, the preceding relation will acquire the form

$$\Delta^e = a_0 \Delta + a_1 \cdot \Delta_k, \tag{50}$$

where a_1 denotes the matrix by which the function describing how finite elements are connected in the structure is supplemented by kinematically indeterminate quantities Δ_k. The forces F^e act at the location and in the direction of the displacement Δ^e. These forces must act on the overall structure because of the existence of node displacements of the overall structure Δ, and they are

$$F^e = K^e \Delta^e = K^e a_0 \Delta + K^e a_1 \Delta_k. \tag{51}$$

Using the principle of virtual displacements and considering the equations of variation of the unknown and independent kinematically indeterminate quantities Δ_k for the given constant displacements Δ, one obtains

$$a_1^T F^e = a_1^T K^e a_0 \Delta + a_1^T K^e a_1 \Delta_k = 0, \tag{52}$$

because the work of external forces is realized only on the given displacements Δ. From this we have

$$\Delta_k = -(a_1^T K^e a_1)^{-1} a_1^T K^e a_0 \Delta, \tag{53}$$

so that the relationship between the node displacements Δ^e and predetermined displacements Δ for the statically indeterminate (kinematically indeterminate) system is

$$\Delta^e = a\Delta, \quad a = a_0 - (a_1^T K^e a_1)^{-1} a_1^T K^e a_0. \tag{54}$$

Taking into account that the node forces F^Δ acting in the direction of displacement Δ are related to the forces F^e via the transformation matrix b (it can be determined by the force method)

$$F^\Delta = b F^e, \tag{55}$$

the total deformation energy in the deformed elastic structure generated from the moment of the beginning of deformation $\Delta^e = 0$ to the final deformation state, caused by the given displacements Δ and resulting node displacements Δ^e, is determined by

$$
\begin{aligned}
A_d &= \frac{1}{2}(F^e)^T \Delta^e = \frac{1}{2}(F^e)^T a \Delta = \frac{1}{2}\Delta^T a^T K^e a \Delta \\
&= \frac{1}{2}\Delta^T a^T K^e a \Delta = \frac{1}{2}\Delta^T K^\Delta \Delta, \quad K^\Delta = a^T K^e a,
\end{aligned}
\tag{56}
$$

where K^Δ is the stiffness matrix of the overall elastic structure defined in the global coordinate frame Δ. The work of the node forces F^Δ on the displacements Δ is determined by

$$
A_F = \frac{1}{2}(F^\Delta)^T \Delta = \frac{1}{2}(F^e)^T b^T \Delta.
\tag{57}
$$

As the works of the forces are the same, then $a = b^T$ and

$$
\frac{1}{2}(F^\Delta)^T \Delta = \frac{1}{2}\Delta^T a^T K^e a \Delta, \quad \Rightarrow F^\Delta = a^T K^e a \Delta = K^\Delta \Delta.
\tag{58}
$$

If the work expressions are known, i.e. if the stiffness matrices and node displacements are determined, the forces F^e along the displacement Δ^e and the forces F^Δ along the displacement Δ are

$$
F^e = \frac{\partial A_d}{\partial \Delta^e} = K^e \Delta^e, \quad F^\Delta = \frac{\partial A_d}{\partial \Delta} = 2\frac{\partial A_F}{\partial \Delta} = a^T K^e a \Delta = K^\Delta \Delta.
\tag{59}
$$

The matrix K^e is a constant diagonal block matrix. The matrix $K^\Delta = a^T K^e a$ is also a constant matrix because the elastic system deformation is considered with respect to the immobile non-deformed state of the system. This means that if the stiffness matrix of the system is known, the forces along the given values of the displacement can be uniquely determined. Also, if the forces along the given displacements are known, the work of internal forces (as a scalar) is uniquely determined. As the strains of elastic structures are considered with respect to the immobile unloaded state, it is customary to measure the components of the vector Δ in the global coordinate frame attached to that state. It is also assumed that the states of the elastic structure before and after the deformation are kinematically determined (the positions of the supports and their displacements are known).

The above discussion has been related to the modeling of an elastic system for the immobile unloaded state 0. Detailed standardized procedures of forming a stiffness matrix for concrete elastic systems can be found in the literature concerning the theory of elastic systems [6, 7]. If the displacements Δ are expressed in external coordinates of the cooperative system, then the results of the theory of elasticity can be used without any alteration to model those phases of the co-operative system's motion in which the unloaded state 0 of the elastic system is immobile. These phases are the gripping and releasing of the immobile object.

In the case of the immobile unloaded state 0, it is necessary to find an expression for the deformation energy in the system of absolute coordinates, which are global coordinates for the elastic system.

In cooperative manipulation, the distances of the nodes $\|\rho_{ij0}\|$, $i, j = 0, 1, \ldots, m$ are known, as well as the relative orientation of the bodies with the MC at the nodes $\|\mathcal{A}_{ij0}\|$, $i, j = 0, 1, \ldots, m$ of the elastic system before deformation and absolute coordinates of the nodes after deformation Y. Let the relation between the given displacements Δ and these quantities be defined as

$$\Delta = \Delta(Y, \|\rho_{ij0}\|_{i,j=0,\ldots,m}, \|\mathcal{A}_{ij0}\|_{i,j=0,\ldots,m}) = \Delta(Y),$$

$$\|\rho_{ij0}\| = \text{const}, \quad \|\mathcal{A}\|_{ij0} = \text{const}, \quad i, j = 0, \ldots, m. \tag{60}$$

Deformation work of the elastic system determined in the coordinate frame Δ by the relation (56), in the new coordinate frame Y, will be

$$A_d = \frac{1}{2}(F^\Delta)^T \Delta = \frac{1}{2}\Delta^T(Y)K^\Delta\Delta(Y) = A_d(\Delta(Y)) = A_d(Y). \tag{61}$$

As the derivative of the scalar function A_d with respect to the vector argument, Y is the vector of the function Δ as an argument of the scalar function A_d determined by

$$\frac{\partial A_d(\Delta(Y))}{\partial Y} = \left(\frac{\partial \Delta}{\partial Y}\right)^T \frac{\partial A_d}{\partial \Delta}, \tag{62}$$

the resulting node forces F^Y along the displacement Y are

$$F^Y = \frac{\partial A_d}{\partial Y} = \left(\frac{\partial \Delta}{\partial Y}\right)^T F^\Delta. \tag{63}$$

If the coordinates Y are expressed as a function of the coordinates Δ and transformation matrix $c(Y)$ by the relation

$$\Delta = c(Y) \cdot Y \quad \Rightarrow \quad \left(\frac{\partial \Delta}{\partial Y}\right)^T = \left(\frac{\partial c(Y)}{\partial Y}Y\right)^T + c^T(Y), \tag{64}$$

the deformation work will be

$$A_d(Y) = \frac{1}{2}Y^T c^T(Y)K^\Delta c(Y)Y = \frac{1}{2}Y^T \pi(Y)Y, \tag{65}$$

where $\pi(Y) = c^T(Y)K^\Delta c(Y)$.

The force F^Y in the direction of the displacement Y is obtained by introducing (64) and (59) into (63) or by differentiating (65)

$$
\begin{aligned}
F^Y &= \left[\left(\frac{\partial c(Y)}{\partial Y}Y\right)^T + c^T(Y)\right]K^\Delta c(Y)Y \\[2mm]
&= \frac{1}{2}Y^T\frac{\partial \pi(Y)}{\partial Y}K^\Delta c(Y)Y + c^T(Y)K^\Delta c(Y)Y \\[2mm]
&= \frac{1}{2}Y^T\frac{\partial}{\partial Y}(c^T(Y)K^\Delta c(Y))Y + c^T(Y)K^\Delta c(Y)Y \\[2mm]
&= \frac{1}{2}Y^T\frac{\partial \pi(Y)}{\partial Y}Y + \pi(Y)Y
\end{aligned}
\tag{66}
$$

or, in a shorter form,

$$
F^Y = \frac{\partial A_d(Y)}{\partial Y} = K(Y)\cdot Y, \quad K(Y) = \left[\left(\frac{\partial c(Y)}{\partial Y}Y\right)^T + c^T(Y)\right]K^\Delta c(Y), \tag{67}
$$

where $K(Y)$ is the generalized stiffness matrix, dependent of the generalized coordinates Y.

Therefore, to form the expression for deformation work (65) or for the resulting node forces of the elastic system in the absolute coordinates (66), it is necessary to determine the relationship between the elastic displacements of the elastic system nodes and absolute coordinates (60), as well as the stiffness matrix of the assembled system K^Δ.

The stiffness matrix of assembled system K^Δ is identical to the stiffness matrix of the elastic system considered with respect to the immobile unloaded state. This matrix is determined by the usual methods of the theory of elasticity. If the method of finite elements is used, it is necessary to divide the elastic system into characteristic finite elements, choose for each of them the local representative displacements Δ^{ei}, determine individual stiffness matrices K^{ei}, and finally, determine the relation $\Delta^e = a\Delta$ for connecting the finite elements into a unique elastic system. The procedure can be carried out only for a concrete known structure of the elastic system. This problem in cooperative manipulation can be overcome if the elastic properties are assigned to the tips of the manipulator grippers only. Then, it is possible to choose in advance the suitable forms of the elastic tips of the grippers as finite elements and determine the matrices of individual stiffness for them in advance. The synthesis of the stiffness matrix of the composed system would reduce to forming efficient on-line algorithms for connecting such finite elements and manipulated object into a unique whole, which can be the subject of future research.

It is not simple to establish a relationship between the elastic displacements of elastic system nodes and absolute coordinates (60). If that relation is of the form (64), the deformation work and forces at the nodes will be of the form (65), (66), (67).

It is necessary to describe the method of forming deformation work (65). The basic goal of the methods of the theory of elasticity is to establish a relationship between the known load of the elastic system and unknown elastic displacements, or between the known elastic displacements and unknown acting load of the elastic system. After establishing these relations, the internal strain and support reactions are determined.

To model the general motion of the cooperative system, it is necessary to express deformation work and/or node forces as a function of absolute coordinates of the loaded elastic system in the form of (65), (66), (67). This can be done without finding the transformations (60) or (64).

The basic idea in describing deformation energy with the aid of absolute coordinates is the following. On the basis of the known instantaneous positions of the nodes of a loaded elastic system and positions of the nodes at the moment of object gripping, instantaneous relative displacements of nodes are found. For the known values of instantaneous relative displacements of nodes, internal forces between them are determined. The deformation energy of the elastic system is determined as one-half of the sum of the products of internal forces and the corresponding relative displacements of the nodes.

Namely, deformation work is the work of the internal forces (strains) and is a function of the relative displacements of nodes as known quantities (see relations (27), (28), (29) and (31)).

$$A_d = \frac{1}{2}\epsilon^T F_{\text{int}} = \frac{1}{2}\epsilon^T \pi_\epsilon \epsilon, \tag{68}$$

where $F_{\text{int}} = F_{\text{int}}(Y)$ and $\epsilon = \epsilon(Y)$ are the vectors of internal forces and relative displacements of nodes (deformation) of the elastic system, and π_ϵ is the constant diagonal matrix in the direction of the action of the internal forces.

Members of the matrix π_ϵ are uniquely calculated as a function of the stiffness matrix K' and spatial characteristics of the unloaded elastic system. Characteristics of the elastic system depend on the type of contact, geometric configuration of contact points, and elastic properties of the object and tips of the manipulators. To each different elastic system corresponds a different stiffness matrix and, consequently, a different matrix π_ϵ.

The procedure to calculate stiffness members of the matrix π_ϵ as a function of the members of the stiffness matrix K is the subject of the theory of elasticity. In deriving the model of cooperative system dynamics, it is essential that this relationship is unique and that stiffness members of the matrix π_ϵ are constant for a concrete elastic system. To illustrate the modeling procedure the adopted members of the matrix π_ϵ (given in Appendix B) represent the stiffness of linear and torsion springs between any two nodes of the elastic system, without determining their values by the procedures of the theory of elasticity for the concrete elastic system.

The relationship between the mutual displacements of elastic system nodes ϵ and their absolute coordinates Y are relatively easily established (see (27)). As a result, one obtains the expression for the deformation energy whose mathematical form is identical to expression (65) (see (33), i.e. the same effect is achieved as in determining the transformation of coordinates (60).

4.4   Elastic System Deformations as a Function of Absolute Coordinates

Let the elastic system be driven out of the state 0 and let the corresponding displacements of the nodes y_i be given as (Figure 13)

$$y_i = \begin{pmatrix} \delta_i \\ \mathcal{A}_i \end{pmatrix} = \begin{pmatrix} r_{ia} - r_{i0} \\ \mathcal{A}_{ia} - \mathcal{A}_{i0} \end{pmatrix} \in R^{6\times 1}, \quad i = 0, 1, \dots, m, \tag{69}$$

where r_{ia} and r_{i0} are the respective position vectors of the instantaneous MC and the MC in the state 0 of the ith body in the three-dimensional Cartesian space, while $\mathcal{A}_{ia}$ and $\mathcal{A}_{i0}$ are the orientation vectors of the instantaneous state and state 0 of the ith body measured by the angular displacement of the coordinate frame with the origin at the MC and axes directed along the main inertia axes with respect to the Cartesian coordinate frame. The subscript $i = 0$ relates to the rigid object handled by the manipulators, whereas the subscripts $i = 1, \dots, m$ refer to the elastic interconnection.

According to (69), it is obvious that there are two different ways of deforming an elastic system:

- deforming the elastic system around its immobile unloaded state

$$\begin{matrix} r_{i0} = \text{const} \\ \mathcal{A}_{i0} = \text{const} \end{matrix} \quad \Rightarrow \quad \begin{matrix} \dot{r}_{i0} = 0 \\ \dot{\mathcal{A}}_{i0} = 0, \end{matrix} \tag{70}$$

- deforming the elastic system around its mobile unloaded state

$$\begin{matrix} r_{i0} \neq \text{const} \\ \mathcal{A}_{i0} \neq \text{const} \end{matrix} \quad \Rightarrow \quad \begin{matrix} \dot{r}_{i0} \neq 0 \\ \dot{\mathcal{A}}_{i0} \neq 0. \end{matrix} \tag{71}$$

The state 0 can be represented by only one coordinate frame which is adopted as the $O_0x_0y_0z_0$ frame, with the origin attached to the manipulated object MC, whose orientation at the given moment is given by the vector $\mathcal{A}^0 = \mathcal{A}^0_0 = \mathcal{A}_{00} = \mathrm{col}(\psi_{00}\ \theta_{00}\ \phi_{00})$, and the position of the coordinate frame origin is r_{00}. The other coordinate frames $O_{i0}x_{i0}y_{i0}z_{i0}$, whose origins are placed at the MCs of elastic interconnections, i.e. at contact points, are rotated with respect to $O_0x_0y_0z_0$ by a constant value of orientation $\mathcal{A}^0_i$ during all the time of the motion of the unloaded state 0 (Figure 14). This means that the coordinate frames $O_{i0}x_{i0}y_{i0}z_{i0}$ for the unloaded state of the elastic system may have a different orientation $\mathcal{A}^0_0, \mathcal{A}^0_0 + \mathcal{A}^0_1, \ldots, \mathcal{A}^0_0 + \mathcal{A}^0_m$.

According to (69) and Figures 13 and 14, the following vector relations hold

$$r_{ia} = r_{i0} + \delta_i, \qquad \dot{r}_{ia} = \dot{r}_{i0} + \dot{\delta}_i, \quad i = 0, 1, \ldots, m,$$

$$\mathcal{A}_{ia} = \mathcal{A}_{i0} + \mathcal{A}_i, \qquad \dot{\mathcal{A}}_{ia} = \dot{\mathcal{A}}_{i0} + \dot{\mathcal{A}}_i, \quad i = 0, 1, \ldots, m, \tag{72}$$

where r_{i0}, r_{ia} and $\mathcal{A}_{i0}$, $\mathcal{A}_{ia}$ are the MC positions and orientations of the ith body in the unloaded and loaded states; δ_i and $\mathcal{A}_i$ are the MC displacements and change of orientation of the ith body in the case of deformation. The dot over a quantity denotes the time derivative of that quantity.

For the unloaded state we have

$$r_{i0} = r_{00} + \rho_{i0}, \qquad \dot{r}_{i0} = \dot{r}_0 + \dot{\rho}_{i0}, \quad i = 1, \ldots, m,$$

$$\mathcal{A}_{i0} = \mathcal{A}^0_0 + \mathcal{A}^0_i = \mathcal{A}^0|_{\mathcal{A}^0_i=0} = \mathcal{A}_{00}|_{\mathcal{A}^0_i=0},$$

$$\dot{\mathcal{A}}_{i0} = \dot{\mathcal{A}}^0_0 + \dot{\mathcal{A}}^0_i = \dot{\mathcal{A}}^0|_{\mathcal{A}^0_i=0} = \dot{\mathcal{A}}_{00}|_{\mathcal{A}^0_i=0}, \tag{73}$$

where ρ_{i0} is the position vector of the MC of the ith object of the elastic system in the unloaded state 0, given with respect to the coordinate origin of the space $O_0x_0y_0z_0$. From this follow the relations for the coordinates and their derivatives of the state 0:

$$r_{00} = r_{0a} - \rho_{00} - \delta_0 = r_{1a} - \rho_{10} - \delta_1 = \ldots = r_{ma} - \rho_{m0} - \delta_m, \quad \rho_{00} = 0,$$

$$\mathcal{A}^0 = \mathcal{A}_{0a} - \mathcal{A}_0 = \ldots = \mathcal{A}_{ia} - \mathcal{A}^0_i - \mathcal{A}_i = \ldots = \mathcal{A}_{ma} - \mathcal{A}^0_m - \mathcal{A}_m,$$

$$\dot{r}_{00} = \dot{r}_{0a} - \dot{\delta}_0 = \dot{r}_{1a} - \dot{\rho}_{10} - \dot{\delta}_1 = \ldots = \dot{r}_{ma} - \dot{\rho}_{m0} - \dot{\delta}_m,$$

$$\dot{\mathcal{A}}^0 = \dot{\mathcal{A}}_{0a} - \dot{\mathcal{A}}_0 = \ldots = \dot{\mathcal{A}}_{ia} - \dot{\mathcal{A}}^0_i - \dot{\mathcal{A}}_i = \ldots = \dot{\mathcal{A}}_{ma} - \dot{\mathcal{A}}^0_m - \dot{\mathcal{A}}_m. \tag{74}$$

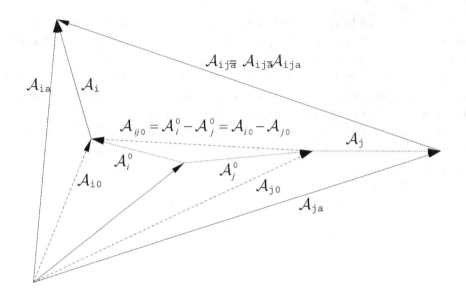

Figure 14. Angular displacements of the elastic system

Relative angular displacement of two arbitrary bodies is defined by the difference of absolute values of their orientation (Figure 14). Let the bodies before elastic system deformation have the same orientation $\mathcal{A}^0$ for $\mathcal{A}_i^0 = 0$, $i = 1, \ldots, m$, or let their orientations differ by a constant value $\mathcal{A}_{ij}^0 = \mathcal{A}_i^0 - \mathcal{A}_j^0, = \mathcal{A}_{i0} - \mathcal{A}_{j0}$, $\mathcal{A}_i^0 \neq \mathcal{A}_j^0$. Starting from the state with the orientation $\mathcal{A}_{i0}$, three (and four if $\mathcal{A}_{ij}^0 \neq 0$), successive changes of orientation yield the same state. Let the initial state have the orientation $\mathcal{A}_{i0}$. By changing the orientation to $\mathcal{A}_i = \mathcal{A}_{ia} - \mathcal{A}_{i0}$, the state with the absolute orientation $\mathcal{A}_{ia} = \mathcal{A}_{i0} + (\mathcal{A}_{ia} - \mathcal{A}_{i0})$ is attained. By changing the orientation to $-\mathcal{A}_{ija} = -(\mathcal{A}_{ia} - \mathcal{A}_{ja})$, the attained orientation of the jth body is $\mathcal{A}_{ja} = \mathcal{A}_{i0} + (\mathcal{A}_{ia} - \mathcal{A}_{i0}) - (\mathcal{A}_{ia} - \mathcal{A}_{ja})$. After a further change of orientation to $-\mathcal{A}_j = -(\mathcal{A}_{ja} - \mathcal{A}_{j0})$ and $\mathcal{A}_{ij0} = (\mathcal{A}_{i0} - \mathcal{A}_{j0}) = (\mathcal{A}_i^0 - \mathcal{A}_j^0)$, the resulting states will be of the orientation $\mathcal{A}_{j0} = \mathcal{A}_{i0} + (\mathcal{A}_{ia} - \mathcal{A}_{i0}) - (\mathcal{A}_{ia} - \mathcal{A}_{ja}) - (\mathcal{A}_{ja} - \mathcal{A}_{j0})$ and $\mathcal{A}_{fin} = \mathcal{A}_{i0} + (\mathcal{A}_{ia} - \mathcal{A}_{i0}) - (\mathcal{A}_{ia} - \mathcal{A}_{ja}) - (\mathcal{A}_{ja} - \mathcal{A}_{j0}) + (\mathcal{A}_{i0} - \mathcal{A}_{j0})$, respectively. Simple adding gives $\mathcal{A}_{fin} = \mathcal{A}_{i0}$, i.e. we return to the state with the initial value of orientation. Hence, it comes out that the change of relative orientation of two arbitrary bodies attained in the loaded state is defined by the difference of the absolute values of their orientations $\mathcal{A}_{ij} = \mathcal{A}_i - \mathcal{A}_j = \mathcal{A}_{ia} - \mathcal{A}_{i0} - (\mathcal{A}_{ja} - \mathcal{A}_{j0}) = \mathcal{A}_{ia} - \mathcal{A}_{ja}$ $-(\mathcal{A}_{i0} - \mathcal{A}_{j0}) = \mathcal{A}_{ia} - \mathcal{A}_{ja} - \mathcal{A}_{ij0} = \mathcal{A}_{ia} - \mathcal{A}_{ja}|_{\mathcal{A}_i^0 = \mathcal{A}_j^0}$. To achieve a more legible presentation, we assume that all coordinate frames $O_{i0}x_{i0}y_{i0}z_{i0}$ have the same orientation for the unloaded state of the elastic system, $\mathcal{A}_i^0 = 0$, $i = 1, \ldots, m$.

Relative displacements of the points of a loaded elastic system are defined by the difference of displacements of the points with respect to the unloaded state

$$
\begin{aligned}
\delta_{ij} &= -\delta_{ji} = \delta_i - \delta_j = r_{ia} - r_{i0} - (r_{ja} - r_{j0}) \\
&= r_{ia} - r_{ja} - (r_{i0} - r_{j0}) = \rho_{ija} - \rho_{ij0} \\
\mathcal{A}_{ij} &= -\mathcal{A}_{ji} = \mathcal{A}_{ia} - \mathcal{A}_{ja} = \mathcal{A}_i - \mathcal{A}_j|_{\mathcal{A}_i^0=0,\,\mathcal{A}_j^0=0}, \quad i,j = 0,1,\ldots,m, \quad (75)
\end{aligned}
$$

where ρ_{ija} and ρ_{ij0} are the relative position vectors of the ith and jth object MC in the loaded and unloaded state of the elastic system, given with respect to the coordinate frame attached to the MC of the manipulated object.

The rate of displacement of the points of the loaded elastic system with respect to the unloaded state is

$$
\dot{\delta}_i = \dot{r}_{ia} - \dot{r}_{i0} = \dot{r}_{ia} - \dot{r}_0 - \omega_0 \times \rho_{i0},
$$

$$
\dot{\mathcal{A}}_i = \dot{\mathcal{A}}_{ia} - \dot{\mathcal{A}}_{i0} = \dot{\mathcal{A}}_{ia} - \dot{\mathcal{A}}^0|_{\mathcal{A}_i^0=0}, \quad (76)
$$

where ω_0 is the angular velocity of the figure formed by the elastic system nodes in the state 0.

As there is no force system acting in the unloaded state, the state 0 moves as a rigid body, so that

$$
\|\rho_{ij0}\| = \|\rho_{i0} - \rho_{j0}\| = \text{const}, \quad \dot{\rho}_{ij0} = \omega_0 \times \rho_{ij0} = -\rho_{ij0} \times \omega_0, \quad (77)
$$

i.e., the vector ρ_{ij0} has a constant intensity and changeable direction. In the case of a translatory motion of the unloaded state 0, we have $\omega_0 = 0$ and $\rho_{ij0} = \text{const}$, and in the case when the MCs of the ith and jth body coincide in the state 0, then $\rho_{ij0} = 0$.

The rate of change of the distance between the loaded state point is

$$
\begin{aligned}
\dot{\delta}_{ij} &= \dot{\delta}_i - \dot{\delta}_j = \dot{r}_{ia} - \dot{r}_{ja} - \dot{\rho}_{ij0} \\
&= \dot{r}_{ia} - \dot{r}_{ja} - \omega_0 \times \rho_{ij0} = \dot{r}_{ia} - \dot{r}_{ja} + \rho_{ij0} \times \omega_0, \\
\dot{\mathcal{A}}_{ij} &= \dot{\mathcal{A}}_{ia} - \dot{\mathcal{A}}_{ja}, \quad (78)
\end{aligned}
$$

By joining (75) and (78), we obtain the relation for displacements and displacement rates between the points of arbitrary MCs of the elastic interconnections and manipulated object (Figure 15).

$$
\begin{aligned}
y_{ij} &= \begin{bmatrix} \delta_{ij} \\ \mathcal{A}_{ij} \end{bmatrix} = \begin{bmatrix} (\delta_{ij}^D + \delta_{ij}^R) \\ \mathcal{A}_{ij} \end{bmatrix} \\
&= \begin{bmatrix} r_{ia} - r_{ja} - \rho_{ij0} \\ \mathcal{A}_{ia} - \mathcal{A}_{ja} \end{bmatrix} = \begin{bmatrix} \rho_{ija} - \rho_{ij0} \\ \mathcal{A}_{ia} - \mathcal{A}_{ja} \end{bmatrix}, \quad (79)
\end{aligned}
$$

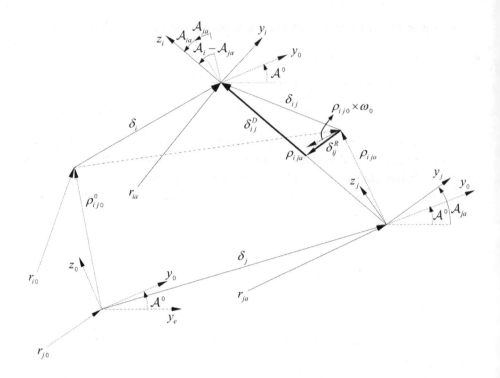

Figure 15. Displacements of the elastic system

$$\dot{y}_{ij} = \begin{bmatrix} \dot{\delta}_{ij} \\ \mathcal{A}_{ij} \end{bmatrix} = \begin{bmatrix} (\dot{\delta}_{ij}^{D} + \dot{\delta}_{ij}^{R}) \\ \mathcal{A}_{ij} \end{bmatrix} = \begin{bmatrix} \dot{r}_{ia} - \dot{r}_{ja} + \rho_{ij0} \times \omega_0 \\ \mathcal{A}_{ia} - \mathcal{A}_{ja} \end{bmatrix}. \qquad (80)$$

In the general case, in relations (79) and (80), the vector ρ_{ij0} and its derivative are unknown quantities. It is known that the intensity of this vector does not change with respect to the one at the moment of formation of contact between the manipulators and the object. In other words, if the stored configuration of contact points in the beginning of the gripping phase is ρ_{ij0}^{0}, then the intensities of the initial radius vectors of mutual displacements of node positions are known. The vector ρ_{ij0} is obtained by rotating the vector ρ_{ij0}^{0}, corresponding to the beginning of the gripping phase for an instantaneous value of orientation $\mathcal{A}^{0}$ of the unloaded state 0. This means that the only unknown quantities are the instantaneous orientation $\mathcal{A}^{0}$ and angular velocity ω_0 of the unloaded state 0.

The vector δ_{ij} can be decomposed into the component δ_{ij}^{D} that is colinear to the vector $r_{ia} - r_{ja} = \rho_{ija}$ and the component δ_{ij}^{R} non-colinear to this vector (Figure 15). The component δ_{ij}^{D} reflects the linear change of the distance between the nodes, whereas the component δ_{ij}^{R} reflects the rotation of the deformed state, i.e.

the curvilinear coordinates of the loaded state. From the point of view of vector calculus, the component δ_{ij}^R reflects the rotation of the vector ρ_{ij0} with respect to the direction defined by the vector ρ_{ija}.

In accordance with the above, the vector y_{ij} is decomposed into the components

$$y_{ij} = y_{ij}^D + y_{ij}^R = \begin{bmatrix} \delta_{ij}^D \\ \mathcal{A}_{ij} \end{bmatrix} + \begin{bmatrix} \delta_{ij}^R \\ 0 \end{bmatrix}. \tag{81}$$

The vector y_{ij} is used to define the potential energy of the elastic system, i.e. its deformation work, and it should be expressed as a function of absolute coordinates only.

As the intensity $\|\rho_{ij0}\|$ does not change, the component y_{ij}^D is defined by

$$
\begin{aligned}
y_{ij}^D &= y_{ij}^D(Y) = \begin{bmatrix} \rho_{ija} - \|\rho_{ij0}\| \dfrac{\rho_{ija}}{\|\rho_{ija}\|} \\ \mathcal{A}_{ia} - \mathcal{A}_{ja} \end{bmatrix} = \begin{bmatrix} \rho_{ija} - \|\rho_{ij0}\| \dfrac{r_{ia} - r_{ja}}{\|r_{ia} - r_{ja}\|} \\ \mathcal{A}_{ia} - \mathcal{A}_{ja} \end{bmatrix} \\
&= \begin{bmatrix} (1 - \dfrac{\|\rho_{ij0}\|}{\|r_{ia} - r_{ja}\|})(r_{ia} - r_{ja}) \\ \mathcal{A}_{ia} - \mathcal{A}_{ja} \end{bmatrix},
\end{aligned} \tag{82}
$$

i.e., by

$$y_{ij}^D = \Lambda_{ij}(Y_i, Y_j)(Y_i - Y_j), \tag{83}$$

where $Y_i = \mathrm{col}(r_{ia}, \mathcal{A}_{ia})$, $i = 0, 1, \ldots, m$ and

$$\Lambda_{ij}(Y_i, Y_j)) = \begin{bmatrix} 1 - \frac{\|\rho_{ij0}\|}{\|r_{ia}-r_{ja}\|} & 0 & 0 & 0 & 0 & 0 \\ 0 & 1 - \frac{\|\rho_{ij0}\|}{\|r_{ia}-r_{ja}\|} & 0 & 0 & 0 & 0 \\ 0 & 0 & 1 - \frac{\|\rho_{ij0}\|}{\|r_{ia}-r_{ja}\|} & 0 & 0 & 0 \\ 0 & 0 & 0 & 1 & 0 & 0 \\ 0 & 0 & 0 & 0 & 1 & 0 \\ 0 & 0 & 0 & 0 & 0 & 1 \end{bmatrix},$$

$$\det \Lambda_{ij}(Y_i, Y_j)) \neq 0. \tag{84}$$

The vector y_{ij}^D (i.e., its component δ_{ij}^D), determined by (82), becomes indeterminate for $\|\rho_{ija}\| = 0$. That case corresponds to the coincidence of the MCs of the ith and the jth body. As all the bodies are by assumption solid and rigid, that case is not possible. For each passage of the elastic structure through the unloaded state 0, the vector y_{ij}^D becomes zero. In that case $\|\rho_{ij0}\| = \|\rho_{ija}\|$ and $\mathcal{A}_{ia} = \mathcal{A}_{ja} = \mathcal{A}^0$, from which $(1 - \|\rho_{ij0}\|/\|\rho_{ija}\|) = 0$ and $\mathcal{A}_{ia} - \mathcal{A}_{ja} = \mathcal{A}^0 - \mathcal{A}^0 = 0$.

The component y_{ij}^R contains only the term δ_{ij}^R. By combining (82) and (79), we obtain

$$\delta_{ij}^R = \delta_{ij} - \delta_{ij}^D = \|\rho_{ij0}\|\frac{\rho_{ija}}{\|\rho_{ija}\|} - \rho_{ij0}$$

$$y_{ij}^R = \text{diag}\left(\frac{\rho_{ij0}}{\|\rho_{ija}\|}, \frac{\rho_{ij0}}{\|\rho_{ija}\|}, \frac{\rho_{ij0}}{\|\rho_{ija}\|}, 0, 0, 0\right)(Y_i - Y_j) - \text{col}(\rho_{ij0}, 0_3),$$

(85)

where 0_3 denotes the zero vector with three components. In this relation, the unknown is the orientation of the vector ρ_{ij0}, i.e. the orientation $\mathcal{A}^0$ of the unloaded state 0.

By differentiating (82) and (85), we obtain complex expressions for the derivatives of the vectors δ_{ij}^D and δ_{ij}^R. Simpler expressions for these derivatives can be obtained in the following way.

The translational component $\dot{\delta}_{ij}^D$ of the displacement rate $\dot{\delta}_{ij}$ is equal to the projection of that rate onto the direction of the vector ρ_{ija}, which is described by

$$\dot{\delta}_{ij}^D = \frac{(\dot{\delta}_{ij}^T \cdot \rho_{ija})}{\|\rho_{ija}\|} \cdot \frac{\rho_{ija}}{\|\rho_{ija}\|} = (\dot{\delta}_{ij}^T \cdot (r_{ia} - r_{ja})) \cdot \frac{r_{ia} - r_{ja}}{\|r_{ia} - r_{ja}\|^2}$$

$$= \frac{(r_{ia} - r_{ja}) \cdot (r_{ia} - r_{ja})}{(r_{ia} - r_{ja})^T \cdot (r_{ia} - r_{ja})} \cdot \dot{\delta}_{ij}.$$

(86)

In the case of the colinearity of $\dot{\delta}_{ij}$ and ρ_{ija}, $\dot{\delta}_{ij}^R = \rho_{ij0} \times \omega_0 = 0$ and $(\dot{\delta}_{ij}^T \cdot \rho_{ija})/\|\rho_{ija}\| = \|\dot{\delta}_{ij}\|$ and, as the unit vectors are identical, $\dot{\delta}_{ij}/\|\dot{\delta}_{ij}\| = \rho_{ija}/\|\rho_{ija}\|$, then $\dot{\delta}_{ij}^D = \dot{\delta}_{ij} = \dot{r}_{ia} - \dot{r}_{ja}$. For small displacements of the elastic system, the vectors $\rho_{ija} = r_{ia} - r_{ja}$ and $\rho_{ij0} \times \omega_0$ are approximately normal (Figure 15), so that $(\rho_{ij0} \times \omega_0)^T \cdot (r_{ia} - r_{ja}) \approx 0$, wherefrom

$$\dot{\delta}_{ij}^D \simeq \frac{(\dot{r}_{ia} - \dot{r}_{ja})^T \cdot (r_{ia} - r_{ja})}{\|r_{ia} - r_{ja}\|^2} \cdot (r_{ia} - r_{ja})$$

$$= \frac{(r_{ia} - r_{ja}) \cdot (r_{ia} - r_{ja})}{(r_{ia} - r_{ja})^T \cdot (r_{ia} - r_{ja})} \cdot (\dot{r}_{ia} - \dot{r}_{ja}),$$

(87)

i.e.

$$\dot{\delta}_{ij}^D \simeq \mathcal{G}_{ija}(r_{ia}, r_{ja})(\dot{r}_{ia} - \dot{r}_{ja}).$$

(88)

The rotation component is given by

$$\dot{\delta}_{ij}^R = \dot{\delta}_{ij} - \dot{\delta}_{ij}^D = (I - \mathcal{G}_{ija}(r_{ia}, r_{ja}))(\dot{r}_{ia} - \dot{r}_{ja}) + \rho_{ij0} \times \omega_0,$$

(89)

where

$$\mathcal{G}_{ija}(r_{ia}, r_{ja}) = \mathcal{G}_{ija}^T(r_{ia}, r_{ja}) = \frac{(r_{ia} - r_{ja}) \cdot (r_{ia} - r_{ja})}{(r_{ia} - r_{ja})^T \cdot (r_{ia} - r_{ja})} \in R^{3 \times 3}, \qquad (90)$$

$$\det \mathcal{G}_{ija}(r_{ia}, r_{ja}) = \frac{\Gamma(r_{ia} - r_{ja})}{\|r_{ia} - r_{ja}\|^2} = 0, \quad r_{ia} \neq r_{ja}, \quad \text{rank}\,\mathcal{G}_{ija}(r_{ia}, r_{ja}) = 1 \quad (91)$$

and $\Gamma(r_{ia} - r_{ja})$ is the determinant of the matrix $(r_{ia} - r_{ja}) \cdot (r_{ia} - r_{ja})$. In the translational motion of the state 0, we have $\omega_0 = 0$, and if the vectors r_{i0}, r_{j0}, r_{ia}, r_{ja}, i.e. ρ_{ij0} and ρ_{ija} are colinear, then $\mathcal{G}_{ija}(r_{ia}, r_{ja})(\dot{r}_{ia} - \dot{r}_{ja}) = \|\rho_{ij0}\|(\dot{r}_{ia} - \dot{r}_{ja})$ and $\dot{\rho}_{ij0} = \rho_{ij0} \times \omega_0 = 0$, so that $\dot{\delta}_{ij}^R = 0$.

In a developed form, $\mathcal{G}_{ija}(r_{ia}, r_{ja})$ is

$$\mathcal{G}_{ija}(r_{ia}, r_{ja}) = \frac{1}{(x_{ia} - x_{ja})^2 + (y_{ia} - y_{ja})^2 + (z_{ia} - z_{ja})^2} \times \qquad (92)$$

$$\times \begin{bmatrix} (x_{ia} - x_{ja})^2 & (x_{ia} - x_{ja})(y_{ia} - y_{ja}) & (x_{ia} - x_{ja})(z_{ia} - z_{ja}) \\ (x_{ia} - x_{ja})(y_{ia} - y_{ja}) & (y_{ia} - y_{ja})^2 & (y_{ia} - y_{ja})(z_{ia} - z_{ja}) \\ (x_{ia} - x_{ja})(z_{ia} - z_{ja}) & (y_{ia} - y_{ja})(z_{ia} - z_{ja}) & (z_{ia} - z_{ja})^2 \end{bmatrix}.$$

In (89), the unknown quantities are the instantaneous orientation A^0 and angular velocity ω_0 of the vector ρ_{ij0}, i.e. the orientation and angular velocity of the unloaded state 0. The instantaneous orientation and angular velocity of the unloaded state 0 are used to calculate only the component δ_{ij}^R and its derivative $\dot{\delta}_{ij}^R$.

This problem can be overcome in several ways.

These components can be neglected, which yields an insufficiently exact description of the deformation work. In this way, part of the load is left out of consideration.

A smaller error is made if the instantaneous orientation and angular velocity of the unloaded state 0 are taken as approximate values. In this way, part of the load is not left out of consideration, but this does not mean that it is taken as accurate. Namely, the elastic displacements are small compared to macro displacements. A sufficiently good approximation of the instantaneous orientation and angular velocity of the unloaded state 0 is obtained by adopting their values as the mean values of instantaneous orientation and angular velocity of the loaded elastic system as of a block.

The problem can also be solved by using the types of contacts and elastic system characteristics that do not produce bending loads, i.e. for which $\delta_{ij}^R = 0$.

Finally, it should be noticed that the expression (82) for the component y_{ij}^D of the vector of relative displacement of elastic system nodes has the same mathematical form as the expression (27) for relative displacements of nodes of the 'linear' elastic system. By its form, the expression (85) for the component y_{ij}^R differs only by the term $\text{col}(\rho_{ij0}, 0_3)$.

4.5   Model of Elastic System Dynamics for the Immobile Unloaded State

In this section, all the models of dynamics are derived using Lagrange's equations

$$\frac{\mathrm{d}}{\mathrm{d}t}\frac{\partial T}{\partial \dot{g}_i} - \frac{\partial T}{\partial g_i} - \frac{\partial \mathcal{D}}{\partial \dot{g}_i} + \frac{\partial \Pi}{\partial g_i} = Q_i, \quad i = 0, 1, \ldots, m, \tag{93}$$

where T is the kinetic energy; Π is the potential energy; $\mathcal{D}$ is the dissipation energy; Q_i are the generalized forces, and g_i, $\dot{g}_i$ are the generalized coordinates and their derivatives.

The complete procedure of deriving elastic system dynamics for an immobile unloaded state is given in Appendix A. Here we give only a summary of the obtained results.

In describing elastic system dynamics for the immobile unloaded state, the deviations y_i ($g_i = y_i$) of the loaded state of the elastic system from its unloaded state 0, are used as generalized coordinates. Such choice of coordinates is correct as the unloaded state 0 is immobile (see (69) and (70)). Also, it is possible to use the results of the theory of elasticity without any alteration, provided the coordinates y_i are used as global coordinates of the elastic system when defining the stiffness matrix, which actually has been assumed.

It is assumed that in the elastic system composed of $m + 1$ elastically interconnected rigid bodies the relations of dimensions of the particular bodies and overall elastic system are such that the continuity of the first derivative of elastic hyper-surface is preserved, i.e. one part of the smooth continual elastic hyper-surface (part of the elastic line of a linear body) can be replaced with a hyper-chord that is sufficiently close to the hyper-tangent to the elastic hyper-surface. Displacements of the elastic system due to the deformation with respect to the state 0 form an elastic hyper-surface. To each point of the elastic hyper-surface corresponds the deflection and slope angle of the hyper-tangent. For the case of the introduced assumption, the elastic system deflection at the MC of a concrete object corresponds to the translation of the object MC, whereas the slope of the hyper-tangent to the elastic hyper-surface at the MC corresponds to the object rotation with respect to the state it had in the state 0 (Figure 16) measured by the angle of rotation of the coordinate frames at the object MC with the axes directed along the main inertia axes. Such an arrangement shows that the choice of the coordinates y_i is also valid for the description of dynamics of the rigid bodies placed at the elastic system nodes.

The effect of the gravitation field is not accounted for via its potential energy but is associated with the external forces acting on the elastic system in the direction y_i^z. The total potential energy Π of the elastic system is only its deformation

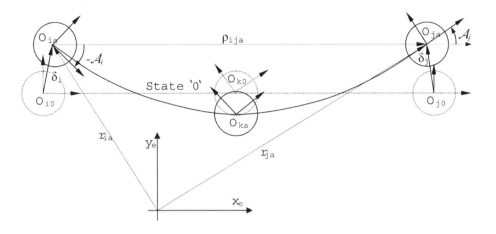

Figure 16. Planar deformation of the elastic system

energy, and it is determined by the expression

$$2\Pi = y^T F_e = y^T K y = (\delta^T \mathcal{A}^T) K \begin{pmatrix} \delta \\ \mathcal{A} \end{pmatrix},$$

$$K = K^T \in R^{(6m+6)\times(6m+6)}, \quad \text{rank } K \le 6m. \qquad (94)$$

If the elastic system is to have displacements of nodes y, then the elastic forces F_e (in this case it is the result of contact and dissipative forces and weight) must act at its nodes O_{ai}, $i = 0, 1, \ldots, m$. In the domain of linear relationship between the strain and dilatation, these forces are equal to the derivative of the potential energy (deformation work) with respect to the displacements

$$F_e = \partial \Pi / \partial y = K y \in R^{6m+6}, \quad F_{ei} = \partial \Pi / \partial y_i = K_i y \in R^{6\times1}, \qquad (95)$$

where $F_e = \text{col}(F_{e0}, F_{e1}, \ldots, F_{em}) = \text{col}(F_{e0}, F_{ec}) \in R^{6m+6}$ are the expanded vector of generalized forces (forces and moments); $y = \text{col}(y_0, y_1, \ldots, y_m) = \text{col}(y_0, y_c) \in R^{6m+6}$ is the expanded vector of displacements; $K \in R^{(6m+6)\times(6m+6)}$ is the stiffness matrix; y_0 and F_{e0} are the vectors of displacements and forces at the manipulated object MC, whereas y_c and F_{ec} are the expanded vectors of displacements and forces at the MCs of the elastic interconnection at the contact. Relation (95) implies, from the principle of potential energy, a minimum for variable displacements (i.e. from the principle of virtual displacements of an elastic body and Castigliano's principles, see (15), (16), (17)). This principle states that, of all the possible displacements allowed on the outer surface, only those will appear for which potential energy of the system has a minimal value.

The stiffness matrix K is determined by some of the methods for static deformation of elastic system. Its rank can be at most rank $K \leq 6m$. The elastic system is kinematically unstable (mobile, contains all the modes of motion of a solid body), so that $6m + 6 -$ rank K displacements should be adopted as known (displacements of the supports or predetermined displacements of the leader) in order to have the system position uniquely determined in space as a function of these displacements and acting forces.

In view of the assumption that all the mass is placed at the grid nodes, the total kinetic energy is only a function of the node coordinates y_i and their derivatives $\dot{y}_i = v_i$, and is defined as a sum of the kinetic energies of the bodies at the elastic system nodes $T = T_0 + T_1 + \cdots + T_i + \cdots + T_m$. After arranging, the total kinetic energy of the elastic system is obtained in the form

$$2T = \sum_{i=0}^{m}\left(\sum_{j=1}^{\infty} dm_j v_j\right) = \sum_{i=0}^{m} m_i \delta_i^2 + \sum_{i=0}^{m} I_i \omega_i^2$$

$$= \sum_{i=0}^{m} v_i^T M_i v_i = v^T M v = \dot{y}^T L_v^T(y) M L_v(y)\dot{y}, \qquad (96)$$

where (see Appendix A)

$$M = \mathrm{diag}(M_0, M_1, \ldots, M_n) \in R^{(6m+6)\times(6m+6)},$$

$$M_i = \mathrm{diag}(m_i, m_i, m_i, A_i, B_i, C_i) \in R^{6\times6}$$

$$v = \mathrm{col}(v_0, v_1, \ldots, v_m) \in R^{6m+6},$$

$$v_i = \mathrm{col}(\dot{\delta}_i, \omega_i(A_i)) = L_{vi}(y_i)\dot{y}_i, \quad \omega_i = L_{\omega i}(A_i)\dot{A}_i \in R^3$$

$$L_{vi}(y_i) = \mathrm{diag}(I_{3\times3}, L_{\omega_i}(A_i)) \in R^{6\times6},$$

$$L_v(y) = \mathrm{diag}(L_{v0}, L_{v1}\ldots.L_{vm}) \in R^{(6m+6)\times(6m+6)} \qquad (97)$$

If the elastic interconnections have dissipative properties, the Rayleigh function (dissipation energy) can be expressed as

$$2\mathcal{D} = -\dot{y}^T D\dot{y}, \quad D = D^T \geq 0, \quad D \in R^{(6m+6)\times(6m+6)}, \qquad (98)$$

where $D \in R^{(6m+6)\times(6m+6)}$ is the matrix of damping coefficients by the corresponding velocity.

Generalized forces Q_i are determined by

$$Q_i = \text{col}(Q_i^1, \ldots, Q_i^6) = G_i(m_i g) + F_{ci}, \quad i = 0, 1, \ldots, m,$$

$$Q_i^j = \sum_{k=0}^{m} \frac{\partial y_k^T}{\partial y_i^j} Q_k = \sum_{k=0}^{m} \frac{\partial y_k^T}{\partial y_i^j} (G_k(m_k g) + F_{ck}), \quad j = 1, \ldots, 6, \quad (99)$$

where G_i is the weight vector; F_{ci} is the vector of external forces, and y_i^j, $j = 1, \ldots, 6$ are the individual components of the vector $y_i = \text{col}(y_i^1, y_i^2, y_i^3, y_i^4, y_i^5, y_i^6) = \text{col}(\delta_i^x \delta_i^y \delta_i^z \psi_i \theta_i \varphi_i)$ at the point i.

By introducing (94), (96), (98) and (99) into the Lagrange equations (93) and uniting all $6m + 6$ equations for the elastic system, which, under the action of the system of external contact forces F_c, performs the motion around the immobile unloaded state 0, one obtains the general form of the model

$$W(y)\ddot{y} + w(y, \dot{y}) = F, \quad (100)$$

where (see Appendix A)

$$W(y) = \text{diag}(W_0(y_0), \ldots, W_m(y_m)) \in R^{(6m+6) \times (6m+6)},$$

$$W(y) = W^T(y), \quad \det W(y) \neq 0,$$

$$w(y, \dot{y}) = \text{col}(w_o(y, \dot{y}), \ldots, w_m(y, \dot{y})) \in R^{6m+6},$$

$$F = \text{col}(0, F_c), \quad F_0 = 0. \quad (101)$$

Of the $6m + 6$ equations (100) the number of independent equations is exactly equal to the rank of the stiffness matrix (rank K).

Equation (100) can be presented in such a way that the description of the motions of elastic interconnections and manipulated object are separated

$$W_c(y_c)\ddot{y}_c + w_c(y, \dot{y}) = F_c,$$

$$W_0(y_0)\ddot{y}_0 + w_0(y, \dot{y}) = 0, \quad (102)$$

where the subscript c denotes the quantities related to the contact points and the subscript 0 stands for the quantities related to the manipulated object. At that (see

Appendix A)

$$y_c = \text{col}(y_1, y_2, \ldots, y_m) \in R^{6m}, \; y_o \in R^6,$$

$$F_c = \text{col}(F_1, F_2, \ldots, F_m) \in R^{6m}, \; F_0 = 0 \in R^6,$$

$$W_c(y_c) = \text{diag}(W_1(y_1), \ldots, W_m(y_m)) \in R^{6m \times 6m},$$

$$W_c(y_c) = W_c^T(y_c), \; \det W_c(y_c) \neq 0,$$

$$w_c(y, \dot{y}) = [\text{col}(w_1(y, \dot{y}), \ldots, w_m(y, \dot{y}))]^T \in R^{6m}, \tag{103}$$

where y_c denotes the expanded vector of positions of contacts in a $6m$-dimensional space and F_c is the expanded vector of contact forces associated with this vector.

It should be noticed that there is no contact force acting directly at the manipulated object MC, so that $F_0 = 0$. Equations (100) and (102) represent the final form of the equations of motion of the elastic system that under the action of the external contact forces F_c performs a general motion around the immobile unloaded state 0.

4.6   Model of Elastic System Dynamics for a Mobile Unloaded State

The complete procedure of deriving the model of elastic system dynamics for the mobile unloaded state is given in Appendix B. In this section we give a brief account of the derivation results for the case when $y_{ij} = y_{ij}^D$.

Let the unloaded state 0 of the elastic system be mobile. The strain of the elastic system takes place under the same principle as if the state 0 is at rest. In other words, the elastic system strain is still considered only with respect to the state 0, which is now mobile (see (69) and (71)). The absolute coordinates of elastic system nodes Y are chosen as generalized coordinates ($g_i = Y_i$). Kinetic, potential, and dissipation energies of the elastic system connections are defined with the aid of the coordinates Y.

The kinetic energy is defined by the absolute velocities as

$$\begin{aligned}
2T_a &= \sum_{i=0}^{m} m_i \dot{r}_{ia}^2 + \sum_{i=0}^{m} I_i \omega_{ia}^2 = \sum_{i=0}^{m} v_{ia}^T M_i v_{ia} \\
&= \dot{Y}^T L_{va}^T(Y) M L_{va}(Y) \dot{Y} = \dot{Y}^T W_a(Y) \dot{Y} m, \tag{104}
\end{aligned}$$

where (see Appendix B) M is the inertial matrix and

$$Y_i = \mathrm{col}(r_{ia}, \mathcal{A}_{ia}) \in R^{6\times 1}, \qquad Y = \mathrm{col}(Y_0, \dots, Y_m) \in R^{6m+6}$$

$$L_{va}(Y) = \mathrm{diag}(L_{v0a}, \dots, L_{vma}) \in R^{(6m+6)\times(6m+6)},$$

$$L_{via} = \mathrm{diag}(I_{3\times 3}, L_{\omega ia}(\mathcal{A}_{ia})) \in R^{6\times 6},$$

$$v_{ia} = \mathrm{col}(\dot{r}_{ia}, \omega_{ia}(\mathcal{A}_{ia})) = L_{via}(Y_i)\dot{Y}_i \in R^{6\times 1},$$

$$\omega_{ia} = L_{\omega ia}(\mathcal{A}_{ia})\dot{\mathcal{A}}_{ia} \in R^{3\times 1},$$

$$Wa(Y) = \mathrm{diag}(W_{0a}, \dots, W_{ma}) = L_{va}^T(Y)ML_{va}(Y) \in R^{(6m+6)\times(6m+6)}. \quad (105)$$

The overall potential energy due to linear and rotational displacements of the body is defined by

$$\Pi_a = \sum_{i=0}^{m}\sum_{j=i+1}^{m} \frac{1}{2}y_{ij}^T K_{ija} y_{ij} = \sum_{i=0}^{m}\sum_{j=i+1}^{m} \frac{1}{2}(Y_i - Y_j)^T \pi_{ij}(Y_i - Y_j)|_{y_{ij}=y_{ij}^D}, \quad (106)$$

where (see Appendix B) $\det \pi_{ij} \neq 0$ and

$$\pi_{ij} = \pi_{ji} = \Lambda_{ij}(Y_i, Y_j))K_{ija}\Lambda_{ij}(Y_i, Y_j)) \qquad (107)$$

$$= \mathrm{diag}(c_{ij}^x \beta, c_{ij}^y \beta, c_{ij}^z \beta, c_{ij}^\psi, c_{ij}^\zeta, c_{ij}^\varphi) \in R^{6\times 6}, \quad \beta = \left(1 - \frac{\|\rho_{ij0}\|}{\|r_{ia} - r_{ja}\|}\right)^2.$$

In combined form, we have

$$2\Pi_a = Y^T \pi_a(Y)Y, \qquad \det \pi_a = 0, \qquad \mathrm{rank}\,\pi_a = 6m, \qquad (108)$$

where $\pi_a(Y)$ is a symmetric matrix, $\pi_a(Y) = \pi_a^T(Y)$, defined by

$$\pi_a(Y) = \begin{bmatrix} \sum_{k=0,k\neq 0}^{m} \pi_{0k} & -\pi_{01} & -\pi_{02} & \cdots & -\pi_{0m} \\ -\pi_{01} & \sum_{k=0,k\neq 1}^{m} \pi_{1k} & -\pi_{12} & \cdots & -\pi_{1m} \\ \cdots & \cdots & \cdots & \cdots & \cdots \\ -\pi_{0m} & -\pi_{1m} & -\pi_{2m} & \cdots & \sum_{k=0,k\neq m}^{m} \pi_{km} \end{bmatrix} \in R^{(6m+6)\times(6m+6)}.$$

$$(109)$$

The overall dissipation energy exchanged in the course of linear and rotational displacements of the body is defined by

$$\mathcal{D}_a = -\sum_{i=0}^{m}\sum_{j=i+1}^{m}\frac{1}{2}(\dot{Y}_i - \dot{Y}_j)^T D_{ij}(\dot{Y}_i - \dot{Y}_j) = -\frac{1}{2}\dot{Y}^T D_a(Y)\dot{Y}, \qquad (110)$$

where

$$D_{ij} = D_{ij}^T = D_{ji} = \mathrm{diag}(\mathcal{G}_{ija}(r_{ia}, r_{ja})D_{ij}^\delta \mathcal{G}_{ija}(r_{ia}, r_{ja}), D_{ij}^A) \in R^{6\times 6}, D_{ij}^\delta, D_{ij}^A)$$

is the damping matrix of elastic interconnections between the ith and jth nodes, and, as $D_{ij} = D_{ji}$

$$D_a(Y) = D_a^T(Y) \qquad (111)$$

$$= \begin{bmatrix} \sum_{k=0,k\neq 0}^{n} D_{0k} & -D_{01} & -D_{02} & \cdots & -D_{0m} \\ -D_{01} & \sum_{k=0,k\neq 1}^{n} D_{1k} & -D_{12} & \cdots & -D_{1m} \\ \cdots & \cdots & \cdots & \cdots & \cdots \\ -D_{0m} & -D_{1m} & -D_{2m} & \cdots & \sum_{k=0,k\neq m}^{n} D_{km} \end{bmatrix}.$$

Generalized forces for the individual components Y_i^j of the vector Y_i are

$$Q_{ia}^j = \sum_{k=0}^{m}\frac{\partial Y_k}{\partial Y_i^j}(G_k(m_k g) + F_{ck})$$

$$= G_i^j(m_i g) + F_{ci}^j, \quad i = 0, 1, \ldots, m, \quad j = 1, \ldots, 6. \qquad (112)$$

By introducing (104), (108), (110) and (112) into the Lagrange equations (93) and after uniting all $6m + 6$ equations, the general form of the model of the elastic system that under the action of the system of external contact forces F_c, performs a macro motion (the mobile unloaded state 0) will be

$$W_a(Y)\ddot{Y} + w_a(Y, \dot{Y}) = F, \qquad (113)$$

where (see Appendix B)

$$W_a(Y) = W_a^T(Y) = \mathrm{diag}(W_{0a}(Y_0), \ldots, W_{ma}(Y_m)) \in R^{(6m+6)\times(6m+6)},$$

$$\det W_a(Y) \neq 0,$$

$$w_a(Y, \dot{Y}) = \mathrm{col}(w_{0a}(Y, \dot{Y}), \ldots, w_{ma}(Y, \dot{Y})) \in R^{6m+6}. \qquad (114)$$

Of the $6m + 6$ equations (113), only rank K equations are independent.

Equation (113) can be presented so that the description of the motion of connections and manipulated object are separated

$$W_{ca}(Y_c)\ddot{Y}_c + w_{ca}(Y, \dot{Y}) = F_c,$$

$$W_{0a}(Y_0)\ddot{Y}_0 + w_{0a}(Y, \dot{Y}) = 0, \tag{115}$$

where the subscript c denotes the quantities related to the contact points, and the subscript 0 denotes the quantities related to the manipulated object. At that (see Appendix B),

$$Y_c = \mathrm{col}(Y_1, \ldots, Y_m) \in R^{6m}, \quad Y_0 \in R^{6 \times 1},$$

$$F_c = \mathrm{col}(F_1, \ldots, F_m) \in R^{6m}, \quad F_0 = 0 \in R^6,$$

$$W_{ca}(Y_c) = W_{ca}^T(Y_c) = \mathrm{diag}(W_{1a}(Y_1), \ldots, W_{ma}(Y_m)) \in R^{6m \times 6m},$$

$$\det W_{ca}(Y_c) \neq 0,$$

$$w_{ca}(Y, \dot{Y}) = (w_{1a}^T(Y, \dot{Y}) \ldots w_{ma}^T(Y, \dot{Y}))^T \in R^{6m \times 1}, \tag{116}$$

where Y_c denotes the expanded vector of contact position in the $6m$-dimensional space and F_c stands for the expanded vector of contact forces acting at the contact points. It should be noticed that no force is acting at the manipulated object MC, so that $F_0 = 0$. Equations (113) and (115) represent the final form of equations describing the behavior of the elastic system that under the action of the system of external forces F_c, performs a general motion around the unloaded state 0, which also performs a general motion.

4.7 Properties of the Potential Energy and Elasticity Force of the Elastic System

Denote by $\mathcal{S}_y \in R^{(6m+6) \times (6m+6)}$ the coordinate frame whose unit vectors coincide with the unit vectors of the generalized coordinates $Y = \mathrm{col}(Y_0, \ldots, Y_m)$, $Y_i = \mathrm{col}(r_{ia} \mathcal{A}_{ia}) \in R^{6 \times 1}$, $i = 0, \ldots, m$ of the manipulated object, MC position, and elastic interconnections.

Assume that with respect to the state characterized by the absence of any displacement ($y_{ij} = 0$), a certain displacement of the nodes, defined by (79), takes place (i.e. $y_{ij} \neq 0$) and, to this new position of elastic system nodes (henceforth, loaded state) let correspond the coordinates Y. Let this displacement be kept constant. It is necessary to determine the properties of the potential energy and force of the elastic system with the change of nodes coordinates in the adopted coordinate frame, i.e. at the translation and rotation of the loaded state without relative displacements of the nodes.

We will briefly repeat the relations needed for the analysis. According to (108), for $y_{ij}^R = 0$, the potential (i.e. deformation) energy of the elastic system at an arbitrary point Y of the system $\mathcal{S}_y$ is

$$2\Pi_a(Y) = Y^T \pi_a(Y) Y \in R^1, \qquad \det \pi_a = 0, \qquad \text{rank } \pi_a = 6m, \qquad (117)$$

where $\pi_a(Y) = \pi_a^T(Y)$ is given by

$$\pi_a(Y) = \begin{bmatrix} \displaystyle\sum_{k=0,k\neq 0}^{n} \pi_{0k} & -\pi_{01} & -\pi_{02} & \cdots & -\pi_{0m} \\[2ex] -\pi_{01} & \displaystyle\sum_{k=0,k\neq 1}^{n} \pi_{1k} & -\pi_{12} & \cdots & -\pi_{1m} \\[2ex] \cdots & \cdots & \cdots & \cdots & \cdots \\[2ex] -\pi_{0m} & -\pi_{1m} & -\pi_{2m} & \cdots & \displaystyle\sum_{k=0,k\neq m}^{n} \pi_{km} \end{bmatrix} \in R^{6(m+1)\times 6(m+1)},$$

$$(118)$$

since

$$\pi_{ij} = \begin{bmatrix} c_{ij}(1 - \frac{\|\rho_{ij0}\|}{\|r_{ia}-r_{ja}\|})^2 & 0 & 0 & 0 & 0 & 0 \\ 0 & c_{ij}(1 - \frac{\|\rho_{ij0}\|}{\|r_{ia}-r_{ja}\|})^2 & 0 & 0 & 0 & 0 \\ 0 & 0 & c_{ij}(1 - \frac{\|\rho_{ij0}\|}{\|r_{ia}-r_{ja}\|})^2 & 0 & 0 & 0 \\ 0 & 0 & 0 & c_{ij}^{\psi} & 0 & 0 \\ 0 & 0 & 0 & 0 & c_{ij}^{\theta} & 0 \\ 0 & 0 & 0 & 0 & 0 & c_{ij}^{\varphi} \end{bmatrix}$$

$$= \pi_{ij}(Y_i, Y_j) = \pi_{ji}(Y_j, Y_i)$$

$$= \text{diag}\left(c_{ij}(1 - \frac{\|\rho_{ij0}\|}{\|r_{ia} - r_{ja}\|})^2 I_{3\times 3}, c_{ij}^{\psi}, c_{ij}^{\theta}, c_{ij}^{\varphi}\right), \qquad \det \pi_{ij} \neq 0. \qquad (119)$$

The potential energy derivative with respect to the coordinate defines the elasticity force decomposed along the coordinates Y of the system $\mathcal{S}_y$ by

$$F_e(Y) = \frac{\partial \Pi_a(Y)}{\partial Y} = \frac{1}{2}\frac{\partial}{\partial Y}(Y^T \bar{\pi}_a(Y) Y)$$

$$= \frac{1}{2} \frac{\partial Y^T \bar{\pi}_a(Y)Y}{\partial Y} + \pi_a(Y)Y = K(Y) \cdot Y \in R^{(6m+6)\times 1}. \quad (120)$$

An arbitrary component F_{ei} of the assembled vector F_e is defined by

$$F_{ei}(Y) = \frac{\partial \Pi_a}{\partial Y_i} = \frac{1}{2} \frac{\partial}{\partial Y_i} (Y^T \bar{\pi}_a(Y)Y)$$

$$= \frac{1}{2} \frac{\partial Y^T \bar{\pi}_a(Y)Y}{\partial Y_i} + \pi_{ia}(Y)Y \in R^{6\times 1}, \quad (121)$$

where $\pi_{ia}(Y) \in R^{6\times(6m+6)}$ are the submatrices composed of the rows starting from the $(6i + 1)$th to $(6i + 6)$th row inclusive, of the matrix $\pi_a(Y)$, and $\partial (Y^T \bar{\pi}_a Y)/\partial Y_i$ is the vector of the derivative of the quadratic form (scalar) $Y^T \pi_a Y$ with respect to the vector Y_i, whereby the macron denotes that partial derivation is carried out over the matrix π_a.

It should be noticed that the potential energy of the elastic system is equal to the sum of the internal forces works. When deriving the expression for potential energy in the adopted generalized coordinates Y, a linear relationship between nodes displacements and elasticity force has been implicitly built in. Studies of the properties of potential energy and elastic system elasticity force in the loaded state motion will be reduced to the study of the behavior of the displacement vector of nodes of the elastic system connected to the mobile loaded state in the fixed frame of adopted coordinates Y.

4.7.1 Properties of potential energy and elasticity force of the elastic system in the loaded state translation

Let the elastic system be translated from the point Y by the vector η, defined by the expression

$$\eta = \begin{bmatrix} \eta_0 \\ \cdots \\ \eta_m \end{bmatrix} = \begin{bmatrix} \bar{\eta} \\ \cdots \\ \bar{\eta} \end{bmatrix} \in R^{(6m+6)\times 1},$$

$$\bar{\eta} = \eta_i = \begin{bmatrix} \bar{\eta}_1 \\ \bar{\eta}_2 \\ \bar{\eta}_3 \\ 0 \\ 0 \\ 0 \end{bmatrix} = \begin{bmatrix} \hat{\eta} \\ 0 \end{bmatrix} \in R^{6\times 1}, \quad i = 0, \dots, m. \quad (122)$$

Let us define the potential energy and elastic force at a point

$$Y_I = Y + \eta \in R^{(6m+6)\times 1},$$

$$Y_{Ii} = \begin{bmatrix} r_{ia} \\ \mathcal{A}_{ia} \end{bmatrix} + \begin{bmatrix} \hat{\eta} \\ 0 \end{bmatrix} = Y_i + \eta_i = Y_i + \bar{\eta}, \ i = 0, \ldots, m. \quad (123)$$

The overall potential energy at that point is

$$2\Pi_a(Y_I) = 2\Pi_a(Y + \eta) = (Y + \eta)^T \cdot \pi_a(Y + \eta) \cdot (Y + \eta), \quad (124)$$

whence

$$\begin{aligned} 2\Pi_a(Y_I) &= Y^T \pi_a(Y + \eta)Y + Y^T \pi_a(Y + \eta)\eta \\ &\quad + \eta^T \pi_a(Y + \eta)Y + \eta^T \pi_a(Y + \eta)\eta. \end{aligned} \quad (125)$$

The product $\pi_a(\#) \cdot \eta$ is

$$\pi_a(\#) \cdot \eta = \begin{bmatrix} \left(\displaystyle\sum_{k=0, k\neq0}^{m} \pi_{0k} - \pi_{01} - \ldots - \pi_{0m} \right) \bar{\eta} \\ \cdots \\ \left(\displaystyle\sum_{k=0, k\neq m}^{m} \pi_{km} - \pi_{0m} - \ldots - \pi_{(m-1)m} \right) \bar{\eta} \end{bmatrix}$$

$$= \begin{bmatrix} 0 \\ \cdots \\ 0 \end{bmatrix} = 0 \in R^{(6m+6)\times1}, \quad (126)$$

so that the quadratic form is $(\#\#) \cdot \pi_a(\#)\eta = 0$, which gives

$$2\Pi_a(Y_I) = Y^T \pi_a(Y + \eta)Y. \quad (127)$$

The matrix $\pi_a(Y + \eta)$ is a function of the submatrices π_{ij}:

$$\begin{aligned} \pi_a(Y + \eta) &= \pi_a(\pi_{01}(Y_0 + \bar{\eta}, Y_1 + \bar{\eta}), \ldots, \\ &\quad \pi_{ij}(Y_i + \bar{\eta}, Y_j + \bar{\eta}), \ldots, \pi_{(m-1)m}(Y_{m-1} + \bar{\eta}, Y_m + \bar{\eta}). \end{aligned} \quad (128)$$

Having in mind the expressions for $\pi_{ij}(Y_i, Y_j)$, Y_{Ii} and Y_{Ij}, the stiffness matrix $\pi_{ij}(Y_{Ii}, Y_{Ij})$ between these nodes is

$$\begin{aligned} \pi_{ij}(Y_{Ii}, Y_{Ij}) &= \text{diag}\left(c_{ij}\left(1 - \frac{\|\rho_{ij0}\|}{\|r_{ia} + \hat{\eta} - (r_{ja} + \hat{\eta})\|}\right)^2 I_{3\times3}, c_{ij}^{\psi}, c_{ij}^{\theta}, c_{ij}^{\varphi} \right) \\ &= \text{diag}\left(c_{ij}\left(1 - \frac{\|\rho_{ij0}\|}{\|r_{ia} - r_{ja}\|}\right)^2 I_{3\times3}, c_{ij}^{\psi}, c_{ij}^{\theta}, c_{ij}^{\varphi} \right) \\ &= \pi_{ij}(Y_i, Y_j) \end{aligned} \quad (129)$$

or, in a shorter form,

$$\pi_{ij}(Y_i + \bar{\eta}, Y_j + \bar{\eta}) = \pi_{ij}(Y_i, Y_j), \tag{130}$$

so that

$$\pi_a(Y + \eta) = \pi_a(\pi_{ij}(Y_i + \bar{\eta}, Y_j + \bar{\eta})) = \pi_a(\pi_{ij}(Y_i, Y_j)) = \pi_a(Y), \tag{131}$$

whereas the overall potential energy is

$$\Pi_a(Y + \eta) = \frac{1}{2} Y^T \pi_a(Y + \eta) Y = \frac{1}{2} Y^T \pi_a(Y) Y = \Pi_a(Y). \tag{132}$$

Let us conclude that the overall potential energy does not change in the course of the parallel displacement of the elastic system, and that this regularity in the system of selected coordinates is described by (132).

The elasticity force $F_e(Y_I)$ at the point Y_I, analogously to (120), is defined by the expression

$$F_e(Y_I) = \frac{\partial \Pi_a(Y_I)}{\partial Y_I}. \tag{133}$$

Since

$$\frac{\partial \Pi_a(Y_I)}{\partial Y} = \left(\frac{\partial Y_I}{\partial Y^T}\right)^T \cdot \frac{\partial \Pi_a(Y_I)}{\partial Y_I} = \frac{\partial Y_I^T}{\partial Y} \cdot \frac{\partial \Pi_a(Y_I)}{\partial Y_I}, \tag{134}$$

then

$$\frac{\partial \Pi_a(Y_I)}{\partial Y_I} = \left(\frac{\partial Y_I^T}{\partial Y}\right)^{-1} \cdot \frac{\partial \Pi_a(Y_I)}{\partial Y}. \tag{135}$$

Substituting the last expression into (133), in view of (132) and (120), we obtain

$$F_e(Y_I) = \left(\frac{\partial Y_I^T}{\partial Y}\right)^{-1} \cdot \frac{\partial \Pi_a(Y)}{\partial Y} = \left(\frac{\partial Y_I^T}{\partial Y}\right)^{-1} \cdot F_e(Y). \tag{136}$$

From expression (123), for Y_I we have $\partial Y_I^T / \partial Y = (\partial Y_I^T / \partial Y)^{-1} = I_{3\times 3}$, so that we can finally conclude that

$$F_e(Y + \eta) = F_e(Y), \tag{137}$$

i.e. in the parallel displacement of the elastic system by the vector η defined by (122), elasticity force does not change.

4.7.2 Properties of potential energy and elasticity force of the elastic system during its rotation in the loaded state

We seek the form of relations that hold in the system of adopted generalized co-ordinates Y during the rotation of the loaded elastic system without a change of relative distances of the nodes.

It is known [31] that every orthogonal transformation of three-dimensional space coordinates

$$\mu = R\zeta, \qquad R^T R = RR^T = I, \qquad R^T = R^{-1}, \tag{138}$$

retains

- the vector module,

- the angle between the vectors,

and if also det $R = 1$, then all basic vectors in the coordinates transformation (138) preserve their mutual orientation (orientation of coordinate frames, vector product and mixed vector product) and such a transformation is called a characteristic rotation. For example, to describe rotation in terms of Euler angles, it is possible to have 12 systems of angles and six variants of the matrix R. For the case of a certain choice of the angles β_1, β_2, β_3, the matrix R will be obtained as a product of three matrices that describe three successive rotations by the selected angle

$$R = R(\beta_1, \beta_2, \beta_3) = R(\beta_1) \cdot R(\beta_2) \cdot R(\beta_3).$$

Let the loaded state before the rotation be at the point Y of $(6m + 6)$-dimensional space. Let the orientation of each body with MC_i, $i = 0, 1, \ldots, m$, i.e. of the overall loaded state, change by the rotation

$$a = (a^\psi \ a^\theta \ a^\varphi)^T \in R^{3 \times 1}. \tag{139}$$

The new orientation of an arbitrary body with the MC_i will be

$$\mathcal{A}_{lia} = \mathcal{A}_{ia} + a, \tag{140}$$

but the mutual orientation of the bodies i and j will remain unchanged,

$$\mathcal{A}_{lij} = \mathcal{A}_{lia} - \mathcal{A}_{lja} = \mathcal{A}_{ia} + a - (\mathcal{A}_{ja} + a) = \mathcal{A}_{ia} - \mathcal{A}_{ja} = \mathcal{A}_{ij}, \tag{141}$$

i.e. in the orientation subsystem of six-dimensional space, the change of orientation by a constant vector a means the translation of the coordinates of this subsystem.

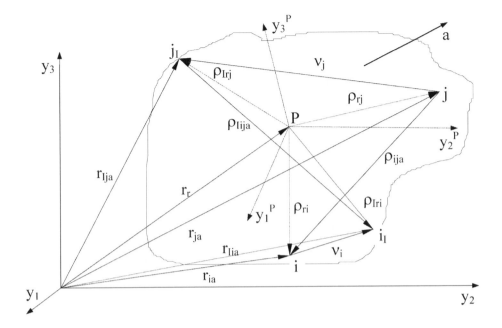

Figure 17. Rotation of the loaded elastic system

Let the loaded state after rotation by the orientation (139) be at the point Y_l (Figure 17). Let P be the instantaneous pole of rotation. Since the loaded state moves as a rigid body, the following relations will hold:

$$r_{lja} = r_{ja} + v_j, \qquad j = 0, 1, \ldots, m. \tag{142}$$

In view of (138)

$$\rho_{lrj} = \rho_{rj} + v_j,$$

$$\rho_{lrj} = A(a)\rho_{rj}, \quad \Rightarrow \quad v_j = \rho_{lrj} - \rho_{rj} = (A(a) - I_{3\times 3})\rho_{rj}, \tag{143}$$

after the substitution, one gets

$$r_{lja} = r_{ja} + (A(a) - I_{3\times 3})\rho_{rj} = r_r + A(a)\rho_{rj} \tag{144}$$

where r_r is the vector of the instantaneous position of the rotation pole; ρ_{*r*} is the position vector of the points of instantaneous rotation pole P whose positions are not known, and $A(a)$ is the matrix of coordinates transformation in the rotation. Since $\rho_{ija} = \rho_{ri} - \rho_{rj} = r_{ia} - r_{ja}$, then from

$$\rho_{lija} = r_{lia} - r_{lja} = r_{ia} - r_{ja} + (A(a) - I_{3\times 3})(\rho_{ri} - \rho_{rj}) = A(a)\rho_{ija}, \tag{145}$$

one gets

$$r_{Ija} = r_{Iia} - A(a)\rho_{ija}. \tag{146}$$

If the positions of all the nodes r_{Ija}, $j = 0, 1, \ldots, m$, are expressed as a function of the positions r_{Iia} of the point i and distance ρ_{ija} of these points from the point i ($\rho_{iia} = 0$), then, in view of (140) and (141), we obtain

$$Y_i = \begin{bmatrix} r_{ia} \\ \mathcal{A}_{ia} \end{bmatrix},$$

$$Y_{Ij} = \begin{bmatrix} r_{Ija} \\ \mathcal{A}_{Ija} \end{bmatrix} = \begin{bmatrix} r_{Iia} - A(a)\rho_{ija} \\ \mathcal{A}_{ia} - \mathcal{A}_{ij} + a \end{bmatrix} = Y_{Ii} - \begin{bmatrix} A(a)\rho_{ija} \\ \mathcal{A}_{ij} \end{bmatrix}. \tag{147}$$

For the case of the absence of rotation, $a = 0$, we have $A(0) = I$ and $A(a)\rho_{ija} = \rho_{ija}$. From (147), we will have

$$Y_{Ij} = Y_{Ii} - \bar{A}_r(a)Y_{ij} \; (= Y_{Ii} - Y_{ij} \mid_{a=0}),$$

$$Y_{ij} = \begin{bmatrix} \rho_{ija} \\ \mathcal{A}_{ij} \end{bmatrix}, \quad \bar{A}_r = \mathrm{diag}(A(a), I_{3\times3}) \in R^{6\times6}. \tag{148}$$

Therefore, for the known absolute coordinates of a node Y_{Ii} and vector of relative positions Y_{ij}, in the absence of rotation and for the known value of the change of orientation a in the rotation space, it is possible to uniquely determine the coordinates of all the nodes of the mobile elastic system in which relative distances of the nodes do not change (fictitious rigid body). The relations (147) and (148) hold for an arbitrary position of the instantaneous rotation pole P and, on the basis of the requirement for, e.g., a manipulated object MC, they will allow the finding of the nominal motion conditions of the other nodes.

If we consider pure rotation about the instantaneous rotation pole (which may also be a node) then, by placing the coordinate frame origin at the instantaneous rotation pole in view of $r_r = 0$, we obtain

$$Y_j = \begin{bmatrix} \rho_{rj} \\ \mathcal{A}_j \end{bmatrix}, \tag{149}$$

$$Y_{Ij} = \begin{bmatrix} \rho_{Irj} \\ \mathcal{A}_{Ij} \end{bmatrix} = \begin{bmatrix} A(a)\rho_{rj} \\ \mathcal{A}_j + a \end{bmatrix} = \bar{A}_r(a)Y_j + \begin{bmatrix} 0 \\ a \end{bmatrix}, \quad j = 0, 1, \ldots, m.$$

After coupling all $m + 1$ relations for pure rotation, the coordinates in the rotated state and derivative with respect to the previous coordinates will be

$$Y_I = A_r(a)Y + a_r(a), \quad \frac{\partial Y_I}{\partial Y} = A_r(a),$$

$$A_r(a) = \mathrm{diag}(\bar{A}_r(a), \ldots, \bar{A}_r(a))$$

$$= \mathrm{diag}(A(a), I_{3\times3}, \ldots, A(a), I_{3\times3}) \in R^{(6m+6)\times(6m+6)},$$

$$a_r(a) = (0_{1\times3}^T \ a^T \ldots 0_{1\times3}^T \ a^T)^T \in R^{(6m+6)\times1}. \tag{150}$$

In view of (138), the vector $y_{ij}^D(Y_i, Y_j)$ (82) in the new deflected position $y_{Iij}^D(Y_{Ii}, Y_{Ij})$, will be

$$y_{Iij}^D(Y_{Ii}, Y_{Ij}) = \left[\begin{array}{c} \left(1 - \dfrac{\|\rho_{ij0}\|}{\|\rho_{Iri} - \rho_{Irj}\|}\right)(\rho_{Iri} - \rho_{Irj}) \\ \mathscr{A}_{Ii} - \mathscr{A}_{Ij} \end{array} \right], \tag{151}$$

and in view of (138), (141) and (143), the norm is

$$\|\rho_{Iri} - \rho_{Irj}\| = \sqrt{(\rho_{Iri} - \rho_{Irj})^T(\rho_{Iri} - \rho_{Irj})} \tag{152}$$

$$= \sqrt{(\rho_{ri} - \rho_{rj})^T A(a)^T A(a)(\rho_{ri} - \rho_{rj})} = \|\rho_{ri} - \rho_{rj}\|.$$

After substituting into the previous relation we get

$$y_{Iij}^D(Y_{Ii}, Y_{Ij}) = \left[\begin{array}{c} \left(1 - \dfrac{\|\rho_{ij0}\|}{\|\rho_{ia} - \rho_{ja}\|}\right) A(a)(\rho_{ri} - \rho_{rj}) \\ \mathscr{A}_{Ii} - \mathscr{A}_{Ij} \end{array} \right]$$

$$= \bar{A}_r y_{ij}^D(Y_i, Y_j). \tag{153}$$

As, analogously to (106), the partial potential energy in the new position is defined by the expression

$$2\Pi_{ija}(Y_{Ii}, Y_{Ij}) = (y_{Iij}^D(Y_{Ii}, Y_{Ij}))^T \cdot K_{ij} \cdot y_{Iij}^D(Y_{Ii}, Y_{Ij}) \tag{154}$$

by introducing (153) into the previous equation and, in view of (82), we obtain

$$2\Pi_{ija}(Y_{Ii}, Y_{Ij}) = \left(\begin{bmatrix} A(a) & 0_{3\times3} \\ 0_{3\times3} & I_{3\times3} \end{bmatrix} \Lambda_{ij} \cdot (Y_i - Y_j) \right)^T$$

$$\times K_{ij} \begin{bmatrix} A(a) & 0_{3\times3} \\ 0_{3\times3} & I_{3\times3} \end{bmatrix} \Lambda_{ij} \cdot (Y_i - Y_j), \tag{155}$$

whereas by introducing Λ_{ij} from (84) and, after transposition, we have

$$2\Pi_{ija}(Y_{Ii}, Y_{Ij}) = (Y_i - Y_j)^T \begin{bmatrix} \left(1 - \dfrac{\|\rho_{ij0}\|}{\|r_{ia} - r_{ja}\|}\right) I_{3\times3} & 0 \\ 0 & I_{3\times3} \end{bmatrix}$$

$$\times \begin{bmatrix} A(a)^T & 0_{3\times3} \\ 0_{3\times3} & I_{3\times3} \end{bmatrix} \begin{bmatrix} c_{ij} I_{3\times3} & 0 \\ 0 & \mathrm{diag}(c_{ij}^\psi, c_{ij}^\theta, c_{ij}^\varphi) \end{bmatrix} \tag{156}$$

$$\times \begin{bmatrix} A(a) & 0_{3\times3} \\ 0_{3\times3} & I_{3\times3} \end{bmatrix} \begin{bmatrix} \left(1 - \dfrac{\|\rho_{ij0}\|}{\|r_{ia} - r_{ja}\|}\right) I_{3\times3} & 0 \\ 0 & I_{3\times3} \end{bmatrix} (Y_i - Y_j),$$

whereas the rearranging gives

$$2\Pi_{ija}(Y_{Ii}, Y_{Ij}) = (Y_i - Y_j)^T \tag{157}$$

$$\times \left[\begin{array}{cc} c_{ij}\left(1 - \dfrac{\|\rho_{ijo}\|}{\|r_{ia} - r_{ja}\|}\right)^2 I_{3\times 3} A(a)^T A(a) & 0 \\ 0 & \mathrm{diag}(c_{ij}^\psi, c_{ij}^\theta, c_{ij}^\varphi) \end{array} \right] (Y_i - Y_j).$$

Since $A(a)^T A(a) = I$, in view of (108) and (107), we get

$$2\Pi_{ija}(Y_{Ii}, Y_{Ij}) = (Y_i - Y_j)^T \tag{158}$$

$$\times \left[\begin{array}{cc} c_{ij}\left(1 - \dfrac{\|\rho_{ijo}\|}{\|r_{ia} - r_{ja}\|}\right)^2 I_{3\times 3} & 0 \\ 0 & \mathrm{diag}(c_{ij}^\psi c_{ij}^\theta c_{ij}^\varphi) \end{array} \right] (Y_i - Y_j) = 2\Pi_{ija}(Y_i, Y_j)$$

or, shorter,

$$\Pi_{ija}(Y_{Ii}, Y_{Ij}) = \Pi_{ija}(Y_i, Y_j). \tag{159}$$

Taking into account (108), along with (150), we conclude that

$$\Pi_a(Y_I) = \Pi_a[A_r(a)Y + a_r(a)] = \Pi_a(Y), \tag{160}$$

i.e. the pure rotation of the loaded state did not change the overall deformation energy.

Using (160), we will find the regularity holding for the elastic force at the rotation of the loaded elastic system. The elasticity force, decomposed along the coordinates Y of the system $\mathcal{S}_y$ in the new position, analogously to (133) and (135), will be determined by the expression

$$\begin{aligned} F_e(Y_I) &= \frac{\partial \Pi_a(Y_I)}{\partial Y_I} = \frac{\partial \Pi_a[A_r(a)Y + a_r(a)]}{\partial(A_r(a)Y + a_r)} \\ &= \left(\frac{\partial[A_r(a)Y + a_r(a)]^T}{\partial Y}\right)^{-1} \frac{\partial \Pi_a(Y)}{\partial Y}. \end{aligned} \tag{161}$$

According to (150) and (138)

$$\left(\frac{\partial[A_r(a)Y + a_r(a)]^T}{\partial Y}\right)^{-1} = A_r(a)^{-1} = \mathrm{diag}(A^{-1}(a)\, I \ldots A^{-1}(a)\, I) \tag{162}$$

$$= \mathrm{diag}(A^T(a)\, I \ldots A^T(a)\, I) = A_r^T(a).$$

By introducing the last expression into (161), and taking (120) into account, we finally obtain a relation between the elasticity forces before and after the rotation of the loaded state in the form

$$F_e(Y_I) = F_e[A_r(a)Y + a_r(a)] = A_r^T(a)F_e(Y), \tag{163}$$

in which, apart from the absolute coordinates of the nodes before and after rotation, is also figuring the change of the orientation a.

In a general motion composed of one translation and one rotation, the force established prior to the rotation is according to (137), equal to the force established at the initial moment of the motion. In other words, if we determine the force in the initial moment and denote it by $F_e(Y)$, and bearing in mind (137) and that, for $a = 0$, there will be $a_r(a) = 0$ and $A_r(a) = I$, then relation (163) can be used as a general expression for determining the force vector at an arbitrary position of the loaded state whose nodes distances do not change with respect to the initial state.

In cooperative manipulation, the nominal conditions are typically given in one of the following two ways:

1. The manipulated object is, in the desired way, transferred from the initial to the final position. During the transfer, the positions of the contact points are registered. Then, on the basis of the object initial position, it is necessary to determine the vectors Y_{ij} and $F_e(Y)$. For an arbitrary position, according to (148) we determine $Y_{Iij} = Y_{Ii} - Y_{Ij} = \bar{A}_r(a)Y_{ij}$. If the object is performing a translatory motion, then $Y_{Iij} = Y_{ij}$ and the contact forces will be determined according to (137) with respect to the initial state. If these two vectors are not identical, then, since the matrix $\bar{A}_r(a)$ has a fixed known structure dependent exactly on three coordinates of the vector a (139), it is necessary to form (148) for the arbitrarily selected nodes and solve it with respect to a and then, using (163), determine the necessary elastic forces in that position.

2. The trajectory of the manipulated object MC r_0 in three-dimensional space and the change of its orientation $\mathcal{A}_0$ are determined first, whereby the rotation is performed about the MC as an instantaneous pole. Then, according to (148), it is necessary to determine the positions of all the nodes Y for the initial instant, as well as the vectors Y_{ij} and $F_e(Y)$. For an arbitrary position on the trajectory, the vector Y_{Ii} is determined by the MC position vector in three-dimensional space r_0 for that point of the trajectory and orientation of the manipulated object from the initial moment (as if it were translated to that position). The value of the vector of the change of orientation is defined by the required orientation of the object $a = \mathcal{A}_0$, so that the positions of the contact points and the forces there are determined according to (148) and (163).

In addition to the source vector Y, with the structure

$$
Y = \begin{bmatrix} Y_0 \\ Y_1 \\ \cdots \\ Y_m \end{bmatrix} = \begin{bmatrix} Y_0 \\ Y_v \\ Y_s \end{bmatrix} = \begin{bmatrix} Y_0 \\ Y_c \end{bmatrix} \in R^{(6m+6)\times 1}, \quad Y_v = Y_1 \in R^{6\times 1},
$$

$$
Y_s = \begin{bmatrix} Y_2 \\ \cdots \\ Y_m \end{bmatrix} \in R^{(6m-6)\times 1},
$$

$$
Y_c = \begin{bmatrix} Y_1 \\ \cdots \\ Y_m \end{bmatrix} = \begin{bmatrix} Y_v \\ Y_s \end{bmatrix} \in R^{6m\times 1}, \tag{164}
$$

use is also made of the vector with a transformed structure

$$
\begin{bmatrix} Y_1 \\ Y_2 \\ \cdots \\ Y_m \\ Y_0 \end{bmatrix} = \begin{bmatrix} Y_v \\ Y_s \\ Y_0 \end{bmatrix} = \begin{bmatrix} Y_c \\ Y_0 \end{bmatrix} \in R^{(6m+6)\times 1}. \tag{165}
$$

For both vectors, the transformation matrix $A_r(a)$ and the vectors $a_r(a)$ and η from (150) and (122) are the same, so that all the previous conclusions, derived for the source vector, also hold for the vector with a transformed structure.

4.8 Model of Manipulator Dynamics

The model of motion of a non-elastic manipulator with six DOFs with non-compliant joints and with the gripper force in the space of internal coordinates, is given by [32–36]

$$
H_i(q_i)\ddot{q}_i + h_i(q_i, \dot{q}_i) = \tau_i + J_i^T f_{ci}, \quad i = 1, \ldots, m, \tag{166}
$$

where $H_i(q_i) \in R^{6\times 6}$ is a positively determined inertia matrix of the ith manipulator; $h_i(q_i, \dot{q}_i) \in R^{6\times 1}$ is the vector taking into account the effect of gravitation, Coriolis acceleration, and friction; $\tau_i \in R^{6\times 1}$ is the vector of joint drives; $J_i \in R^{6\times 6}$ is the transformation matrix of the velocity vector of internal coordinates into velocity vector of the manipulator tip, and $f_{ci} = -F_i \in R^{6\times 1}$ is the contact force at the manipulator tip.

For m manipulators, the model acquires a general form

$$
H(q)\ddot{q} + h(q, \dot{q}) = \tau + J^T f_c \tag{167}
$$

with the following designations:

$$H(q) = \text{blockdiag}(H_1(q_1), \ldots, H_m(q_m)) \in R^{6m \times 6m},$$

$$h(q, \dot{q}) = \text{col}(h_1(q_1, \dot{q}_1), \ldots, h_m(q_m, \dot{q}_m)) \in R^{6m \times 1},$$

$$\tau = \text{col}(\tau_1, \ldots, \tau_m) \in R^{6m \times 1},$$

$$J^T = \text{blockdiag}(J_1^T, \ldots, J_m^T) \in R^{6m \times 6m},$$

$$f_c = \text{col}(f_{c1}, \ldots, f_{cm}) = \text{col}(-F_1, \ldots, -F_m) \in R^{6m \times 1},$$

$$q = \text{col}(q_1, \ldots, q_m) \in R^{6m \times 1},$$

$$\dot{q} = \text{col}(\dot{q}_1, \ldots, \dot{q}_m) \in R^{6m \times 1}. \tag{168}$$

The vector equation (167) determines $6m$ connections.

4.9 Kinematic Relations

By assumption, the contact of the manipulator and object is stiff and rigid. The contact position is determined by the position and orientation of the manipulator gripper tip on the one hand and by the position of the external node of the elastic system on the other. We assume that the work space is six-dimensional and the manipulators are non-redundant so a unique mutual kinematic relation is established through the contact position between the internal coordinates of each manipulator and the position of its contact with the elastic system described by the external (absolute) coordinates.

If the kinematic relation of the internal and absolute coordinates Y of contact points is expressed as

$$Y_i = \Theta_i(q_i) \in R^{6 \times 1}, \quad i = 1, \ldots, m, \tag{169}$$

then the relation between their velocities and accelerations will be [32–34]

$$\dot{Y}_i = \frac{\partial \Theta_i(q_i)}{\partial q_i} \cdot \dot{q}_i = J_i(q_i)\dot{q}_i \in R^{6 \times 1}, \quad i = 1, \ldots, m, \tag{170}$$

$$\ddot{Y}_i = \dot{J}_i(q_i)\dot{q}_i + J_i(q_i)\ddot{q}_i \in R^{6 \times 1}, \quad i = 1, \ldots, m, \tag{171}$$

or in united form

$$Y_c = \Theta(q) \in R^{6m \times 1},$$

$$\dot{Y}_c = J(q)\dot{q} \in R^{6m \times 1},$$

$$\ddot{Y}_c = \dot{J}(q)\dot{q} + J(q)\ddot{q} \in R^{6m \times 1}, \qquad (172)$$

where $Y_c = \text{col}(Y_1, \ldots, Y_m)$.

Kinematic relations of the internal coordinates and the coordinates used to describe the motion around the immobile y and mobile Y unloaded states are of the same form. Relations (171) are given for the coordinates Y whereas kinematic relations are obtained for the coordinates y by introducing into (171) y instead of Y. The concrete dependence $\Theta(q)$ is formed for each concrete choice of the cooperative system structure, i.e. the arrangement of the manipulators and their kinematic characteristics.

4.10   Model of Cooperative System Dynamics for the Immobile Unloaded State

The equations describing the behavior of the elastic system (100) or (102) and manipulators (167) and kinematic relations (172) of the internal coordinates q and elastic system displacements y define the model of cooperative work of m rigid manipulators with six DOFs, handling a rigid object whose general motion in three-dimensional space is unconstrained, whereby the connections between the object and manipulators are elastic and the motion takes place around the immobile unloaded state 0.

The number of inputs into the model is $6m$, whereas the number of independent state quantities (positions and velocities) is $2 \cdot (6m + 6)$, of which $2 \cdot \text{rank } K$ are dictated by the elastic system and $2 \cdot (6m + 6 - \text{rank } K)$ are dictated by the leader's dynamics.

State quantities can be chosen in different ways. For example, possible choices are internal coordinates vector $q \in R^{6m \times 1}$ and position vector of the manipulated object MC $y_0 \in R^{6 \times 1}$ and their derivatives, or the position vector of the MCs of elastic interconnections $y_c \in R^{6m \times 1}$ and of manipulated object $y_0 \in R^{6 \times 1}$ and their derivatives, or the vector of the leader's internal coordinates $q_v \in R^{6 \times 1}$ and position vector of the MCs of the remaining elastic interconnections $y_s \in R^{(6m-6) \times 1}$ and their derivatives, or the corresponding internal coordinates $q_s \in R^{(6m-6) \times 1}$ of the followers and the MC of the manipulated object $y_0 \in R^{6 \times 1}$.

Let the state quantity vector $z = \text{col}(q, y_0) \in R^{6m+6}$ and its derivatives be prescribed.

By introducing kinematic relations between the internal coordinates q and contact points displacements y_c defined by (172) into (102), we obtain

$$W_c(\Theta(q))(\dot{J}(q)\dot{q} + J(q)\ddot{q}) + w_c(\Theta(q), J(q)\dot{q}, y_0, \dot{y}_0) = F_c,$$

$$W_0(y_0)\ddot{y}_0 + w_0(\Theta(q), J(q)\dot{q}, y_0, \dot{y}_0) = 0. \qquad (173)$$

As $F_c = -f_c$, by introducing F_c from the last equation in (167), we get

$$(H(q) + J^T(q)W_c(\Theta(q))J(q))\ddot{q} + h(q, \dot{q}) + J^T(q)W_c(\Theta(q))\dot{J}(q)\dot{q}$$

$$+ J^T(q)w_c(\Theta(q), J(q)\dot{q}, y_0, \dot{y}_0) = \tau,$$

$$W_0(y_0)\ddot{y}_0 + w_0(\Theta(q), J(q)\dot{q}, y_0, \dot{y}_0) = 0, \qquad (174)$$

i.e. in a shorter form,

$$N(q)\ddot{q} + n(q, \dot{q}, y_0, \dot{y}_0) = \tau,$$

$$W(y_0)\ddot{y}_0 + w(q, \dot{q}, y_0, \dot{y}_0) = 0, \qquad (175)$$

where

$$N(q) = H(q) + J^T(q)W_c(\Theta(q))J(q) \in R^{6m \times 6m},$$

$$n(q, \dot{q}, y_0, \dot{y}_0) = h(q, \dot{q}) + J^T(q)W_c(\Theta(q))\dot{J}(q)\dot{q}$$

$$+ J^T(q)w_c(\Theta(q), J(q)\dot{q}, y_0, \dot{y}_0) \in R^{6m \times 1},$$

$$W(y_0) = W_0(y_0),$$

$$w(q, \dot{q}, y_0, \dot{y}_0) = w_0(\Theta(q), J(q)\dot{q}, y_0, \dot{y}_0) \in R^{6 \times 1}. \qquad (176)$$

By introducing the vector $z = \mathrm{col}(q, y_0)$ into (175), a general form of the model of cooperative manipulation is obtained in the form

$$\Phi(z)\ddot{z} + \phi(z, \dot{z}) = \delta_\tau, \quad \det\Phi(z) \neq 0, \qquad (177)$$

where

$$\Phi(z) = \begin{bmatrix} N(q) & 0 \\ 0 & W(y_0) \end{bmatrix} \in R^{(6m+6) \times (6m+6)},$$

$$\phi(z, \dot{z}) = \begin{pmatrix} n(q, \dot{q}, y_0, \dot{y}_0) \\ w(q, \dot{q}, y_0, \dot{y}_0) \end{pmatrix} \in R^{6m+6},$$

$$\delta_\tau = \mathrm{col}(\tau, 0, \ldots, 0) \in R^{6m+6},$$

$$\tau \in R^{6m \times 1}. \qquad (178)$$

4.11 Model of Cooperative System Dynamics for the Mobile Unloaded State

Equations (113) or (115), (167) and (172) define the model of cooperative work of m rigid manipulators with six DOFs handling a rigid object whose general motion in three-dimensional space is unconstrained, whereby the connections between the object and manipulators are elastic and the unloaded state of the elastic system also performs general motion.

As in the model of cooperative work for the motion around the immobile unloaded state of elastic system, in this example too, the number of model inputs is $6m$, whereas the number of state quantities (positions and velocities) is $2 \cdot (6m+6)$.

The possible choices of state quantities are the internal coordinates vector $q \in R^{6m \times 1}$ and position vector of the manipulated object MC $Y_0 \in R^{6 \times 1}$ and their derivatives, or the position vector of the MCs of elastic interconections and of manipulated object $Y \in R^{6m+6}$ and their derivatives, or the internal coordinates vector of the leader $q_v \in R^{6 \times 1}$ and vector of positions of the MCs of the remaining elastic interconnections $Y_s \in R^{(6m-6) \times 1}$ and the MC of the manipulated object $Y_0 \in R^{6 \times 1}$ and their derivatives.

The choice of vectors of the absolute coordinates of contact points $Y_c \in R^{6m \times 1}$ for the state quantities requires finding an inverse function for mapping the manipulator tip position into internal coordinates $q = \Theta^{-1}(Y_c)$. As this task is not easily solvable, to derive a general model of cooperative manipulation it is necessary to adopt the vector of state quantities $z = \text{col}(q, Y_0) \in R^{6m+6}$ and its derivatives.

Replacing (172) into (115) yields

$$W_{ca}(\Theta(q))(\dot{J}(q)\dot{q} + J(q)\ddot{q}) + w_{ca}(\Theta(q), J(q)\dot{q}, Y_0, \dot{Y}_0) = F_c,$$

$$W_{0a}(Y_0)\ddot{Y}_0 + w_{0a}(\Theta(q), J(q)\dot{q}, Y_0, \dot{Y}_0) = 0. \tag{179}$$

Since $F_c = -f_c$, after introducing F_c from the last equation into (167), we get

$$(H(q) + J^T(q)W_{ca}(\Theta(q))J(q))\ddot{q} + h(q, \dot{q}) + J^T(q)W_{ca}(\Theta(q))\dot{J}(q)\dot{q}$$

$$+ J^T(q)w_{ca}(\Theta(q), J(q)\dot{q}, Y_0, \dot{Y}_0) = \tau,$$

$$W_{0a}(Y_0)\ddot{Y}_0 + w_{0a}(\Theta(q), J(q)\dot{q}, Y_0, \dot{Y}_0) = 0 \tag{180}$$

or, in a shorter form,

$$N(q)\ddot{q} + n(q, \dot{q}, Y_0, \dot{Y}_0) = \tau,$$

$$W(Y_0)\ddot{Y}_0 + w(q, \dot{q}, Y_0, \dot{Y}_0) = 0,$$

$$P(q)\ddot{q} + p(q, \dot{q}, Y_0, \dot{Y}_0) = F_c. \tag{181}$$

where

$$N(q) = H(q) + J^T(q)W_{ca}(\Theta(q))J(q) \in R^{6m \times 6m},$$

$$n(q, \dot{q}, Y_0, \dot{Y}_0) = h(q, \dot{q}) + J^T(q)W_{ca}(\Theta(q))\dot{J}(q)\dot{q}$$

$$+ J^T(q)w_{ca}(\Theta(q), J(q)\dot{q}, Y_0, \dot{Y}_0) \in R^{6m \times 1},$$

$$W(Y_0) = W_0(Y_0),$$

$$w(q, \dot{q}, Y_0, \dot{Y}_0) = w_{0a}(\Theta(q), J(q)\dot{q}, Y_0, \dot{Y}_0) \in R^{6 \times 1},$$

$$P(q) = W_c(\Theta(q))J(q) \in R^{6m \times 6m},$$

$$p(q, \dot{q}, Y_0, \dot{Y}_0) = W_c(\Theta(q))\dot{J}(q)\dot{q} + w_c(\Theta(q), J(q)\dot{q}, Y_0, \dot{Y}_0) \in R^{6m \times 1}.$$

$$(182)$$

The first two equations in (181) describe the behavior of the cooperative system, whereas the third equation defines the dependence of the contact forces on internal coordinates. This dependence is of a differential type and is described by the model of elastic system dynamics (179). By their structure, the matrices $H(q)$, $J(q)$, $W_c(\Theta(q))$ from (181) are diagonal block matrices. Each submatrix from the diagonals of these matrices is non-singular and is only a function of the internal coordinates of the manipulator that this submatrix is related to. Hence, the inertia matrices $N(q)$ and $P(q)$ in (181) are also of block-diagonal form.

If the vector $z = \operatorname{col}(q, Y_0)$ is introduced into (181), a general form of the model of cooperative work will be

$$\Phi(z)\ddot{z} + \phi(z, \dot{z}) = \delta_\tau, \quad \det \Phi(z) \neq 0, \tag{183}$$

where

$$\Phi(z) = \begin{bmatrix} N(q) & 0 \\ 0 & W(Y_0) \end{bmatrix} \in R^{(6m+6) \times (6m+6)},$$

$$\phi(z, \dot{z}) = \begin{pmatrix} n(q, \dot{q}, Y_0, \dot{Y}_0) \\ w(q, \dot{q}, Y_0, \dot{Y}_0) \end{pmatrix} \in R^{6m+6},$$

$$\delta_\tau = \operatorname{col}(\tau, 0, \ldots, 0) \in R^{6m+6},$$

$$\tau \in R^{6m \times 1}. \tag{184}$$

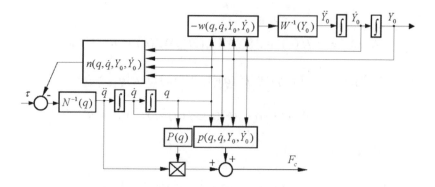

Figure 18. Block diagram of the cooperative system model

Obviously, the two model forms, (177) and (183), are identical. From this form of the model we get the essential characteristic of cooperative work that the number of possible (physical) inputs is smaller than the number of model state quantities (order of the system).

Therefore, the presented modeling procedure based on Lagrange's equations yielded a complete dynamic model of the cooperative work of several manipulators with six DOFs in several mathematical forms for different conditions of cooperative work. The problem of force uncertainty has been completely resolved, and the causes of force uncertainty and kinematic uncertainty in cooperative work have been explained. It was shown that this problem does not physically exist and a procedure was given to solve the force uncertainty problem in accordance with physical phenomena. We modeled the general motion of the system of elastically connected rigid bodies with six DOFs, starting from the assumption that absolute coordinates of the body MC during the motion and distance from the MC of the elastic system before deformation are known.

A block diagram of the cooperative work model (183) is given in Figure 18.

4.12   Forms of the Motion Equations of Cooperative System

First, we recapitulate on the mathematical modeling of cooperative manipulation.

The clue to solving force and position redundancy, described in the introductory Section 3.1 as the problem of cooperative manipulation (as dealt with in the available literature), is the abandoning of the assumption that the object, manipulators, and contacts are rigid. To present the conceived solution of the model of cooperative manipulation, we have chosen to analyze the simplest case of contact. It is the rigid contact of the manipulator tips and object in which no change of contact point on the manipulated object is allowed, whereas the forces and moments are possible in all directions, irrespective of whether they are positive or negative. On the other hand, the manipulator tips and manipulated object form one whole (as if they were glued), so that the elasticity model introduced holds either for the rigid manipulator and elastic object, or for a rigid manipulator with an elastic tip and rigid object, or for an elastic manipulator tip and elastic object. The simplest case is when the tips of the manipulators are elastic, as then all the model parameters can be easily determined.

The system manipulators-manipulated object is decomposed into the elastic system and rigid manipulators. The elastic system is first described approximately by considering the discontinual structure with $6m + 6$ DOFs of motion, which 'communicates' with the manipulators via force and position. Practically, an elastic spatial grid is formed with rigid objects at the nodes, which is a relatively rough picture of reality, but a very practical one for engineering applications and sufficiently correct provided the influential (Maxwell's) coefficients are constant. If the grid is elastic there will be a unique relationship between the grid position and forces acting on the grid. For the selected model of elastic system only the grid nodes are under the influence of forces equal to the sum of inertial, damping, gravitational, and contact forces. The acting forces are balanced by elastic forces both under the static and dynamic conditions. A description of this property has the same form when the unloaded state is either at rest or in the state of motion, and is given by the relations

$$F_e(y) = \frac{\partial \Pi_a}{\partial y} = Ky, \quad K = \text{const} \in R^{(6m+6)\times(6m+6)}, \quad \text{rank } K \leq 6m, \quad (185)$$

$$F_e(Y) = \frac{\partial \Pi_a(Y)}{\partial Y} = K(Y)Y, \quad K(Y) \in R^{(6m+6)\times(6m+6)}, \quad \text{rank } K(Y) \leq 6m.$$

From the Lagrange equations (93) for the immobile and mobile unloaded state, we have

$$\frac{\partial \Pi}{\partial y} \;\Big|\; = \;\Big|\; -\frac{d}{dt}\frac{\partial T}{\partial \dot{y}} \;+\; \frac{\partial T}{\partial y} + \frac{\partial \mathcal{D}}{\partial \dot{y}} \;\Big|\; + Q,$$

$$\frac{\partial \Pi_a}{\partial Y} \;\Big|\; = \;\Big|\; -\frac{d}{dt}\frac{\partial T_a}{\partial \dot{Y}} \;+\; \frac{\partial T_a}{\partial Y} + \frac{\partial \mathcal{D}_a}{\partial \dot{Y}} \;\Big|\; + Q_a$$

$$\Leftrightarrow \quad \frac{\partial \Pi_*}{\partial \heartsuit} \;\Big|\; = \;\Big|\; (\clubsuit\clubsuit\clubsuit)\cdot\ddot{\heartsuit} \;+\; (\spadesuit\spadesuit\spadesuit\spadesuit)\cdot\dot{\heartsuit} \;\Big|\; + G_* + F_*, \quad \heartsuit = y, Y, \; * = -, a,$$

$$F_e \;=\; F_{\ddot{d}} \;+\; F_{\dot{d}} \;+\; G_* + F_*. \quad (186)$$

Evidently, the first and second equations of (185) are of identical form.

If the potential energy (deformation work) is zero, the system of $6m + 6$ equations (185) describes the motion of $m + 1$ rigid bodies. If $m = 0$ too, the system describes the motion of the rigid manipulated object only. If $m = 1$, the system describes the motion of one manipulator in contact with the object.

The first equation in (185), with $F_e = G + F$, is the subject matter of the strength of materials, and the theory of construction or aeroelasticity. The stiffness matrix is constant with the rank dictated by the degree of static uncertainty of the considered construction (for rank $K = 6m$ the construction is statically determined) and is determined with respect to the Cartesian coordinate frame for the unloaded state (for example, by some of the finite-element methods). It should be noticed that the stiffness matrix is here determined for the adopted coordinate frame attached to the unloaded state, and that it is not identical to the stiffness matrices for other coordinate frames, although their properties are the same. The form of Equation (186) illustrates D'Alembert's principle of solving dynamic problems by static methods.

Since, in a mathematical sense, the behavior of all the equations of elastic system are equivalent, for the purpose of this analysis we assume that $\heartsuit = y$. Let us denote by F_d the sum of dynamic forces $F_d = F_{\ddot{j}} + F_{\dot{j}}$ in (186). Taking into account (20), (100), (102), (113), (115), the last two equations of (186) can also be presented in the form (matrix A in the subsequent relations differs from the transformation matrix in the rotation of the loaded elastic system given in the previous section):

$$F_{ec} = F_{dc} + G_c + F_c = Ay_c + by_0,$$

$$F_{e0} = F_{d0} + G_0 + F_0 = cy_c + dy_0, \tag{187}$$

where the subscripts c and 0 refer to the contact points and manipulated object MC, respectively, whereby

$$\hbar_\star = \begin{bmatrix} \hbar_{\star c} \\ \hbar_{\star 0} \end{bmatrix} \in R^{6m+6}, \quad \hbar_{\star c} \in R^{6m \times 1}, \quad \hbar_{\star 0} \in R^{6 \times 1}, \quad \hbar = F, G, y, \quad \star = e, d, -$$

$$K = \begin{bmatrix} A & b \\ c & d \end{bmatrix}, \quad \text{rank } K = 6m, \quad |d| \neq 0. \tag{188}$$

If $F_d = 0$, the equations of behavior describe in full the static conditions of the elastic system, and the following two cases are possible:

1. First, corresponding to the static conditions of the manipulated object on the support, the object being under the action of the contact forces F_c and gravitation forces G, whereby the support reaction is the force F_0. The cor-

responding equations are

$$G_c + F_c = Ay_c + by_0,$$

$$G_0 + F_0 = cy_c + dy_0, \qquad F_d = 0, \quad F_0 \neq 0. \tag{189}$$

2. Second, corresponding to the static conditions of the manipulated object held in space by the manipulators and experiencing the contact forces F_c and gravitation forces G, whereby no other external force is acting at the object MC (internal node of the grid), i.e. $F_0 = 0$. The corresponding equations are

$$G_c + F_c = Ay_c + by_0,$$

$$G_0 \qquad = cy_c + dy_0, \qquad F_d = 0, \quad F_0 = 0. \tag{190}$$

If the equations also encompass the description of dynamics, the corresponding term should be different from zero, i.e. $F_d \neq 0$. Since this term can be split into the inertial and damping terms, $F_d = F_{in} + F_t$, there are a lot of different combinations of forces that might be taken into consideration. We will give several examples that are of interest for cooperative manipulation.

1. The first case corresponds to the immobile manipulated object on the support, experiencing the forces of dynamics F_{dc}, contact, F_c, and gravitation, G, whereby the support reaction is the force F_0. The corresponding equations are

$$F_{dc}+ \quad G_c + F_c = Ay_c + by_0,$$

$$G_0 + F_0 = cy_c + dy_0, \qquad F_{d0} = 0, \quad F_0 \neq 0. \tag{191}$$

This case corresponds to the cooperative work of the manipulators on a robust object (e.g. car body). If the mutual influence of the manipulators via elastic properties can be neglected, the case reduces to the independent work of m manipulators on the same object. In the case of a single manipulator, the first equation represents the 'dynamic environment' of that manipulator, the subject being treated in [35–39]. Then, the dynamic environment describes the dynamics of connection from the point of view of cooperative work (e.g. in grinding, the grinder is immobile so that $F_{d0} = 0$, but the connection manipulator tip–workpiece has a certain dynamics). The overall model is obtained by associating equations of environment dynamics with the equations describing the behavior of the manipulators and kinematic relations between the internal and external coordinates.

2. The second case corresponds to the mobile manipulated object on the support, experiencing the forces of dynamics F_{dc}, contact F_c, and gravitation G, whereby the support reaction is the force F_0 (e.g. in machining with pitch and depth control without controlling the machining force, in some cases of assembly on the conveyer). The corresponding equations are

$$F_{dc} + \ G_c + F_c = Ay_c + by_0,$$
$$F_{d0} + \ G_0 + F_0 = cy_c + dy_0, \qquad F_d \neq 0, \ \ F_0 \neq 0. \qquad (192)$$

3. The third case refers to the immobile manipulated object in space, experiencing the forces of dynamics F_{dc}, contact F_c, and gravitation G, whereby no other external force is acting at the object MC, i.e. $F_0 = 0$ (e.g. assembly involving the fixed casing of the corresponding parts of the assembly block). The corresponding equations are

$$F_{dc} + \ G_c + F_c \ = Ay_c + by_0,$$
$$G_0 \qquad\quad = cy_c + dy_0, \qquad F_{d0} = 0, \ \ F_0 = 0. \qquad (193)$$

4. The fourth case corresponds to the mobile manipulated object in space, experiencing the forces of dynamics F_{dc}, contact F_c, and gravitation G, whereby no other external force is acting at the object MC, i.e. $F_0 = 0$. This case corresponds to the cooperative manipulation and has been dealt with in Sections 4.6 and 4.11. The corresponding equations are

$$F_{dc} + \ G_c + F_c \ = Ay_c + by_0,$$
$$F_{d0} + \ G_0 \qquad = cy_c + dy_0, \qquad F_0 = 0. \qquad (194)$$

This form of equations also holds in the case when the number of manipulators is $m = 1$, i.e. when one manipulator is handling an elastic or rigid object with elastic contact.

For practical applications, the above forms of equations are rather complex, as they are of a high order and hence require a lot of calculation, using powerful (i.e. expensive) hardware. Hence, it is often plausible to make simplifications at the expense of accuracy. The main characteristic of cooperative manipulation that cannot be neglected is its physical feature. However, it is often possible to neglect the mass properties of the elastic interconnections with respect to those of the manipulated object ($F_{cin} = 0$), which means a reduction of the system's order by $6m$. If it is possible then to neglect the damping properties of elastic interconnections ($F_{ct} = 0$), the system's order is further reduced by $6m$, and the overall model acquires the form

$$G_c + F_c = Ay_c + by_0, \qquad F_{dc} = 0,$$
$$F_{d0} + G_0 = cy_c + dy_0, \qquad F_0 = 0, \qquad (195)$$

composed of $6m$ algebraic equations and six second-order differential equations to describe the dynamics of the manipulated object that cannot be avoided when its general motion is concerned.

For the overall description of cooperative manipulation, to one of the selected forms of equations from (189) to (195) one has to associate the equations describing the manipulator's dynamics (167) and equations of kinematic relations (172) between the internal coordinates q and the coordinates y of the coordinate systems fixed at the corresponding contact points.

It is necessary to notice some more characteristic properties of the forms of equations describing cooperative manipulation. We will analyze the general form of equations of the elastic system while taking into account the specific feature of the cooperative work implicitly built-in into these equations. The previous general form (187) can be written as

$$F_{ec} = Ay_c + by_0,$$

$$F_{e0} = cy_c + dy_0, \qquad F_0 = 0, \qquad (196)$$

where $A \in R^{6m \times 6m}$, $b \in R^{6m \times 6}$, $c \in R^{6m \times 6}$ and $d \in R^{6 \times 6}$ are the submatrices of the singular stiffness matrix K.

Matrix $d \in R^{6 \times 6}$ represents elastic properties in the directions of manipulated object DOFs. Hence, it is an irregular matrix, $|d| \neq 0$, otherwise it would contradict the assumptions about the model. The rank of the matrix K is rank $K = 6m$ irrespective of which node is selected to define the elastic system (grid) in space, because equations of external equilibrium of forces and moments must hold for that point, and each of these equations represent the sum of m corresponding rows of the rest of the stiffness matrix. If the characterization in space is based on the selected point y_0, then a unique relation must exist between y_c and the acting forces, i.e. the matrix A has to be non-singular too, $|A| \neq 0$.

It is known [40] that for the non-singular matrices, A and d holds

$$\det K = \det \begin{bmatrix} A & b \\ c & d \end{bmatrix} = |A| \cdot |d - cA^{-1}b|$$
$$= |A - bd^{-1}c| \cdot |d|, \qquad |A| \neq 0, |d| \neq 0. \qquad (197)$$

and since $\det K = 0$ and $|d| \neq 0$ then $|A - bd^{-1}c| = 0$ should be satisfied, i.e. the matrix $A - bd^{-1}c$ must be singular, and because of $|A| \neq 0$ is $|d - cA^{-1}b| = 0$, so that the matrix $d - cA^{-1}b$ is singular too.

What are the repercussions of this statement to cooperative work?

By solving the first equation of (196) with respect to y_c and after rearranging, we have

$$y_c = A^{-1}F_{ec} - A^{-1}by_0,$$

$$F_{e0} = cA^{-1}F_{ec} + (d - cA^{-1}b)y_0, \quad \Rightarrow d = cA^{-1}b. \tag{198}$$

By solving the second equation of (196) with respect to y_0 and after rearranging, one gets

$$F_{ec} = (A - bd^{-1}c)y_c + bd^{-1}F_{e0}, \quad \Rightarrow A = bd^{-1}c,$$

$$y_0 = d^{-1}cy_c + -d^{-1}F_{e0}. \tag{199}$$

The previous equations must be satisfied for all static and dynamic conditions, including the conditions $y_0 = 0$ and $F_{e0} = 0$, irrespective of how long they last. Then, the right-hand sides of Equations (198) and (199) acquire a rectangular form, with the elasticity forces F_{ec} and displacements of the contact points y_c as independent arguments. Equation (198) can establish a unique correspondence between the displacements of contact points and forces acting at them $y_c = y_c(F_{ec}, y_0)$. As in Equation (199) rank $[((A - bd^{-1}c)^T \mid (d_{-1}c)^T)^T] < 6m$, it is not possible to establish a unique correspondence $F_{ec} = F_{ec}(y_c, F_{e0})$. Then it is necessary to seek the part of the newly-formed matrix whose rank is $6m$ and establish for it the correspondence between the quantities on the left and right sides of the equality sign. In that case, the vector on the left-hand side will be composed of the part of the vector of contact forces and displacements of the manipulated object MC. If y_0 and F_{e0} are not equal to zero, they represent independent parameters, and for their concretely selected values, the equations will be fully and uniquely solvable. Thus we described the system's physical nature, which cannot be disturbed. These simple functional relations must be taken into account when introducing control laws, or more precisely, when setting up correct conditions for the control system's requirements. Namely, there may exist two unique combinations of the quantities on the right and left sides of the previous equalities ((198) always exists). It is possible to do two things: either to prescribe the requirements for the independent quantities and select the dependent ones as input or vice versa (provided that the choice is realizable). Let us notice that the equations should not be necessarily solved only with respect to y_0 or y_c, but this choice is logical for cooperative manipulation (a possible alternative would be solving with respect to the position of the leader and follower).

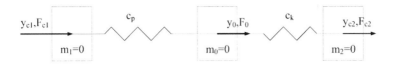

Figure 19. Elastic system of two springs

The above will be illustrated in a simple example. Let us consider again two linear springs placed between zero masses at the nodes of which act the forces as depicted in Figure 19.

The behavior equations are given in the matrix form

$$
\begin{bmatrix} F_{c1} \\ F_{c2} \\ F_0 \end{bmatrix} = \begin{bmatrix} c_p & 0 & -c_p \\ 0 & c_k & -c_k \\ -c_p & -c_k & c_p + c_k \end{bmatrix} \cdot \begin{bmatrix} y_{c1} \\ y_{c2} \\ y_0 \end{bmatrix}.
\tag{200}
$$

Here $A = \mathrm{diag}(c_p, c_k)$, $b = (-c_p - c_k)^T$, $c = (-c_p - c_k)$ and $d = c_p + c_k$. By introducing these quantities into (198) and (199), we obtain from (198)

$$
y_c = \begin{bmatrix} \dfrac{1}{c_p} & 0 \\ 0 & \dfrac{1}{c_k} \end{bmatrix} F_{ec} + \begin{bmatrix} 1 \\ 1 \end{bmatrix} y_0,
$$

$$
F_{e0} = (-1 - 1) F_{ec} + 0 \qquad \Leftrightarrow \sum F = 0
\tag{201}
$$

and from (199)

$$
F_{ec} = \frac{1}{c_p + c_k} \begin{bmatrix} c_p c_k & c_p c_k \\ c_p c_k & c_p c_k \end{bmatrix} y_c + \begin{bmatrix} 1 \\ 1 \end{bmatrix} F_{e0},
$$

$$
y_0 = -\frac{1}{c_p + c_k} (c_p \ c_k) y_c - \frac{1}{c_p + c_k} F_{e0}.
\tag{202}
$$

Obviously, the relation between $F_{ec}(y_c, y_0)$ and $F_{e0}(F_{ec}, y_0) = F_{e0}(F_{ec})$ is unique for an arbitrary y_0. The same conclusion would also hold if, instead of y_0, we chose the position vector of an arbitrary contact point y_{ci}. The relation $F_{ec} = F_{ec}(y_c, F_{e0})$ is not unique, but on the left-hand side we have to choose a new vector $(F_{eci} \ y_0)^T$, to establish a unique correspondence with y_c and F_{e0}, which will exist only when $c_p \neq c_k$.

In agreement with the needs of the concrete task in solving the systems of equations that describe the behavior of the elastic and cooperative system as a whole, it is possible to choose known rank K arbitrary quantities. This implies the need for rearranging the equations of behavior, the corresponding vectors, and matrices, as well as the introduction of designations for the newly-formed quantities.

To make our discussion easier to follow, we will present in brief the adopted way of dividing matrices and vectors and of the system of assigning the subscripts.

The source equations of the elastic system behavior are defined for the immobile unloaded state by (100), and for the mobile unloaded state by (113). The position vector in the $(6m + 6)$-dimensional space for the immobile unloaded state is denoted by y, and for the mobile unloaded state by Y. These vectors define a point in these spaces. Both vectors are composed by the same principle of $(m + 1)$-dimensional vectors y_i, i.e. Y_i, where the subscript takes the values $i = 0, 1, \ldots, m$. The structure of these vectors is the same and has already been given by (164) for the vector Y. The first subscript refers to the name of the part of the system structure for which the values of the coordinates Y or y are given. The second subscript is added only when we want especially to point to the ordinal number of the coordinate of the already indexed vector. The first subscript 'c' is associated to the positions of contact points, 'v' to the leader and 's' to the followers, '0' to the manipulated object MC, whereas the ordinal number of the corresponding manipulator refers to the concrete contact point. For the vector y, instead of Y, we should write only y, i.e.,

$$y = \begin{bmatrix} y_0 \\ y_1 \\ \cdots \\ y_m \end{bmatrix} = \begin{bmatrix} y_0 \\ y_v \\ y_s \end{bmatrix} = \begin{bmatrix} y_0 \\ y_c \end{bmatrix} \in R^{6m+6}, \qquad y_v = y_1 \in R^{6\times 1},$$

$$y_s = \begin{bmatrix} y_2 \\ \cdots \\ y_m \end{bmatrix} \in R^{6(m-1)\times 1}, \qquad y_c = \begin{bmatrix} y_1 \\ \cdots \\ y_m \end{bmatrix} = \begin{bmatrix} y_v \\ y_s \end{bmatrix} \in R^{6m\times 1}. \tag{203}$$

To the position vectors Y and y with such a structure correspond the $(6m + 6)$-dimensional force vectors F and the source $(6m + 6) \times (6m + 6)$-dimensional matrices of stiffness, K^i, and damping D^i. These matrices are constant when the elastic system motion is performed around the immobile unloaded state. The structure and subscripts of the force vectors are identical to those of the position vectors. Force vectors may be of different origin. When we want to emphasize this property we associate another subscript with the force vectors, which are always put in the first place. The subscript 'd' refers to dynamic forces, 'e' to elastic, etc. For example, the source structure of the vector of the elasticity force is obtained by putting F_e instead of Y in (164) (or in the preceding expression). Structures of the matrices of stiffness and damping remain the same. The notations adopted for the stiffness matrix are associated as the subscripts to the corresponding submatrices of the damping matrix. The source structure of the matrix is

$$K^i = \begin{bmatrix} d & c \\ b & A \end{bmatrix} \in R^{(6m+6)\times(6m+6)}, \qquad D^i = \begin{bmatrix} D_d & D_c \\ D_b & D_A \end{bmatrix} \in R^{(6m+6)\times(6m+6)},$$

$$A, D_A \in R^{6m\times6m}, \qquad b, D_b \in R^{6m\times6}, \qquad c, D_c \in R^{6\times6m}, \qquad d, D_d \in R^{6\times6}. \quad (204)$$

Since no force is directly acting at the manipulated object MC, to single out one of the manipulators (for example, the leader), the source equations of elastic system behavior (100) and (113) should be rearranged by writing first the equation of the objects associated with the contact points, starting the enumeration from the contact point of the leader. Thus, we obtain the equations of behavior for the immobile unloaded state (102) and for the mobile unloaded state (115). To these equations correspond the rearranged matrices of stiffness and damping

$$K = \begin{bmatrix} A & b \\ c & d \end{bmatrix} \in R^{(6m+6)\times(6m+6)},$$

$$D = \begin{bmatrix} D_A & D_b \\ D_c & D_d \end{bmatrix} \in R^{(6m+6)\times(6m+6)}. \quad (205)$$

Because of the later application, a more detailed division is performed so that submatrices are obtained which are suitable for multiplying by the rearranged position vectors and their derivatives. The structure of the stiffness matrix is

$$K = \begin{bmatrix} A_{11} & | & A_{12} & \cdots & A_{1m} & | & b_1 \\ --- & | & --- & --- & --- & | & --- \\ A_{21} & | & A_{22} & \cdots & A_{2m} & | & b_2 \\ \cdots & | & \cdots & \cdots & \cdots & | & \cdots \\ A_{m1} & | & A_{m2} & \cdots & A_{mm} & | & b_m \\ --- & | & --- & --- & --- & | & --- \\ c_1 & | & c_2 & \cdots & c_m & | & d \end{bmatrix}$$

$$= \begin{bmatrix} u_v & u_s & u_0 \\ A_v & A_s & A_0 \\ c_v & c_s & c_0 \end{bmatrix} \in R^{(6m+6)\times(6m+6)},$$

$$u_v = A_{11} \in R^{6\times6}, \qquad u_s = (A_{12} \cdots A_{1m}) \in R^{6\times(6m-6)},$$

$$u_0 = b_1 \in R^{6\times6} \quad A_v = \begin{bmatrix} A_{21} \\ \cdots \\ A_{m1} \end{bmatrix}, \qquad A_0 = \begin{bmatrix} b_2 \\ \cdots \\ b_m \end{bmatrix} \in R^{(6m-6)\times6},$$

$$A_s = \begin{bmatrix} A_{22} & \cdots & A_{2m} \\ \cdots & \cdots & \cdots \\ A_{m2} & \cdots & A_{mm} \end{bmatrix} \in R^{(6m-6)\times(6m-6)} c_v = c_1 \in R^{6\times6},$$

$$c_s = (c_2 \cdots c_m) \in R^{6\times(6m-6)}, \qquad c_0 = d \in R^{6\times6}. \tag{206}$$

The structure of the damping matrix is

$$D = \begin{bmatrix} D_{A11} & | & D_{A12} & \cdots & D_{A1m} & | & D_{b1} \\ --- & | & --- & --- & --- & | & --- \\ D_{A21} & | & D_{A22} & \cdots & D_{A2m} & | & D_{b2} \\ \cdots & | & \cdots & \cdots & \cdots & | & \cdots \\ D_{Am1} & | & D_{Am2} & \cdots & D_{Amm} & | & D_{bm} \\ --- & | & --- & --- & --- & | & --- \\ D_{c1} & | & D_{c2} & \cdots & D_{cm} & | & D_d \end{bmatrix}$$

$$= \begin{bmatrix} D_{uv} & D_{us} & D_{u0} \\ D_{Av} & D_{As} & D_{A0} \\ D_{cv} & D_{cs} & D_{c0} \end{bmatrix} \in R^{(6m+6)\times(6m+6)},$$

$$D_{uv} = D_{A11} \in R^{6\times6}, \qquad D_{us} = (D_{A12} \cdots D_{A1m}) \in R^{6\times(6m-6)},$$

$$D_{u0} = D_{b1} \in R^{6\times6},$$

$$D_{Av} = \begin{bmatrix} D_{A21} \\ \cdots \\ D_{Am1} \end{bmatrix}, \quad D_{A0} = \begin{bmatrix} D_{b2} \\ \cdots \\ D_{bm} \end{bmatrix} \in R^{(6m-6)\times6},$$

$$D_{As} = \begin{bmatrix} D_{A22} & \cdots & D_{A2m} \\ \cdots & \cdots & \cdots \\ D_{Am2} & \cdots & A_{Amm} \end{bmatrix} \in R^{(6m-6)\times(6m-6)},$$

$$D_{cv} = D_{c1} \in R^{6\times6}, \qquad D_{cs} = (D_{c2} \cdots D_{cm}) \in R^{6\times(6m-6)},$$

$$D_{c0} = D_d \in R^{6\times6}. \tag{207}$$

The structure of the rearranged vectors of positions and forces that correspond to these matrices is the same as the structure of (165) defined for the position vector y. For the selected vector, instead of y we should put the designation for that vector in this expression. For example, to determine the structure of the vector of dynamic forces, it is necessary to put F_d instead of y in (165), so that we obtain

$$\begin{bmatrix} F_{d1} \\ F_{d2} \\ \cdots \\ F_{dm} \\ F_{d0} \end{bmatrix} = \begin{bmatrix} F_{dv} \\ F_{ds} \\ F_{d0} \end{bmatrix} = \begin{bmatrix} F_{dc} \\ F_{d0} \end{bmatrix} \in R^{6m+6\times 1}. \tag{208}$$

When determining the nominal conditions we also use the following structure of stiffness matrix:

$$\begin{aligned} K &= \begin{bmatrix} u_v & u_s & u_0 \\ A_v & A_s & A_0 \\ C_v & C_s & C_0 \end{bmatrix} = \begin{bmatrix} u_v & u_{s0} \\ A_v & A_{s0} \\ C_v & C_{s0} \end{bmatrix} \\ &= \begin{bmatrix} A_{uv} & A_{us0} \\ C_v & C_{s0} \end{bmatrix} \in R^{(6m+6)\times(6m+6)}, \end{aligned}$$

$$u_{s0} = (A_{12} \cdots A_{1m}\ b_1) = (u_s\ u_0) \in R^{6\times 6m},$$

$$A_{s0} = \begin{bmatrix} A_{22} & \cdots & A_{2m} & b_2 \\ \cdots & \cdots & \cdots & \cdots \\ A_{m2} & \cdots & A_{mm} & b_3 \end{bmatrix} = (A_s\ A_0) \in R^{(6m-6)\times 6m},$$

$$A_{uv} = \begin{bmatrix} u_v \\ A_v \end{bmatrix} \in R^{6m\times 6}, \quad A_{us0} = \begin{bmatrix} u_s & u_0 \\ A_s & A_0 \end{bmatrix} = \begin{bmatrix} u_{s0} \\ A_{s0} \end{bmatrix} \in R^{6m\times 6m},$$

$$C_{s0} = (c_2 \cdots c_m\ d) = (c_s\ d) \in R^{6\times 6m}. \tag{209}$$

To this matrix structure corresponds the structure of vectors of displacements and forces, exemplarily given for the vector y by

$$y = \begin{bmatrix} y_v \\ y_s \\ y_0 \end{bmatrix} = \begin{bmatrix} y_v \\ y_{s0} \end{bmatrix} \in R^{(6m+6)\times 1}, \quad y_{s0} = \begin{bmatrix} y_s \\ y_0 \end{bmatrix} \in R^{6m\times 1}. \tag{210}$$

The enumeration convention will be shown in the model of cooperative manipulation (181): the subscript 1 being for the leader and the subscripts from 2 to m for the followers. Using this convention, Equations (181) and (179) become

$$N_v(q_v)\ddot{q}_v + n_v(q, \dot{q}, Y_0, \dot{Y}_0) = \tau_v,$$

$$N_s(q_s)\ddot{q}_s + n_s(q, \dot{q}, Y_0, \dot{Y}_0) = \tau_s,$$

$$W(Y_0)\ddot{Y}_0 + w(q, \dot{q}, Y_0, \dot{Y}_0) = 0,$$

$$P_v(q_v)\ddot{q}_v + p_v(q, \dot{q}, Y_0, \dot{Y}_0) = F_{cv},$$

$$P_s(q_s)\ddot{q}_s + p_s(q, \dot{q}, Y_0, \dot{Y}_0) = F_{cs}. \tag{211}$$

The first three equations of (211) are the equations of the cooperative system's behavior. The last two differential equations (211) describe the dependence of the contact forces on the internal coordinates. If the control in cooperative manipulation is realized via contact forces, these equations represent equations of the output of the cooperative system. Obviously, in that case the output equations are differential and not algebraic, as is accustomed in theory of control.

By their structure, the matrices $H(q)$, $J(q)$, $W_c(\Theta(q))$, in (181), i.e. in (211) are block-diagonal matrices. Hence, the inertia matrices $N(q)$ and $P(q)$ in (181) and (211) are also of block-diagonal structure. The structures of the matrices are given by

$$N(q) = H(q) + J^T(q)W_c(\Theta(q))J(q) = \mathrm{diag}(N_1(q_1), N_2(q_2), \ldots, N_m(q_m))$$

$$= \mathrm{diag}(N_v(q_v), N_s(q_s)) \in R^{6m \times 6m}, \quad |N(q)| \neq 0,$$

$$N_i(q_i) = H_i(q_i) + J_i^T(q_i)W_{ci}(\Theta_i(q_i))J_i(q_i) \in R^{6 \times 6},$$

$$|N_i(q_i)| \neq 0, i = 1, \ldots, m,$$

$$N_v(q_v) \in R^{6 \times 6}, \quad |N_v(q_v)| \neq 0, \quad N_s(q_s) \in R^{(6m-6) \times (6m-6)}, \quad |N_s(q_s)| \neq 0,$$

$$P(q) = W_c(\Theta(q))J(q) \quad = \quad \mathrm{diag}(P_1(q_1), P_2(q_2), \ldots, P_m(q_m))$$

$$= \quad \mathrm{diag}(P_v(q_v), P_s(q_s)) \in R^{6m \times 6m}, \quad |P(q)| \neq 0,$$

$$P_i(q_i) = W_{ci}(\Theta_i(q_i))J_i(q_i) \in R^{6 \times 6}, \quad |P_i(q_i)| \neq 0, \quad i = 1, \ldots, m,$$

$$P_v(q_v) \in R^{6 \times 6}, |P_v(q_v)| \neq 0, \quad P_s(q_s) \in R^{(6m-6) \times (6m-6)}, |P_s(q_s)| \neq 0. \quad (212)$$

A mathematical model of cooperative manipulation in the form (211) is the basic expression of the model for the analysis and synthesis of nominals and cooperative system control laws.

4.13   Stationary and Equilibrium States of the Cooperative System

To make the problems appearing in the synthesis of nominal motion and control laws more understandable, we will consider the characteristics of the stationary state and motion of the uncontrolled (without feedback loops) cooperative system.

The natural state of a cooperative system is the one when it is at rest. For that state, let the elastic system load $G + \mathrm{col}(F_c, 0)$ be known. The state of the cooperative system at rest is described by the system of differential equations of the stationary state (181) and it is determined from the condition $\dot{q} = \ddot{q} = 0$ ($\dot{Y}_c = \ddot{Y}_c = 0$), $\dot{Y}_0 = \ddot{Y}_0 = 0$, by the equations

$$
\begin{aligned}
F_e = K(Y) \cdot Y &= \frac{1}{2} \frac{\partial Y^T \bar{\pi}_a Y}{\partial Y} + \pi_a(Y)Y = G + \begin{bmatrix} F_c \\ 0 \end{bmatrix} \\
&= \begin{bmatrix} A & b \\ c & d \end{bmatrix} \begin{bmatrix} Y_c \\ Y_0 \end{bmatrix} = \begin{bmatrix} u_v & u_s & u_0 \\ A_v & A_s & A_0 \\ c_v & c_s & c_0 \end{bmatrix} \begin{bmatrix} Y_v \\ Y_s \\ Y_0 \end{bmatrix}, \quad (213)
\end{aligned}
$$

$$
h(q,0) = \tau + J^T f_c \in R^{6m \times 1},
$$

$$
Y_c = \Theta(q) \in R^{6m \times 1},
$$

whereby the partition of the matrix $K(Y)$ is performed in a way that rank $K(Y)$ = rank $A = 6m$, and that the block matrices are congruous for multiplication by the vector $Y = \text{col}(Y_c, Y_0) = \text{col}(Y_v, Y_s, Y_0) = \text{col}(Y_v, Y_{so})$, $Y_c = \text{col}(Y_v, Y_s)$, $Y_{so} = \text{col}(Y_s, Y_0)$. The matrix $c_0 = d \in R^{6 \times 6}$ represents the elastic properties in the directions of DOFs of the independent variable positions chosen for the spatial characterization of the elastic system. Hence, it is a non-singular matrix $|d| \neq 0$. If we consider the cooperative system motion only around the stationary unloaded state of the elastic system and if the model is described by the deviation coordinates y, then the matrices A, b, c and d are constant, and the y coordinates define displacements of the elastic structure nodes.

For non-redundant manipulators under stationary conditions, the last two equations of (213) make the basis for establishing a mutually unique correspondence between the internal coordinates q and driving torques τ, on the one hand, and the positions of contact points Y_c and loads $G_c + F_c$ acting at them on the other. Because $K \in R^{(6m+6) \times (6m+6)}$ and rank $K = 6m$, from the first equation in (213), it is not possible to establish a mutually unique correspondence between the known load $G + \text{col}(F_c, 0)$ at the elastic system nodes and its position Y. It is possible to determine $6m$ components of the positions of the elastic system nodes as a function of the known load, and six components of the independent variable positions chosen for determining the position of the elastic system in space. These six independently variable positions of the elastic system nodes can be chosen arbitrarily. Here we choose that the independently variable positions determine the spatial position of one node only, and that is the manipulated object MC position Y_0. If Y_0 is defined in advance, then there exists a unique relationship between the positions and orientations Y_c of the contacts and contact forces F_c (the matrix A is non-singular, $|A| \neq 0$ and G is known). A unique correspondence exists between the elastic system stress state and nodes displacements relative to the system's unloaded state, or to the forces acting at the nodes [6]. When we define (i.e. select and control in an ideal way) the position of one node in space and magnitude of the desired stress state in the elastic system, the value of the load acting at the node is uniquely determined and, by the same token, so is the stationary state of the cooperative system.

The point coordinates Y_0, selected to determine the elastic system position in space, are independent variables. Therefore, it can be concluded that the cooperative system stationary state can be the whole three-dimensional space within which the manipulated object moves, i.e., at least the part of this space that represents the manipulator work space. In other words, under the stationary conditions, the same elastic forces F_e (i.e. stress state) of the elastic system can be generated according to (213) in any part of the space by the same load $G + \text{col}(F_c, 0)$, which is fully in agreement with the statics of the elastic systems [6, 7].

The equilibrium state of the non-linear system described by the second-order differential equations (181) is determined by the conditions $\tau = 0$, $\dot{q} = \ddot{q} = 0$, $\dot{Y}_0 = \ddot{Y}_0 = 0$. In reality, there is no zero driving torque condition, $\tau = 0$, and therefore there is no reason to determine the equilibrium state. Nevertheless, it is reasonable to analyze the system of differential equations (181), to derive conclusions about the stability properties of the equilibrium state, i.e. about the behavior of the homogeneous part of the solution of these differential equations.

In the case of non-redundant manipulators, there is the mutually unique correspondence (172) between the manipulator internal coordinates and their derivatives on the one hand, and the state quantities of the objects, with the MCs at contact points and their derivatives on the other. The features of the homogenous part of the solution of the differential equations (181) are determined by the properties of the solution of the non-homogenous differential equations (113), which describe the dynamic behavior of the elastic system.

Equations (113) describe the force and moment dynamic equilibrium at each node of the mobile elastic structure, expressed in absolute coordinates. Each equation for a particular node determines the force or moment equilibrium in the selected direction, and is equal to the sum of all other equations defining the equilibrium of forces and moments at the remaining nodes in the same direction. As a consequence, of $6m + 6$ equations only $6m$ are independent, although there are $2 \times (6m + 6)$ independent quantities, necessary and sufficient for the description of the elastic system motion. The difference between the approach in this work and the approaches to the problems of elastic structure statics and dynamics in the available literature is as follows. In [4, 5], the initial assumption is that the elastic displacements needed for the determination of the position of the elastic system in space are not independent variables (state quantities), but they are given in advance as in the case of statics of stationary elastic structures [6, 7]. An implication of this *a priori* assumption is that the position and orientation of the elastic structure's unloaded state are known in advance and the stiffness matrix is non-singular, which is favorable. For determination in space, any point of the elastic structure can be selected, even the contact one. A consequence of this would be that the internal coordinates of the manipulator to which this contact point belongs are given in advance, i.e. they are not state quantities, which is erroneous. In developing the mathematical model in [8], and in this work too, all displacements of the elastic system (i.e. positions of contact points and manipulated object MC) are independent variables, i.e. they are state quantities, necessary and sufficient to describe the cooperative system dynamics.

Further consequences of the singularity of the matrix $K(Y)$ on the character of the solution of Equations (113) can be seen if linearization of the system (113) around the unloaded state of the mobile elastic system is carried out. The position of the unloaded state of the elastic system during the motion is not known. It is known that this state moves as a rigid body because all elastic displacements are zero. The general motion of the elastic system can be considered as a complex motion that consists of the transfer motion of the mobile unloaded elastic system and its relative motion around that state due to elastic displacements. Let the relationship between the elastic system node displacements and its stresses be linear. The application of the usual linearization procedure [31] to Equations (113) yields a linear system with $6m + 6$ second-order differential equations of the form

$$W'\ddot{y}^k + 2D'\dot{y}^k + K'y^k = F', \qquad (214)$$

where W', D', K' are the constant matrices of the order $(6m + 6) \times (6m + 6)$; $F' = \text{col}(F'_c, 0_{6 \times 1})$ is the force vector; $y^k = \text{col}(y^k_c, y^k_0)$, $\dot{y}^k$, $\ddot{y}^k$ are the elastic displacement vector and its derivatives relative to the instantaneous position and orientation of the elastic system unloaded state, defined by the vector $Y^u(t) = \text{col}(Y^u_c, Y^u_0)$, $\text{col}(Y_c, Y_0) = \text{col}(Y^u_c, Y^u_0) + \text{col}(y^k_c, y^k_0)$. The equilibrium state of this system is determined by the solution of the algebraic equation $K'y^k = 0$, rank $K' = \text{rank } A' = 6m$. Therefore, the equilibrium state of the contact points positions can be expressed as a function of the elastic displacements y^k_0 of the manipulated object MC, $y^k_{cr} = -A'^{-1}b'y^k_0$. As the vector y^k_0 is an independent variable, the equilibrium state of the contact points positions is the whole space of the manipulated object MC positions, which is a direct consequence of the elastic system mobility. The same conclusion also holds for the stationary state of differential equation (214), determined as the solution of the algebraic equation $K'y^k = F'$.

The character of the solution of system (214) is governed by the disposition of the system characteristic equation roots

$$|\lambda^2 W' + 2\lambda D' + K'| = 0. \qquad (215)$$

The sufficient conditions needed to establish the relations between the matrices W', D' and K' and sign definiteness and properties of the roots of the characteristic equation (215) are known in the mathematical literature [41]. If the matrices W', D' and K' are non-negative definite and if at least one of the matrices W' and K' is positive definite, the previous characteristic equation does not possess roots with positive real parts. If, in addition to that, the matrices W' and K' are non-negative definite, and the matrix D' is positive definite, then the only zero of the real part is the root $\lambda = 0$. The matrix $W' = \text{diag}(W_c(Y_c^u), W_0(Y_0^u))$ represents the inertia matrix of the elastic system, calculated for its unloaded state, and is always positive definite $W' > 0$ ($\det W' \neq 0$). The damping matrix D' is the result of selecting damping features of the elastic contacts. In engineering applications, elastic contacts are always selected so that this matrix is non-negative definite $D' \geq 0$. The definiteness by the sign of the stiffness matrix K' is governed by the elastic properties and disposition of the elastic system nodes. The object grasping can be always planned in a way that this matrix is non-negative definite, $K' \geq 0$. The planning of the object grasping will not be considered here, but it is assumed that the gripping is performed so that the non-negative definite stiffness matrix is always obtained. Accordingly, the real parts of the characteristic equation roots are non-positive, with potential zero real part roots, whose multiplicity has to be examined.

By introducing the state variables $\xi_1 = y^k$ and $\xi_2 = \dot{y}^k$, Equation (214) is transformed into

$$\dot{\xi} = \begin{pmatrix} 0 & I \\ -W'^{-1}K' & -2W'^{-1}D' \end{pmatrix} \xi - \begin{pmatrix} 0 \\ W'^{-1} \end{pmatrix} F = A_\xi \xi + B_\xi F, \qquad (216)$$

where I and 0 are the unit and square zero-matrix of the order $6m + 6$; $\xi = \text{col}(\xi_1, \xi_2)$, A_ξ and B_ξ, are the matrices accompanying the vectors ξ and $F = \text{col}(0_{6m \times 1}, F')$. As the rank of the matrices K' and D' equals $6m$, the matrix A_ξ rank equals rank $A_\xi = 6m + 6 + 6m = (12m + 12) - 6$. Hence, the characteristic equation $A_\xi - sI = 0$ is of the form $s^6 \cdot f(s) = 0$, where $f(s)$ is the polynomial with the highest degree $12m + 6$, i.e. characteristic equation has six roots with the zero real part. According to the linear system theory, the consequence of the root zero real part multiplicity is the unstable character of the solution of the system (214) describing the relative motion of the elastic system. As the transfer motion of the mobile unloaded elastic system, being a rigid-body motion, is determined, the general motion of the uncontrolled elastic system is of an unstable character. This conclusion can be interpreted in the following manner. Under dynamic conditions, the same system elastic forces $K'y^k$ (i.e. stress state) can be produced in any part of the elastic displacements y^k domain according to (214), by the same resulting load $F' - W'\ddot{y}^k - 2D'\dot{y}^k$.

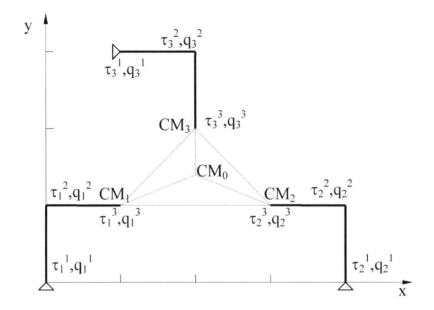

Figure 20. Initial position of the cooperative system

4.14 Example

The presented modeling procedure is illustrated by an example of the cooperative work of three manipulators with three DOFs, handling an object with the mass center CM_0, while contacts are being established at the mass centers of the connections CM_1, CM_2 and CM_3. (In Figure 20 the initial position of the cooperative system is given.) In the example, the state vector is $(Y_0^T q^T)^T = Y_0^x Y_0^y Y_0^\varphi q_1^1 q_1^2 q_1^3 \cdots q_3^1 q_3^2 q_3^3)^T \in R^{12\times 1}$ and the model has the following form (according to (181), (172) and (115):

$$\ddot{q} = N^{-1}(q)(\tau - n(q, \dot{q}, Y_0, \dot{Y}_0)) \ [\text{rad/s}^2] \in R^{9\times 9},$$

$$\ddot{Y}_0 = -W^{-1}w(q, \dot{q}, Y_0, \dot{Y}_0) \ [\text{m/s}^2] \in R^{3\times 1},$$

$$F_c = W_c(\dot{J}(q)\dot{q} + J(q)\ddot{q}) + w_c(q, \dot{q}, Y_0, \dot{Y}_0) \ [\text{N}] \in R^{9\times 1},$$

where

$$N(q) = H(q) + J^T(q)W_c J(q) \in R^{9\times 9},$$

$$n(q, \dot{q}, Y_0, \dot{Y}_0) = h(q, \dot{q}) + J^T(q)W_c \dot{J}(q)\dot{q} + J^T(q)w_c(q, \dot{q}, Y_0, \dot{Y}_0) \in R^{9\times 1},$$

According to (166)

$$H(q) = \text{blockdiag}(H_1(q_1), H_2(q_2), H_3(q_3)) \in R^{9\times 9},$$

$$H_i(q_i) = \begin{bmatrix} a_{11}(q_i^2) & a_{12}(q_i^2) & a_{13} \\ a_{12}(q_i^2) & a_{22} & a_{23} \\ a_{13} & a_{23} & a_{33} \end{bmatrix}$$

and, in the selected example, the coefficients are

$$
\begin{aligned}
a_{11}(q_i^2) &= I_i^1 + (a_i^1)^2 m_i^1 + I_i^2 + ((L_i^2)^2 + (a_i^2)^2 + 2L_i^1 a_i^2 \cos q_i^2) m_i^2 \\
&\quad + I_i^3 + ((L_i^1)^2 + (L_i^2)^2 + +2L_i^1 L_i^2 \cos q_i^2) m_i^3, \\
a_{12}(q_i^2) &= I_i^2 + ((a_i^2)^2 + L_i^1 a_i^2 \cos q_i^2) m_i^2 + I_i^3 + ((L_i^2)^2 + L_i^1 L_i^2 \cos q_i^2) m_i^3, \\
a_{22} &= I_i^2 + (a_i^2)^2 m_i^2 + I_i^3 + (L_i^2)^2 m_i^3, \\
a_{13} &= a_{23} = a_{33} = I_i^3,
\end{aligned}
$$

where $L_i^j = 1$ [m], $i = 1, 2, 3$, $j = 1, 2$ are the lengths of manipulators links, $a_i^j = 0.25$ [m], $i = 1, 2, 3$, $j = 1, 2$, are distances of the links MCs from the rotation points; $m_i = (3, 3, 2)$ [kg], $i = 1, 2, 3$ and $I_i = (0.25, 0.25, 0.01)$ [kg m^2], $i = 1, 2, 3$ are the vectors of masses and moments of inertia of the manipulators links.

Vector components

$$h(q, \dot{q}) = (h_1^T(q_1, \dot{q}_1) \ h_2^T(q_2, \dot{q}_2 \ h_3^T(q_3, \dot{q}_3))^T \in R^{9 \times 1}$$

are given by

$$h_i(q_i, \dot{q}_i) = \begin{bmatrix} -L_i^1(a_i^2 m_i^2 + L_i^2 m_i^3)\dot{q}_i^2(\dot{q}_i^2 + 2\dot{q}_i^1)\sin q_i^2 \\ -L_i^1(a_i^2 m_i^2 + L_i^2 m_i^3)\dot{q}_i^1(\dot{q}_i^1 + 2\dot{q}_i^2)\sin q_i^2 \\ 0 \end{bmatrix} \in R^{3 \times 1}.$$

Matrices W_c and W are constant diagonal matrices

$$
\begin{aligned}
W_c &= \operatorname{diag}(m_1, m_1, A_1, m_2, m_2, A_2, m_3, m_3, A_3) \\
&= \operatorname{diag}(1, 1, 0.005, 1, 1, 0.005, 1, 1, 0.005) \in R^{9 \times 9},
\end{aligned}
$$

$$W = W_0 = \operatorname{diag}(m_0, m_0, A_0) = \operatorname{diag}(5, 5, 0.1) \in R^{3 \times 3},$$

where $m_0 = 5$ [kg], $m_i = 1$ [kg] and $A_0 = 0.1$ [kg m], $A_i = 0.005$ [kg m], $i = 1, 2, 3$ are the masses and moments of inertia of the body CM_0 and of the elastic interconnections CM_1, CM_2, and CM_3. The relation (169) is defined by

$$Y_i = \Theta_i(q_i) = \begin{bmatrix} L_i^1 \cos q_i^1 + L_i^2 \cos(q_i^1 + q_i^2) + Xb_i \\ L_i^1 \sin q_i^1 + L_i^2 \sin(q_i^1 + q_i^2) + Yb_i \\ q_i^1 + q_i^2 + q_i^3 \end{bmatrix} \in R^{3 \times 1}, \quad i = 1, 2, 3,$$

where $(Xb_i, Yb_i) = (0, 0)$, $(4, 0)$ and $(1, 3)$ [m, m] are the coordinates of the ith manipulator support, so that the corresponding Jacobian matrices and their derivatives are

$$J_i(q_i) = \begin{bmatrix} -L_i^1 \sin q_i^1 - L_i^2 \sin(q_i^1 + q_i^2) & -L_i^2 \sin(q_i^1 + q_i^2) & 0 \\ L_i^1 \cos q_i^1 + L_i^2 \cos(q_i^1 + q_i^2) & L_i^2 \cos(q_i^1 + q_i^2) & 0 \\ 1 & 1 & 1 \end{bmatrix} \in R^{3 \times 3}, \quad i = 1, 2, 3,$$

$$\dot{J}_i(q_i) = \begin{bmatrix} -L_i^1 \dot{q}_i^1 \cos q_i^1 - L_i^2 (\dot{q}_i^1 + \dot{q}_i^2) \cos(q_i^1 + q_i^2) & -L_i^2 (\dot{q}_i^1 + \dot{q}_i^2) \cos(q_i^1 + q_i^2) & 0 \\ -L_i^1 \dot{q}_i^1 \sin q_i^1 - L_i^2 (\dot{q}_i^1 + \dot{q}_i^2) \sin(q_i^1 + q_i^2) & -L_i^2 (\dot{q}_i^1 + \dot{q}_i^2) \sin(q_i^1 + q_i^2) & 0 \\ 0 & 0 & 0 \end{bmatrix} \in R^{3 \times 3}.$$

The generalized Jacobian matrix and its derivative are given by

$$J(q) = \begin{bmatrix} J_1(q_1) & 0 & 0 \\ 0 & J_1(q_1) & 0 \\ 0 & 0 & J_1(q_1) \end{bmatrix} \in R^{9 \times 9},$$

$$\dot{J}(q) = \begin{bmatrix} \dot{J}_1(q_1) & 0 & 0 \\ 0 & \dot{J}_1(q_1) & 0 \\ 0 & 0 & \dot{J}_1(q_1) \end{bmatrix} \in R^{9 \times 9}.$$

The vector $w_a \in R^{12 \times 1}$ from (113) is composed of the vectors $w \in R^{3 \times 1}$ and $w_c \in R^{9 \times 1}$

$$w_a = \begin{pmatrix} w \\ w_c \end{pmatrix} = D_a \dot{Y} + \pi_a Y + \frac{1}{2} \frac{\partial Y^T \bar{\pi}_a Y}{\partial Y} \in R^{12 \times 1}.$$

The damping matrices D_a, of stiffness π_a, and $\partial \pi_a / \partial Y_i^j$ have the form

$$*_a = \begin{bmatrix} *_{01} + *_{02} + *_{03} & -*_{01} & -*_{02} & -*_{03} \\ -*_{01} & *_{01} + *_{12} + *_{13} & -*_{12} & -*_{13} \\ -*_{02} & -*_{12} & *_{02} + *_{12} + *_{23} & -*_{23} \\ -*_{03} & -*_{13} & -*_{23} & *_{03} + *_{13} + *_{23} \end{bmatrix}$$
$$\in R^{12 \times 12},$$

where $* = D, \pi, \partial \pi / \partial Y_i^j$. The damping submatrices are

$$D_{ij} = \begin{pmatrix} \mathcal{G}_{ij} D_{ij}^\delta \mathcal{G}_{ij} & 0 \\ 0 & d_{ij}^\varphi \end{pmatrix} \in R^{3 \times 3}, \quad ij = 01, 02, 03, 12, 13, 23,$$

where $r_{ia} = (x_{ia} y_{ia})^T, i = 1, 2, 3,$

$$\mathcal{G}_{ij} = \frac{1}{\|r_{ia} - r_{ja}\|^2} \begin{pmatrix} (x_{ia} - x_{ja})^2 & (x_{ia} - x_{ja})(y_{ia} - y_{ja}) \\ (x_{ia} - x_{ja}^x)(y_{ia} - y_{ja}) & (y_{ia} - y_{ja})^2 \end{pmatrix} \in R^{2 \times 2},$$

whereby the vector norm is $\|\rho_{ija}\| = \|r_{ia} - r_{ja}\|^2 = ((x_{ia} - x_{ja})^2 + (y_{ia} - y_{ja})^2)^{0.5}$.

The constant damping matrix D_{ij} is formed by the damping of the linear motion, given in the diagonal matrix $D_{ij}^{\delta} = \mathrm{diag}(d_{ij}^{x}, d_{ij}^{y}) \in R^{2 \times 2}$ and the damping of the rotational motion d_{ij}^{φ}. The adopted values are $d_{ij}^{x} = d_{ij}^{y} = 20$ [N/(m/s)] and $d_{ij}^{\varphi} = 15$ [Nm/(rad/s)], $ij = 01, 02, 03, 12, 13, 23$.

Submatrices of the stiffness matrix are defined by

$$\pi_{ij} = \begin{bmatrix} c_{ij}^{x} \left(1 - \dfrac{\|\rho_{ij0}\|}{\|r_{ia} - r_{ja}\|}\right)^2 & 0 & 0 \\[2ex] 0 & c_{ij}^{y} \left(1 - \dfrac{\|\rho_{ij0}\|}{\|r_{ia} - r_{ja}\|}\right)^2 & 0 \\[2ex] ds0 & 0 & c_{ij}^{\varphi} \end{bmatrix} \in R^{3 \times 3},$$

where the coefficients $c_{ij}^{x} = c_{ij}^{y} = 1225$ [N/m] and $c_{ij}^{\varphi} = 25$ [Nm/rad], $ij = 01, 02, 03, 12, 13, 23$. The derivatives of the stiffness submatrices are determined by

$$\frac{\partial \pi_{ij}}{\partial *_{\mathrm{ind}}} = \begin{cases} 0, \quad *_{\mathrm{ind}} \neq x_{ia}, x_{ja}, y_{ia}, y_{ja} \\[2ex] 2(*_{\mathrm{ind}} - *_{ja}) \left(1 - \dfrac{\|\rho_{ij0}\|}{\|r_{ia} - r_{ja}\|}\right) \dfrac{\|\rho_{ij0}\|}{\|r_{ia} - r_{ja}\|^3} \mathrm{diag}(c_{ij}^{x}, c_{ij}^{y}, 0), \\[1ex] \qquad * = x, y, \quad \mathrm{ind} = ia, \\[2ex] -2(*_{ia} - *_{\mathrm{ind}}) \left(1 - \dfrac{\|\rho_{ij0}\|}{\|r_{ia} - r_{ja}\|}\right) \dfrac{\|\rho_{ij0}\|}{\|r_{ia} - r_{ja}\|^3} \mathrm{diag}(c_{ij}^{x}, c_{ij}^{y}, 0), \\[1ex] \qquad * = x, y, \quad \mathrm{ind} = ja, \end{cases}$$

where $\|\rho_{ij0}\| = ((x_{i0} - x_{j0})^2 + (y_{i0} - y_{j0})^2)^{0.5}$ and $ij = 01, 02, 03, 12, 13, 23$.

Simulation results are presented in Figures 21 and 22.

In Figures 21a and 21b simulation results are given for all driving torques equal to zero. In Figures 22a to 22g the results obtained for the driving torques $\tau_1^1 = 50$ [Nm] and $\tau_2^1 = -50$ [Nm] are presented, while all the other driving torques are zero (Figure 22g). The components of the realized contact forces at the contact points CM_1 and CM_2 (Figure 20) along the x-axis, $F_1^x = -50$ [N] and $F_2^x = 50$ [N] (Figure 22f) can serve as control quantities.

In all the diagrams, the independent variable (the abscissa) is the simulation time given in seconds, while the dependent variables are the inputs and simulation results. Each diagram is supplied with an explanation giving first the independent variable (T) and then one or more corresponding dependent variables along with their dimensions. All symbols are given by capital letters and numbers, the letter

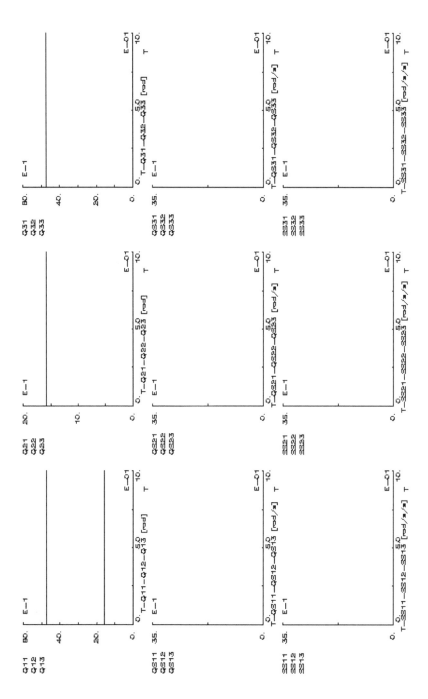

Figure 21a. Simulation results for $\tau_i^j = 0$, $i, j = 1, 2, 3$

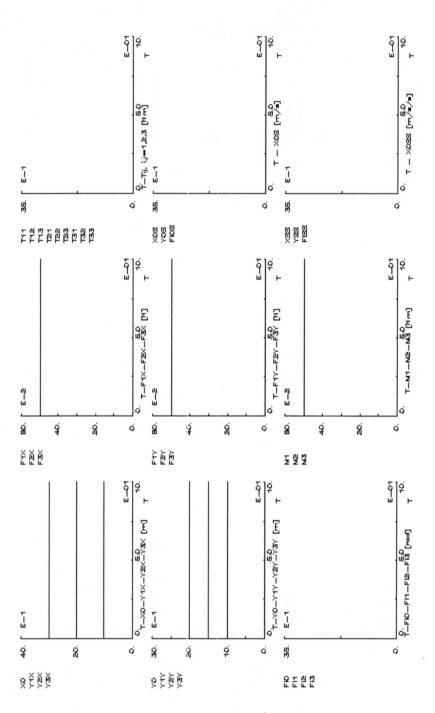

Figure 21b. Simulation results for $\tau_i^j = 0$, $i, j = 1, 2, 3$

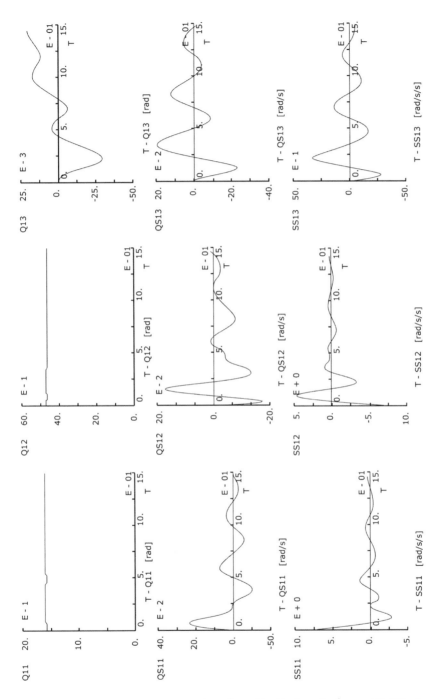

Figure 22a. Simulation results for $\tau_1^1 = 50$ [Nm] and $\tau_2^1 = -50$ [Nm]

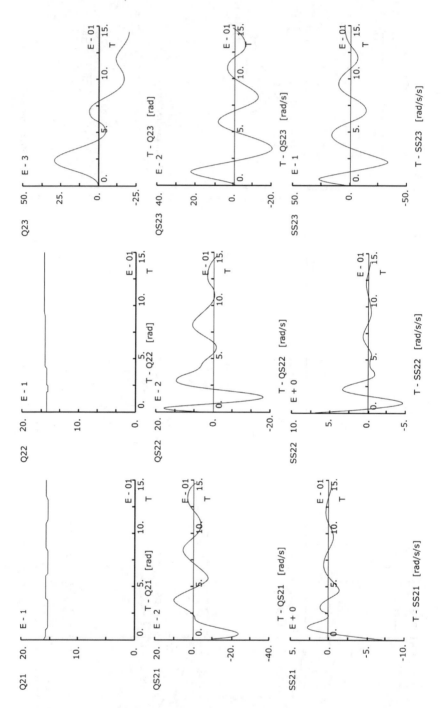

Figure 22b. Simulation results for $\tau_1^1 = 50$ [Nm] and $\tau_2^1 = -50$ [Nm]

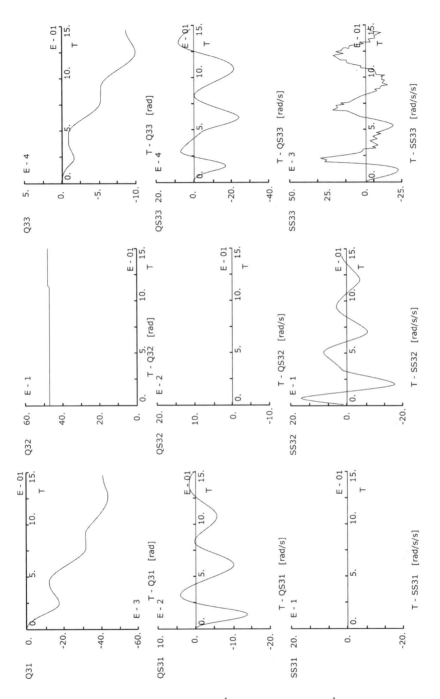

Figure 22c. Simulation results for $\tau_1^1 = 50$ [Nm] and $\tau_2^1 = -50$ [Nm]

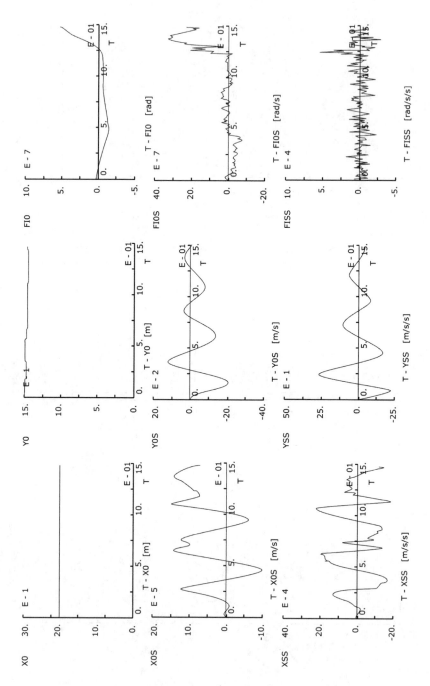

Figure 22d. Simulation results for $\tau_1^1 = 50$ [Nm] and $\tau_2^1 = -50$ [Nm]

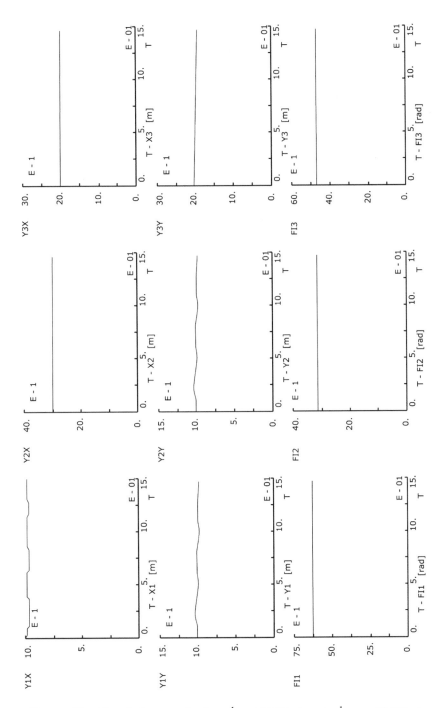

Figure 22e. Simulation results for $\tau_1^1 = 50$ [Nm] and $\tau_2^1 = -50$ [Nm]

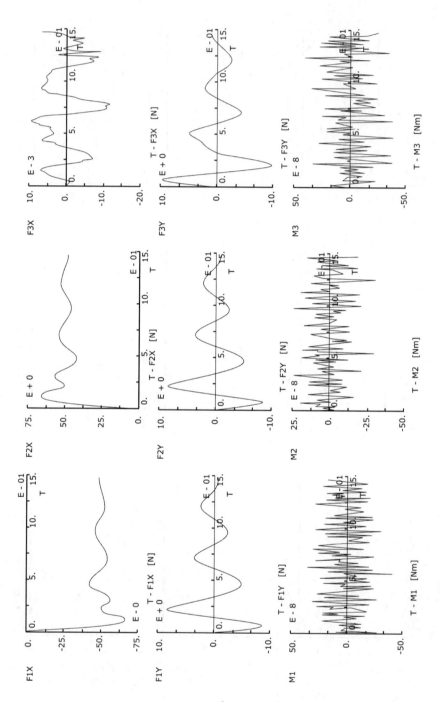

Figure 22f. Simulation results for $\tau_1^1 = 50$ [Nm] and $\tau_2^1 = -50$ [Nm]

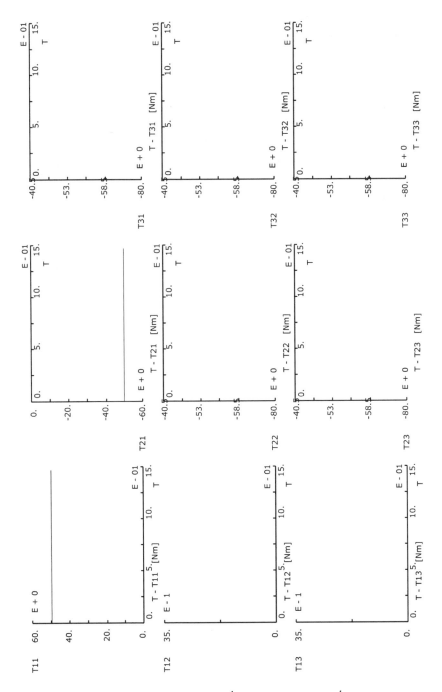

Figure 22g. Simulation results for $\tau_1^1 = 50$ [Nm] and $\tau_2^1 = -50$ [Nm]

part referring to the physical quantity used in modeling and the number indicating the ordinal number of the physical quantity vector. Thus, $Q13$ is the symbol for the internal coordinate q_1^3, whereas $QS13$ and $SS13$ are the symbols for its first and second derivatives $\dot{q}_1^3$ and $\ddot{q}_1^3$. Diagrams of the dependent variable and its derivatives are always given one below the other. The symbol Tij, $i, j = 1, 2, 3$ is associated to the driving moments τ_i^j. The symbols of the quantities at the manipulated object MC $*0$, $*0S$, $*SS$, $* = X$, Y, FI and at contact points $\&i\#$, $\& = Y$, FI, F, M, $\# = X$, Y, denote respectively the linear and angular displacements of the manipulated object MC $*_0$, $\dot{*}_0$, $\ddot{*}_0$, $* = X$, Y, φ, linear and angular displacements of the contact points $\&_i^\#$, $\& = Y$, $\# = X$, Y and φ_i, $i = 1, 2, 3$, are the forces and moments at the contact points $\&_i^\#$, $\& = F$, $\# = X$, Y and M_i, $i = 1, 2, 3$. For example, $Y1X$, $Y1Y$ and $FI1$ are the symbols of the displacement components Y_1^x, Y_1^y and φ_1 of the first contact point, while $F1X$, $F1Y$ and $M1$ are the symbols of the components of the forces F_1^x and F_1^y and moment M_1 in the direction of the displacements Y_1^x, Y_1^y and φ_1 of the first contact point.

5 SYNTHESIS OF NOMINALS

We understand *cooperative system trajectory* as a line described by the state vector in the state space during cooperative system motion, or the image of this line in some other space of the same dimension. From the point of view of mathematics, the trajectory represents a hodograph of the time-dependent vector, characterized by the number of coordinates corresponding to the state space dimension. Each motion of the cooperative system that takes place without external perturbations is *unperturbed motion*. The trajectory described in the state space during the unperturbed motion is called an *unperturbed trajectory*. *Nominal motion of the cooperative system* is any of its unperturbed motions satisfying a certain set of conditions. The maximum number of independent conditions that can be imposed on the nominal motion of a cooperative system is equal to the number of its independent inputs. Constraints can be imposed either on the input or states of the cooperative system. A *nominal trajectory* is an unperturbed trajectory that is realized by the cooperative system during its nominal motion. The *nominal input* is the vector of external actions under which the nominal motion is performed. The input to a cooperative system is represented by the vector of the manipulator driving torques. By the *nominal of a cooperative system* is understood the nominal input and its corresponding nominal trajectory. From the mathematical point of view, a nominal trajectory is the solution of the system of differential equations describing the cooperative system dynamics which is obtained by the action of the nominal input. The problem of determining the nominal motion considered in this section is to define a procedure for the synthesis of the nominal vector, i.e. the vector of the nominal trajectories and the vector of nominal inputs, so to ensure that the differential equations describing the cooperative system dynamics are identically satisfied, provided part of the nominal vector has been given in advance. Such an approach ensures that the determined nominals are realizable under the condition that the mathematical model describes well enough the system dynamics.

In this chapter we present a procedure for the synthesis of the cooperative

137

system nominals. The procedure comes from the solution of the problem of co-ordinated motion of an elastic structure, taking into account the specific features of cooperative manipulation. The procedure has been defined on the basis of the mathematical model of the dynamics of the cooperative manipulation of the object by the non-redundant manipulators with six DOFs, in which the problem of force uncertainty is solved by introducing elastic properties into part of the cooperative system.

5.1 Introduction – Problem Definition

Part of the cooperative system nominals represent the inputs to its control. Hence, the first step in solving the task of cooperative system control is to determine its nominal motion, be it considered as a system of either rigid or elastic bodies.

Generally, the task of the synthesis can be formulated as follows. The cooperative system is primarily used to manipulate objects. The user chooses and defines the manipulation object and its characteristics (dimensions, shape and maximal allowed gripping intensity). The desired object motion is defined by defining the trajectory of a chosen point on the object (e.g. the MC or some other reference point). Starting from the data thus defined, the problem is how to determine the driving torques that are to be introduced at the manipulator's joints to ideally realize the preset requirement in the case of the absence of any disturbance. At that, the stress of the object and manipulator must be within the allowed limits.

When the problem of cooperative system operation is approached from the point of view of the mechanics of a rigid body, which is the usual procedure in the available literature, there appears the problem of force uncertainty. The problem of determining nominals for a rigid cooperative system has not been discussed in the literature. The problem has been reduced to determining the nominals of driving torques to drive rigid manipulators. As we know the kinematic relations between the internal and external coordinates for a known load and tip position of non-redundant manipulators, the problem is easily solvable. The problem of planning (optimal) trajectories of the cooperative system in the work space with or without obstacles has been treated in a number of works [26, 42–46].

As the models used do not faithfully describe the cooperative system's dynamics, the nominal motion of cooperative systems in the available literature has not been determined as its realizable motion, even when the maximum possible number of preset requirements (equal to the number of driving torques) has been ideally realized, irrespective of whether the fulfillment of these conditions results in its optimal or non-optimal motion.

The problem of force uncertainty is solved by considering the cooperative system as an elastic system. In Section 3.3, we showed that the problem of force uncertainty in a cooperative system can be solved by introducing the assumption of the elastic properties of the entire cooperative system or part of it. The cooperative system was approximated by *m* non-redundant manipulators with six DOFs of motion in contact with a body that can move in three-dimensional space without any constraint (Figure 1). The manipulators and object are all assumed to be rigid apart from the neighborhoods of the contact points, whereby the resulting manipulator-object contact is elastic and the manipulator tip cannot move over the object surface. The manipulated object and the neighborhood of its contact points with the manipulators are approximated by an elastic system of *m* + 1 elastically interconnected solid rigid bodies. Each body is allowed to have six DOFs. For the elastic system, gravitational and contact forces are the external forces acting at the MCs of these bodies. By contact forces is understood the six-dimensional vector of generalized force formed from the three-dimensional vector of axial force (dimension [N]) and three-dimensional vector of torques (dimension [Nm]).

The dynamics of a cooperative system thus defined is modeled as a general motion of an elastic structure. Such expansion yields a complex mathematical model, but without force uncertainty. This model faithfully describes the dynamics and statics of the cooperative system. The model of a rigid manipulated object is expanded by equations of elastic connections. This yields a dynamic model of the separated elastic system, composed of a model of rigid body dynamics and a set of equations to describe the elastic interconnections. Depending on the introduced assumptions on the characteristics of elastic connections, these equations are differential (if neither mass nor damping are neglected) or algebraic (see Section 4.12). Nominal motion is determined on the basis of the model given by Equations (102) and (175) for gripping, and by Equations (115) and (181) for the general motion in the form (211). The model characteristics presented in Sections 4.12 and 4.13 show that there is a functional dependence between the kinematic configuration and elastic system load. This property makes the problem of the synthesis of nominals of the elastic cooperative system essentially more complex.

The problem of determining the nominal motion of an elastic cooperative system can be interpreted in the following way. In the cooperative system's motion, the nominal trajectory and its derivatives of the MC or some other reference point of the manipulated object (one node of the elastic system), is prescribed. In this way, six kinematic conditions for describing an elastic system in space are defined. It is assumed that in the course of the cooperative system nominal motion the prescribed trajectory is realized in an ideal way. The mathematical model of cooperative manipulation establishes a functional dependence between the kinematic configuration and the elastic system load. The model of the elastic system establishes a relation between $6m$ active forces and $6m+6$ kinematic quantities and their derivatives. As only six dynamic conditions are defined, the problem is how to define the rest $6m + 6m$ quantities in order to get the desired nominal motion of the cooperative system. At that, any motion of the cooperative system can be chosen as the nominal motion, including the one that corresponds to the resonance states of the elastic system.

From a mathematical point of view, after introducing the desired quantities (as if they were ideally fulfilled) the mathematical model is transformed into a non-homogeneous system of differential equations involving differential constraints. Such a system is solved by taking the left-hand side of the equality being given and seeking the right-hand one, or vice versa, or by giving additional conditions until the task becomes closed in a mathematical sense. The problem is how to set out the conditions that are given in advance and, when these conditions are being fulfilled, how to find the solution of the obtained system of equations.

In cooperative manipulation, one cannot simultaneously prescribe the arbitrary trajectories of the object (6 quantities) and manipulator ($6m$ quantities) and seek active forces ($6m$ quantities), as there can appear excessive contacts and internal stress of the object and manipulators. On the other hand, active forces (contact forces or driving torques) in the course of cooperative system's motion are not known. However, even if the values of contact forces F_c ($6m$ quantities) are known, because of the singularity of the elastic system stiffness matrix, it is not possible to simply give and solve the system of equations and obtain the remaining $6m$ unique nominal trajectories. Such an approach does not ensure a unique description of the cooperative system in space.

A consistent solution of the cooperative system nominals assumes a solution that ensures its unique position in space and a unique load of the elastic structure (object), or the values of contact forces in that position.

The problem of determining nominal motion can be solved by introducing additional conditions, specific to the cooperative manipulation. Namely, in the case when the cooperative system's kinematic configuration represents a copy of some existing natural kinematic configuration, it is possible to record the nominal trajectories of all the links of the natural cooperative system and, on the basis of these records, determine the nominal contact forces and check the system stresses. If this is not possible, it is necessary to determine the contact forces first and then, based on them as driving torques and known position in space, by solving the differential equations that describe elastic system dynamics (102) or (115), determine the nominal trajectories of all the elastic system nodes. After that, the determination of the nominals of rigid non-redundant manipulators with the aid of (166) is a simple and uniquely solvable problem.

Several approaches can yield the solution of the values of contact forces appearing during the motion.

1. A first approach starts from the *a priori* definition of the optimal motion, yielding the extremal values of the acting load [26].

2. The second approach starts from the condition that the vector of contact forces always remains inside the friction cone and the force intensity ensures permanent contact of the manipulators and the manipulated object. The stress state of the elastic system is determined by its total load, obtained as a resultant of the inertial, damping, contact, and gravitation forces. Hence, the fulfillment of the condition for contact force does not guarantee non-violation of the system's permitted stresses.

3. The third approach starts from the condition required of the elastic system stress state (equivalent to the elastic force condition), without taking explicit care of the contact maintaining conditions. Several versions can be distinguished in the scope of this approach.

   - The first version is based on the requirement that there is a certain relationship between the ratio and magnitude of the elastic forces and displacement (as with a pilot, see [47]). Because the work phase schedule in cooperative manipulation is known, the remaining variants are based on the requirement that the motions of the object and manipulators are coordinated.

   - The second version relies upon the possibility that the coordination is achieved by presetting the motion conditions either to the MC or to one contact point of the manipulated object and permitting elastic displacements of the elastic system nodes after the gripping step, due to a change of dynamic forces, orientation during the motion and, possibly, the required changes in gripping conditions.

   - The third version starts from the assumption that the coordination is achieved by setting the motion conditions to one contact point and preserving in the motion the shape of the geometric figure formed by the contact points at the end of the gripping phase.

In this chapter we consider the second and third variants of determining the nominal coordinated motion of the cooperative system on the basis of the conditions of manipulated object MC and one contact point. In the proposed procedure, we first analyze the nominal motion of the separated elastic system and then, on the basis of this analysis, determine the nominal motion of the manipulators. The nominals are determined only for the phases of object gripping and manipulation. The result is a set of nominal quantities (states and inputs) defining different nominal motions of the cooperative system. From this set, the desired inputs to the cooperative system control are selected. Thus, the quantities are selected that are directly tracked, i.e. the quantities that close the feedback loops of the control system of the cooperative system.

5.2   Elastic System Nominals

Let the nominal trajectory of one node of the elastic system be given in advance and let a coordinated motion of the cooperative system be required. Under these assumptions, we define a procedure for the synthesis of separated elastic system in the phases of gripping and general motion. We consider two cases: the case of the object MC trajectory given in advance (the whole $\mathrm{col}(Y_0^0, \dot{Y}_0^0) \in R^{12}$ or only the position part $Y_0^0 \in R^6$), and the case of the prescribed trajectory of one contact point (contact point of the 'leader' $\mathrm{col}(Y_v^0, \dot{Y}_v^0) \in R^{12}$ or $Y_v^0 \in R^6$). The proposed procedure allows the synthesis of the nominal contact force F_c^0 and nominal (not given in advance) trajectories of all the nodes of the elastic system.

5.2.1   Nominal gripping of the elastic system

Let the manipulated object be immobile on its base in the initial moment t_0 (Figure 23). The nominal motion in the instant t_0 is defined as the motion of the elastic system's immobile unloaded state, determined by the coordinates Y_c^u and Y_0^u, with the manipulator tips at contact points, the contact and gravitational forces being zero ($F_{ec}^u = 0$, $F_{e0}^u = 0$). Let the object orientation at the elastic system nodes be the same, with the value $\mathcal{A}_0$. Due to the acting forces, the cooperative system will be displaced to a new equilibrium state. Accordingly, the completed nominal gripping is the cooperative system state in which elastic forces have attained the desired value F_e^s.

Such a description of the gripping phase is consistent because it establishes a full correspondence between the elastic forces, elastic system's nodes displacement, and stresses consequently occurring in the system. It is necessary to know the elasticity forces and displacements of the elastic system's nodes in order to determine the conditions for the manipulators, whereas knowing the stresses is needed to determine the limits of the forces and displacements that will not produce undesired states on the manipulators and manipulated object (e.g. undesired deformation of the object or tips of the manipulators).

Completion of the gripping is not necessarily determined by the stationary conditions of the cooperative system, but, after the stress state has been attained, the motion can be continued without stopping. In order to obtain a better insight into the features originating from the cooperative system elastic properties, let us as-

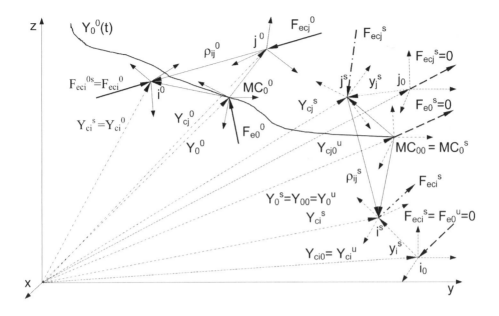

Figure 23. Nominal trajectory of the object MC

sume that the completion of the gripping phase is determined by the stationary state. According to this assumption, the stationary conditions (213) have to hold, and they are expressed by

$$F_{ec}^s = G_c(y_c^s) + F_c^s = A(y^s)y_c^s + b(y^s)y_0^s,$$

$$F_{e0}^s = G_0(y^s) + F_0^s = c(y^s)y_c^s + d(y^s)y_0^s, \tag{217}$$

where the superscript '*s*' denotes the values of the quantities at the end of gripping, and F_0^s represents the resistance of the manipulated object support.

It is essential to note that the attained elastic force of gripping F_e^s, which provides high-quality conditions for the object under static conditions, is distributed onto the gravitational vector G and contact forces F_c. Under the dynamic conditions, the elasticity force, apart from the gravitation and contact forces, is also distributed over dynamic forces.

Since it has been adopted that the initial state is the immobile unloaded state, the elastic displacements of nodes y_*^s at the end of the gripping phase can be measured just from that state. In that case, the stiffness matrix K is constant, i.e., the matrices A, b, c, d, are constant, and can be determined by the finite-element method [6, 7]. The addition of these displacements to the absolute coordinates at the beginning of the gripping yields the absolute coordinates of the cooperative system at the end of the gripping phase.

Further, it is assumed that the displacement of some of the elastic system nodes at the end of gripping phase is known. For example, it can be assumed that the position of the manipulated object MC remains unchanged during the gripping $y_0^s = 0$, or that it changes by the preset value $y_0^s \neq 0$ or, that the displacement of the leader's contact point ($y_v^s \neq 0$) is known.

Let it be required that $y_0^s \neq 0$. Then, from the first equality of (217), the values of the necessary contact displacements y_c^s that ensure realization of the required elastic forces F_{ec}^s can be explicitly calculated

$$y_c^s = A^{-1}F_{ec}^s - A^{-1}by_0^s = A^{-1}(G_c + F_c^s) - A^{-1}by_0^s. \tag{218}$$

It is obvious that there exists a mutually unique correspondence $F_c^s = F_{ec}^s - G_c$ between the required elastic force F_{ec}^s and contact force F_c^s. In the gripping phase, ending with static conditions, it is all the same which force is required as the nominal: the elastic F_{ec}^s or the contact F_c^s. The second expression in (217) represents the equilibrium condition of the forces and moments at the manipulated object MC. Only in the special case of the appropriate choice of F_{ec}^s or F_c^s from (217), an exactly determined value of the reaction force F_0^s of the manipulated object MC can be obtained, such as $F_{e0}^s = G_0(y^s) + F_0^s = cA^{-1}F_{ec}^s + (d - cA^{-1}b)y_0^s = cA^{-1}(G_c + F_c^s)|_{y_0^s=0}$.

If it is required that some exactly determined force is attained under nominal conditions at the manipulated object MC, $F_{e0}^s \neq 0$ (e.g. $F_{e0}^s = G_0$, $F_0^s = 0$ for a hovering object), then the displacement of an arbitrary contact point must be realized as a function of the displacements of the other contact points. To explain this property, it is assumed, for example, that the leader's contact point displacement is a function of the displacements of the other points.

Let the coordinates of that point be denoted by $y_v^s = y_1^s \in R^{6 \times 1}$ and the coordinates of other contact points of the followers by $y_s^s = \mathrm{col}(y_2^s, y_3^s, \ldots, y_m^s) \in R^{(6m-6) \times 1}$. Let us introduce the notations $c = (c_v \ c_{s1} \ldots c_{s(m-1)}) = (c_v \ c_s)$, where the submatrices are $c_* \in R^{6 \times 6}$, $* = v, s1, \ldots, s(m-1)$, $c_s \in R^{6 \times (6m-6)}$.

For the known $y_0^s \neq 0$, from the second expression in (217), it follows that

$$F_{e0}^s = c_v y_v^s + c_{s1}y_{s1}^s + \ldots c_{s(m-1)}y_{s(m-1)}^s + dy_0^s = c_v y_v^s + c_s y_s^s + dy_0^s$$

$$\Rightarrow \quad y_v^s = c_v^{-1}F_{e0}^s - c_v^{-1}c_s y_s^s - c_v^{-1}dy_0^s. \tag{219}$$

By introducing the matrix A in the form of the block matrix $A_{ij} \in R^{6 \times 6}$ and elasticity force by $F_{ec}^s = \mathrm{col}(F_{ev}^s, F_{es}^s) \in R^{6m \times 1}$, $F_{es}^s \in R^{(6m-6) \times 1}$ into the first equation of (217), we obtain

$$
Ay_c^s = \begin{bmatrix} A_{11} & A_{12} & \cdots & A_{1m} \\ \cdots & \cdots & \cdots & \cdots \\ A_{m1} & A_{m2} & \cdots & A_{mm} \end{bmatrix} \begin{bmatrix} c_v^{-1}F_{e0}^s - c_v^{-1}c_s y_s^s - c_v^{-1}\,\mathrm{d}y_0^s \\ y_s^s \end{bmatrix}
$$

$$
= \begin{bmatrix} F_{ev}^s \\ F_{es}^s \end{bmatrix} = F_{ec}^s, \tag{220}
$$

from which follows

$$
\begin{aligned}
F_{ev}^s &= (-A_{11}c_v^{-1}c_s + (A_{12}\ldots A_{1m}))y_s^s - A_{11}c_v^{-1}\,\mathrm{d}y_0^s + A_{11}c_v^{-1}F_{e0}^s \\
&= a_y y_s^s + a_{y0}y_0^s + a_{fe0}F_{e0}^s,
\end{aligned}
$$

$$
\begin{aligned}
F_{es}^s &= \begin{bmatrix} -A_{21}c_v^{-1}c_s + (A_{22}\ldots A_{2m})) \\ \cdots\cdots \\ -A_{m1}c_v^{-1}c_s + (A_{m2}\ldots A_{mm})) \end{bmatrix} y_s^s + \begin{bmatrix} A_{21}c_v^{-1} \\ \cdots \\ A_{m1}c_v^{-1} \end{bmatrix}(F_{e0}^s - \mathrm{d}y_0^s) \\
&= A_y y_s^s + A_{fe0}(F_{e0}^s - \mathrm{d}y_0^s), \tag{221}
\end{aligned}
$$

where $a_y = -A_{11}c_v^{-1}c_s + (A_{12}\ldots A_{1m}) \in R^{6\times(6m-6)}$, $a_{y0} = -A_{11}c_v^{-1}d$, $a_{fe0} = A_{11}c_v^{-1} \in R^{6\times6}$, $A_y = -[A_{i1}]_{i=2\ldots m}c_v^{-1}c_s + [A_{ij}]_{i=2\ldots m, j=2\ldots m} \in R^{(6m-6)\times(6m-6)}$ and $A_{fe0} = [A_{i1}]_{i=2\ldots m}c_v^{-1} \in R^{(6m-6)\times6}$. It is evident that the mathematical form of these relations is $F_{ev}^s = F_{ev}^s(y_0^s, y_s^s, F_{e0}^s)$ and $F_{es}^s = F_{es}^s(y_0^s, y_s^s, F_{e0}^s)$.

By calculating y_s^s as a function of F_{es}^s, from the last equation we can to express the node displacements y_v^s and y_s^s and forces at the leader's node F_{ev}^s as a function of the displacement y_0^s and forces $F_{e0}^s = G_0$ and F_{es}^s. For example, for the case $y_0^s = 0$, these relations are

$$
y_v^s = -c_v^{-1}c_s A_y^{-1}F_{es}^s + c_v^{-1}(c_s A_y^{-1}A_{fe0} + I_{6\times6})F_{e0}^s,
$$

$$
y_s^s = A_y^{-1}F_{es}^s - A_y^{-1}A_{fe0}F_{e0}^s,
$$

$$
F_{ev}^s = a_y A_y^{-1}F_{es}^s + (a_{fe0} - a_y A_y^{-1}A_{fe0})F_{e0}^s,
$$

$$
F_{es}^s = F_{es}^s. \tag{222}
$$

From the previous relations, it is possible to conclude the following: if the manipulated object MC displacement y_0^s and an exact force F_{e0}^s acting at the MC are selected under the nominal gripping conditions, then the displacement of one node y_v^s and the forces F_{ec}^s at all nodes must be expressed as a function of the displacements y_s^s of the remaining $m - 1$ contact points. The relations defined in

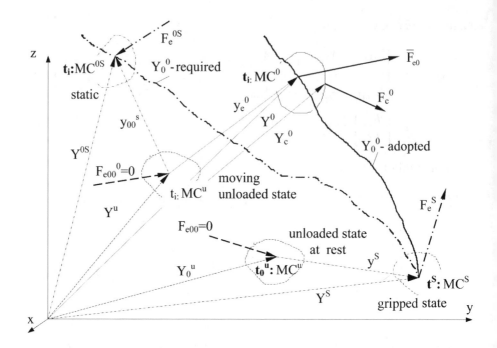

Figure 24. Elastic deviations from the nominal trajectory

(219), (221) and (222), expressed as a function of the elasticity forces F_{es}^s at the contact points y_s^s, must also be satisfied.

Let us consider the approaching and gripping phases in which not an exact but only an approximate position of the manipulated object MC is required. During the approaching, one manipulator comes first and establishes contact with the object, without changing its position. Further, the other manipulators also form contacts, and gripping is performed without significant change in the object's position. Namely, the position of the tip of one manipulator, that is of one contact point, will be given. Let it be the coordinates of the first contact point $y_v = y_1$. Using (209), Equation (217) can be written in the form

$$
F_{ec}^s = \begin{bmatrix} F_{ev}^s \\ F_{es}^s \end{bmatrix} = \begin{bmatrix} u_v & u_s & u_0 \\ A_v & A_s & A_0 \end{bmatrix} \begin{bmatrix} y_v \\ y_s \\ y_0 \end{bmatrix}
$$

$$
= \begin{bmatrix} u_v & u_{s0} \\ A_v & A_{s0} \end{bmatrix} \begin{bmatrix} y_v \\ y_{s0} \end{bmatrix} = A_{us0} y_{s0}^s + A_{uv} y_v^s,
$$

$$
F_{e0}^s = c_s y_s^s + c_0 y_0^s + c_v y_v^s = c_{s0} y_{s0}^s + c_v y_v^s, \tag{223}
$$

where $A_{us0} = ([A_{ij}]_{i,j=2...m} \mid b) \in R^{6m \times 6m}$, $A_{uv} = [A_{i1}]_{i=1...m} \in R^{6m \times 6}$, $c_{s0} =$

$([c_i]_{i=2...m} \mid d) \in R^{6 \times 6m}$ and $c_v = c_1 \in R^{6 \times 6}$. For the given y_v and F^s_{ec}, the position vector of the other nodes y^s_{s0} is calculated from (223) as

$$y^s_{s0} = \begin{bmatrix} y^s_s \\ y^s_0 \end{bmatrix} = A^{-1}_{us0} F^s_{ec} - A^{-1}_{us0} A_{uv} y^s_v \tag{224}$$

and, consequently, the force at the manipulated object MC at the end of the gripping phase will be

$$F^s_{e0} = c_{s0} A^{-1}_{us0} F^s_{ec} + (c_v - c_{s0} A^{-1}_{us0} A_{uv}) y^s_v. \tag{225}$$

It should be noticed that in the case of nominal gripping, it is not necessary to give the overall vector of elasticity force F^s_e, but only the part associated to the contact points F^s_{ec}, which is equivalent to prescribing the vector of contact forces F^s_c. Expressions (224) and (225) can be interpreted in the following way: to determine all the characteristics of the elastic system at the end of gripping phase, it suffices to know the position of one contact point and forces at the other contact points. In other words, it is not necessary to know the properties of the manipulated object in order to be able to reach a conclusion about the elastic system position. Moreover, on the basis of knowing the position of one contact point and forces at the other contact points, it is possible to determine the displacement and forces at the manipulated object MC. Namely, (224) determines $y_{s0} = (y^T_s \ y^T_0)^T$. In this way the object MC y_0 is uniquely determined and, by replacing it into (223), one can calculate the force F^s_{e0} at the object's MC.

If, however, exact elasticity force at the object MC $F^s_{e0} (= G_0)$ is required, then, as in the previous case of nominal gripping, the displacement of another node, different from the contact point of the leader, must be in agreement with the preset force requirement, leader's displacement (generally different from zero) and with the state at other contact points. Namely, as $\det K = 0$, then according to (197), $\det(c_v - c_{s0} A^{-1}_{us0} A_{uv}) = 0$, so that on the basis of the known forces F^s_{e0} and F^s_{ec} from (225) one cannot calculate the necessary leader's displacement y^s_v. It is necessary to first fix the elastic system in space by giving, e.g., the leader's displacements y^s_v as independent variables and then, on the basis of the requirement for the force F^s_{e0} at the manipulated object MC, determine the displacement of one node as a function of the displacements of the other nodes and required force. Hence, it is necessary to start from another equation (223), which can be written in the form

$$c_s y^s_s + c_0 y^s_0 = \sum_{i=1}^{m-1} c_{si} y^s_{si} + c_0 y^s_0 = F^s_{e0} - c_v y^s_v. \tag{226}$$

The vector $F^s_{e0} - c_v y^s_v$ is a known quantity, so that one of the vectors of the followers' displacement y^s_{si}, $i = 1, m - 1$ or displacement of the object's MC y^s_0

can be calculated as a function of non-selected vectors of the followers' displacements and known vector. Let the displacement of the manipulated object MC y_0^s be calculated. From (226) we get that this displacement can be expressed in the form

$$y_0^s = -c_0^{-1}c_s y_s^s - c_0^{-1}c_v y_v^s + c_0^{-1}F_{e0}^s = y_0^s(y_s^s, y_v^s, F_{e0}^s) \tag{227}$$

as a function of the required force at the manipulated object MC, F_{e0}^s, given the leader's displacement y_v^s, and the state of the followers' displacements y_s^s. Under these conditions, the forces acting at contact points

$$
\begin{aligned}
F_{ev}^s &= (u_s - u_0 c_0^{-1} c_s) y_s^s + (u_v - u_0 c_0^{-1} c_v) y_v^s + u_0 c_0^{-1} F_{e0}^s \\
&= F_{ev}^s(y_s^s, y_v^s, F_{e0}^s), \\
F_{es}^s &= (A_s - A_0 c_0^{-1} c_s) y_s^s + (A_v - A_0 c_0^{-1} c_v) y_v^s + A_0 c_0^{-1} F_{e0}^s \\
&= F_{es}^s(y_s^s, y_v^s, F_{e0}^s), \tag{228}
\end{aligned}
$$

will be calculated for the known displacements of the contact points and known (required) force at the object's MC.

Since the matrix $A_s - A_0 c_0^{-1} c_s$ is non-singular, the followers' displacements y_s^s and displacement of the manipulated object MC y_0^s can be determined as a function of forces at the followers' contact points F_{es}^s, displacements of the leader's contact points y_v^s, and of the sought force at the object's MC F_{e0}^s from the expressions

$$
\begin{aligned}
y_s^s &= (A_s - A_0 c_0^{-1} c_s)^{-1} F_{es}^s \\
&\quad -(A_s - A_0 c_0^{-1} c_s)^{-1}(A_v - A_0 c_0^{-1} c_v) y_v^s \\
&\quad -(A_s - A_0 c_0^{-1} c_s)^{-1} A_0 c_0^{-1} F_{e0}^s = y_s^s(F_{es}^s, y_v^s, F_{e0}^s), \\
y_0^s &= -c_0^{-1} c_s (A_s - A_0 c_0^{-1} c_s)^{-1} F_{es}^s - (A_v - A_0 c_0^{-1} c_v) y_v^s \\
&\quad +[c_0^{-1} c_s (A_s - A_0 c_0^{-1} c_s)^{-1}(A_v - A_0 c_0^{-1} c_v) - c_0^{-1} c_v] y_v^s \\
&\quad +[c_0^{-1} c_s (A_s - A_0 c_0^{-1} c_s)^{-1} A_0 c_0^{-1} + c_0^{-1}] F_{e0}^s \\
&= y_0^s(F_{es}^s, y_v^s, F_{e0}^s), \tag{229}
\end{aligned}
$$

whereas the force at the leader's contact point will be determined by the relation

$$
\begin{aligned}
F_{ev}^s &= (u_s - u_0 c_0^{-1} c_s)(A_s - A_0 c_0^{-1} c_s)^{-1} F_{es}^s \\
&\quad + [u_v - u_0 c_0^{-1} c_v - (u_s - u_0 c_0^{-1} c_s)(A_s - A_0 c_0^{-1} c_s)^{-1}(A_v - A_0 c_0^{-1} c_v)] y_v^s
\end{aligned}
$$

$$+ [u_0 c_0^{-1} - (u_s - u_0 c_0^{-1} c_s)(A_s - A_0 c_0^{-1} c_s)^{-1} A_0 c_0^{-1}] F_{e0}^s$$

$$= F_{ev}^s (F_{es}^s, y_v^s, F_{e0}^s). \tag{230}$$

The difference between the nominal conditions given via the object's MC and the connection's MC at the contact point is in the number of requirements to be met by the manipulated object. In the former case, there are two requirements and only one in the latter. Hence, although one starts from the same expression for the force at the manipulated object MC, the requirements concerning node displacements and node force are not the same. By assigning the nominal gripping conditions via the manipulated object MC, one obtains a functional dependence between the displacement of the leader's contact point and the displacements of the other contact points. The assigning of nominal conditions via the contact permits an arbitrary value of the object MC displacement y_0^s, determined by (227) as a function of $y_0^s(y_s^s, y_v^s, F_{e0}^s)$, or by (229) as a function of $y_0^s(F_{es}^s, y_v^s, F_{e0}^s)$. As the object must remain within the geometric figure determined by the contact points, then, although the displacements y_0^s are arbitrary, the object's position after gripping cannot be essentially changed.

Nominal displacements in the gripping phase can be also considered starting from the state acquired by the elastic system as a consequence of the previous action of the contact forces or gravitation forces. In determining the initial position of the elastic system due to gravitational forces, three cases may appear, viz.

- *The object is rigid and the manipulators' tips are elastic.* The position of the object MC is not a function of elastic properties but is determined as the rigid body MC, so that the initial displacement of this point is zero $y_0^g = 0$. Positions of the manipulators' tips are functions of the weight of elastic interconnections $y_c^g = A^{-1} G_c$, obtained using (217).

- *The object is elastic and the manipulators' tips are rigid.* In that case, the theory of elasticity is applied to calculate the displacements due to the action of concentrated gravitation forces at the elastic system's nodes, the supports position of which is known [6, 7]. Namely, expression (217) is expanded by the number of support displacements (which are zero if the object lies on the support surface), whose position in space is known, and is solved with respect to the sought displacements of the connections and manipulated object MC.

- *Both the object and manipulators' tips are elastic.* Then the initial position is calculated as for the elastic object, whereby the masses of connections are equal to the sum of the masses of elastic parts of the manipulators and object associated with the connections.

As a result, we obtain the initial position of the elastic system (displacements of the nodes) due to the gravitational forces. If some contact forces already exist, then, by an analogous procedure, we can find the displacements of the nodes due to their action. If the absolute coordinates of the nodes, defining gravitational and contact loads, are known, then, by subtracting initial displacements from them, we obtain the absolute coordinates of the unloaded state 0 in which the displacements of the nodes are zero. Further, it is possible to apply the procedure of nominal gripping, from the initial state with zero displacements already defined.

Nominal quantities for the beginning and end of gripping, which is ended by static conditions, are defined by the relations (219) and (221) or (222) when assigning nominal conditions to the manipulated object MC and relations (227) and (228), or by (229) and (230) when assigning the conditions to a selected contact point. It remains to define the nominal quantities during the motion in the gripping phase. This practically means that the forces balancing the elastic forces should be supplemented by dynamic forces, so that the solution of nominal conditions will not be determined by the solution of the system of algebraic but of differential equations. All the conditions that are valid for the system of algebraic equations must be fully satisfied for the solution of the differential equations too. When the transition process is completed, the solution of the system of differential equations becomes identical to that of the system of algebraic equations.

The dynamic behavior of the elastic system in the gripping phase can be most simply described either by (100) or (102), given for the immobile state, to which the system would return when the action of the introduced forces stopped.

For the nominal gripping defined by the requirements for the manipulated object MC it is necessary to put in Equations (100) or (102) $y_0 = \dot{y}_0 = \ddot{y}_0 = 0$ and introduce the driving forces at contacts, F_c. As the gripping is the introductory step to the motion, it is assumed that the object at the end of gripping is hovering in space, i.e. $F_{e0}^s = G_0$.

Forces have to be defined as a $6m$-dimensional vector of contact forces defined for the followers as an independent variable vector, and for the leader as a dependent variable vector. The change of contact forces in time, from an initial to the end value, may be an arbitrary monotonous (usually linear) function. However, to the components of each of these forces upon termination of the transition phase (after a certain period of time, the same for all forces) should be assigned a nominal value equal to $F_c = F_e - G$, where the values of F_e are calculated from (222) or (228). After introducing the adopted nominal conditions into (102), we obtain the following system of differential equations:

$$W_c(y_c)\ddot{y}_c + w_c(y_c, \dot{y}_c) = F_c,$$

$$w_0(y_c, \dot{y}_c) = 0. \tag{231}$$

For the first $6m$ differential equations the last six equations represent non-holonomic constraints. The developed form of these equations is

$$W_c(y_c)\ddot{y}_c + F_{bc}(y_c, \dot{y}_c) + D_A \dot{y}_c + A y_c = G_c + F_c,$$

$$D_c \dot{y}_c + c y_c = G_0, \tag{232}$$

where $F_{bc}(y_c, \dot{y}_c) \in R^{6m \times 6m}$ are the force vectors whose components $F_{bi} = \dot{W}_i(y_i)\dot{y}_i - \partial T_i(y_i, \dot{y}_i)/\partial y_i,\ i = 1, \ldots, m$, D_A and D_c are parts of the constant damping matrix D associated to the vector y_c in the same way as the submatrices A and c of the stiffness matrix K were assigned. After differentiating the equations of connections and after introducing the subscripts for the leader v ($y_v = y_1$) and s for the followers, and having in mind the notations (206), (207) and (203) for the structure of matrices and vectors defined at the end of Section 4.12 the last equation obtains the form

$$W_v(y_v)\ddot{y}_v + F_{bv}(y_v, \dot{y}_v) + D_{uvs}\dot{y}_c + u_{vs} y_c = G_v + F_v,$$

$$W_s(y_s)\ddot{y}_s + F_{bs}(y_s, \dot{y}_s) + D_{Avs}\dot{y}_c + A_{vs} y_c = G_s + F_s,$$

$$D_{cv}\ddot{y}_v + D_{cs}\ddot{y}_s + c\dot{y}_c = 0, \tag{233}$$

where

$$
\begin{aligned}
W_v(y_v) &= W_1(y_1) \in R^{6 \times 6}, \\
F_{bv}(y_v, \dot{y}_v) &= F_{b1}(y_1, \dot{y}_1) \in R^{6 \times 1}, \\
D_{uvs} &= (D_{uv} \mid D_{us}) = [D_{1i}]_{i=1\ldots m} \in R^{6 \times 6m}, \\
u_{vs} &= (u_v \mid u_s) = [A_{1i}]_{i=1\ldots m} \in R^{6 \times 6m}, \\
G_v &= G_1 \in R^{6 \times 1}, \qquad F_v = F_{c1} \in R^{6 \times 1}, \\
W_s(y_s) &= \text{diag}(W_2(y_2), \ldots, W_m(y_m)) \in R^{(6m-6) \times (6m-6)}, \\
F_{bs} &= \text{col}(F_{b2}(y_2), \ldots, F_{bm}(y_m)) \in R^{(6m-6) \times 1}, \\
D_{Avs} &= (D_{Av} \mid D_{As}) = [D_{ij}]_{i=2\ldots m, j=1\ldots m} \in R^{(6m-6) \times 6m}, \\
A_{vs} &= (A_v \mid A_s) = [A_{ij}]_{i=2\ldots m, j=1\ldots m} \in R^{(6m-6) \times 6m}, \\
D_{cv} &= D_{c1} \in R^{6 \times 6}, \qquad D_{cs} = (D_{c2} \ldots D_{cm}) \in R^{6 \times (6m-6)}. \quad (234)
\end{aligned}
$$

From the equation of connection, one can calculate the leader's acceleration as a function of the acceleration of the followers. It is obvious that the leader may be only that manipulator whose contact point velocity is characterized by the non-singular matrix D_{cv} (det $D_{cv} \neq 0$). By introducing into the first equation of the found acceleration, we obtain the leader's contact force, so that all the quantities sought can be expressed as a function of the acceleration of followers by

$$
\begin{aligned}
\ddot{y}_v &= -D_{cv}^{-1} D_{cs} \ddot{y}_s - D_{cv}^{-1} c \dot{y}_c, \\
F_v &= -W_v(y_v) D_{cv}^{-1} D_{cs} \ddot{y}_s + F_{bv}(y_v, \dot{y}_v) \\
&\quad + (D_{uvs} - W_v(y_v) D_{cv}^{-1} c) \dot{y}_c + u_{vs} y_c - G_v, \\
F_s &= W_s(y_s) \ddot{y}_s + F_{bs}(y_s, \dot{y}_s) + D_{Avs} \dot{y}_c + A_{vs} y_c - G_s. \quad (235)
\end{aligned}
$$

As the inertia matrix $W_s(y_s)$ is always non-singular, the followers' accelerations $\ddot{y}_s$ are uniquely calculated as a function of the followers' contact forces, whose change can be given as the nominal $F_s = F_s^s(t)$. Thus, one obtains

$$
\begin{aligned}
\ddot{y}_v &= -D_{cv}^{-1} D_{cs} W_s(y_s)^{-1} F_s^s \\
&\quad + D_{cv}^{-1} D_{cs} W_s(y_s)^{-1}(F_{bs}(y_s, \dot{y}_s) + D_{Avs} \dot{y}_c + A_{vs} y_c - G_s) - D_{cv}^{-1} c \dot{y}_c, \\
\ddot{y}_s &= W_s(y_s)^{-1} F_s^s - W_s(y_s)^{-1}(F_{bs}(y_s, \dot{y}_s) + D_{Avs} \dot{y}_c + A_{vs} y_c - G_s), \\
F_v &= -W_v(y_v) D_{cv}^{-1} D_{cs} W_s^{-1}(y_s) F_s^s \\
&\quad + W_v(y_v) D_{cv}^{-1} D_{cs} W_s^{-1}(y_s)(F_{bs}(y_s, \dot{y}_s) + D_{Avs} \dot{y}_c + A_{vs} y_c - G_s) \\
&\quad + F_{bv}(y_v, \dot{y}_v) + (D_{uvs} - W_v(y_v) D_{cv}^{-1} c) \dot{y}_c + u_{vs} y_c - G_v, \\
F_s &= F_s^s. \quad (236)
\end{aligned}
$$

The expression for the followers' acceleration $\ddot{y}_s$ defines the full system of $6m - 6$ second-order differential equations, whose solving gives the nominal trajectories $y_s^s(t)$ of the contact points of the followers in the gripping phase. By solving six second-order equations for the leader's acceleration $\ddot{y}_v$ or the last six first-order equations (232) for the leader's velocity, the nominal trajectories $y_v^s(t)$ of the leader's contact points are obtained. The simplest way to obtain such a solution is the simulation with $F_s^s(t)$ as input, whose initial and final values are determined from static conditions. By introducing the obtained values for the leader's contact force F_v into (236), we obtain the nominal value of the leader's contact force $F_v^s(t)$, whereby all the values of the nominal quantities of gripping under the conditions $y_0 = \dot{y}_0 = \ddot{y}_0 = 0$ and $F_{e0}^s = G_0$ are determined.

For the nominal gripping determined by the requirements for the leader's contact point, the conditions $(y_v, \dot{y}_v, \ddot{y}_v)$ and gripping forces (elastic and the contact one) at all contact points of the leader in the beginning and at the end of gripping are known. Assuming that the initial state of all the nodes is known, it is necessary to determine the trajectories and forces at the nodes during the gripping phase. This can be done, as in the previous case, while considering the conditions for the leader and manipulated object as non-holonomic constraints for the rest of the system. More complex expressions would be obtained than by assigning the nominal requirements for the object MC.

To get a more vivid picture of the initial state of the nominal motion, let us recapitulate what we said about the object gripping. The gripping phase was observed beginning from the elastic system state Y_{c0}, Y_{00} to which corresponded a zero values of all the forces ($F_{ec0} = 0$, $F_{e00} = 0$). In that state, the orientations of the object and connections were the same, $\mathcal{A}_0$. Therefore, the gripping to attain the nominal gripping force $F_e^s = \text{col}(F_{ec}^s, F_{e0}^s)$ was performed, and the resulting displacements of the nodes, $y^s = \text{col}(y_c^s, y_0^s)$ were measured from the initial immobile unloaded state. The final state of the nominal gripping at the moment t^s is the initial state of the nominal motion with the absolute coordinates $Y_c^s = Y_c^u + y_c^s$, $Y_0^s = Y_0^u + y_0^s$ in which the elastic forces F_{ec}^s, F_{e0}^s are acting (Figures 23 and 24), realized after the nominal gripping. In the initial state of the nominal motion, the coordinates of an arbitrary contact point and of the object MC are $Y_{ci}^s = \text{col}(r_{ci}^s, \mathcal{A}_i^s)$ $= \text{col}(r_{ci}^s, \mathcal{A}_0 + \mathcal{A}_i) \in R^{6\times1}$ and $Y_0^s = \text{col}(r_0^s, \mathcal{A}_0^s) \in R^{6\times1}$, where r_{ci}^s and r_0^s are the vectors of Cartesian coordinates of the MC and $\mathcal{A}_i$ are the vectors of orientation increments during the gripping.

Further, the contact forces acting during the motion along the required nominal trajectory are to be determined.

5.2.2 Nominal motion of the elastic system

From the above discussion it is possible either to prescribe the forces and seek the kinematic quantities or to prescribe the kinematic quantities and seek for the forces of the elastic system. The problem is how to prescribe some of the mentioned quantities that yield a coordinated motion in space. The procedure proposed for gripping provides the initial and final position under static conditions and the forces corresponding to them. It is implicitly assumed that the elastic system's unloaded state does not move. Also, it is proposed that the change of the gripping force from the initial to end state is a monotonous function. The problem is closed in a mathematical sense, and the desired coordination of motion in gripping is achieved. In the case of the motion along a given trajectory, the unloaded state is mobile, and its position is not known. Even if its position were known, the motion around the mobile state would not proceed as around the unloaded immobile state. The same conclusion would also hold for the solution of the coordinated motion. Hence, a two-stage procedure is proposed to determine the nominal quantities during the motion.

In brief, the procedure to calculate the nominals during the motion can be re-capitulated as follows: It is proposed that during the nominal motion, the problem of determining the contact forces has to be resolved by setting the requirement that the motion in the cooperative manipulation is coordinated. By the coordinated motion of the cooperative system is meant the motion by which the manipulated object is initially gripped to a definite elastic force, and then it continues to perform the general motion, whereby the manipulators move in a way that ensures the gripping conditions are not essentially violated. It is assumed that the elastic displacements are not large and that the positions of elastic system's nodes during the static displacement and at the end of the motion along the trajectory given for the manipulated object, cannot essentially change. A two-stage procedure is proposed. In the first stage, during the coordinated quasi-static motion, the contact forces are calculated as approximate values by applying static methods. From the initial motion state at the instant t^s (end of gripping – the quantities have the superscript 's') the gripped object is statically transferred to the series of selected points on the trajectory (the variables correspond to the instants t_i and bear the superscript '$0s$'), keeping the fictitious action of the forces at the end of gripping in the coordinate system attached to the loaded state, without taking into consideration the actual loads. After canceling the fictitious action of these forces, the unloaded state of the elastic system (the variables have the superscript 'u') in the transferred position is obtained (Figure 24). The loaded state of the elastic system in the trans-ferred position is obtained by the static action of the resultants of the gravitational forces, rotated contact forces from the end of gripping, and dynamic forces at each of the elastic system nodes. Dynamic forces are determined by using the acceler-ations and velocities of the nodes, obtained from the condition that, from the end of gripping on, the elastic system moves as a rigid body. If, in addition to the ma-nipulated object motion along the nominal trajectory, a simultaneous change of the gripping forces is required, then, instead of the rotated contact forces from the end of the gripping step, the sought contact forces are used to calculate the results. For the obtained trajectories, the approximate contact forces needed to bring the elastic system nodes to the calculated positions, are determined. In the second stage, these contact forces are adopted as the nominal forces in the coordinated motion. It is proposed that during the motion between the selected points on the trajectory, the changes of contact forces are monotonous functions. The trajectories that satisfy the motion equations are determined by numerically solving the full system of dif-ferential equations that describes the dynamic contacts of the followers, whereby the nominal forces of the system input are adopted. Nominal conditions at the leader's contact point are dependent on the manipulated object nominal conditions and on the nominal conditions at the contact points of the manipulators-followers.

Let the nominal trajectory of the manipulated object MC be set as the line $Y_0^0(t) = \mathrm{col}(r_0^0(t), \mathcal{A}_0^0(t)) \in R^{6 \times 1}$, to which belongs the point Y_0^s (Figures 23 and 24). Under purely static conditions, to transfer the gripped manipulated object from the position CM_0^s to the position CM_0^{0s} on the trajectory $Y_0^{0s}(t_i)$, it is necessary to make one translation by the vector $r_0^0 - r_0^s$ and one orientation change around CM_0^{0s} for $\mathcal{A}_0^0 - \mathcal{A}_0^s$ of the gripped object (loaded state of the elastic system after gripping being completed on the whole). The absolute coordinates of the elastic system nodes in the transferred position, for the instantaneous rotation pole of the manipulated object MC, are (see Section 4.7, relations (123) and (150))

$$Y^{0s} = \eta + A_r(\mathcal{A}_0^0 - \mathcal{A}_0^s)\rho_0^s + a_r(\mathcal{A}_0^0 - \mathcal{A}_0^s). \tag{237}$$

The forces acting at the nodes are

$$F_e^{0s} = A_r^T(\mathcal{A}_0^0 - \mathcal{A}_0^s)F_e^s = A_r^T(\mathcal{A}_0^0 - \mathcal{A}_0^s)(G + \mathrm{col}(F_c^s, 0)), \tag{238}$$

where

$$A_r(a) = \mathrm{diag}(A(a), I_{3 \times 3}, \ldots A(a), I_{3 \times 3}) \in R^{(6m+6) \times (6m+6)},$$

$$a_r(a) = \mathrm{col}(0_{1 \times 3}, a, \ldots 0_{1 \times 3}, a) \in R^{(6m+6) \times 1},$$

$$a = \mathcal{A}_0^0 - \mathcal{A}_0^s,$$

$A(a)$ is the coordinate transformation matrix at the rotation by the orientation a; F_e^s is the elastic force attained at the end of gripping; $\rho_0^s = \mathrm{col}(\rho_{00}^s, \rho_{01}^s, \ldots, \rho_{0m}^s)$, $\rho_{00}^s = 0$, $\rho_{00}^s = 0$, is the vector of distance of the nodes from the manipulated object MC at the end of gripping, and $\eta = \mathrm{col}(r_0^0 - r_0^s \ 0 \ r_0^0 - r_0^s \ 0 \ \ldots \ r_0^0 - r_0^s \ 0)$ is the expanded vector of absolute coordinates, defining the translation of the elastic system nodes at the end of gripping as if they were rigid body points.

Since gravitational forces do not change the direction of their action, the elastic forces in the rotated position will differ from the F_e^{0s} calculated from the expression (238) by $\Delta G = (I - A_r^T)G$ and, in proportion to that force, some additional displacement of the nodes will take place.

Because of the limited time interval needed for the motion along the trajectory, the trajectory is preset not only as a function of space but also as a function of time, $Y_0^0 = Y_0^0(t) \in R^{6 \times 1}$. A consequence of this is also the appearance of dynamic forces at the nodes that are equal to the sum of inertial and damping forces. The elastic forces at the manipulated object MC are balanced by the gravitation force and produced dynamic force $F_{e0}^0 = G_0 + F_{d0} = G_0 + F_{in0} + F_{t0}$. The key issue of the nominal motion and the later introduction of the control laws is how to realize the dynamic force F_{d0} on account of the additional displacements of the nodes, and especially of the contact points through which energy is introduced into the system. This means that the motion after the gripping phase is not possible without the additional motion of the elastic system's nodes.

The above properties can be described in a simplest way in the case when the elastic system upon gripping, performs only a translatory motion without the action of any damping force. Then, the motion equations will be

$$
\begin{aligned}
F_{d0} + G_0 \quad\quad\;\;\; &= F_{e0}, \\
F_{dc1} + G_1 \quad +F_{c1} &= F_{ec1}, \\
\cdots \quad\quad\quad \cdots \quad\;\; &\;\;\; \cdots \\
F_{dcm} + G_m \quad +F_{cm} &= F_{ecm}.
\end{aligned}
\tag{239}
$$

If there would be no first equation, then the value of any contact force would change by the value of the produced dynamic force $F_{dci} = F_{ini}$, $i = 1, \ldots, m$ and the motion would take place in the desired nominal manner. In the first equation, the force F_{d0} is a function only of the derivative of the object MC coordinates Y_0^0, i.e. $F_{d0} = F_{in0} = F_{in0}(\ddot{Y}_0^0, \dot{Y}_0^0, Y_0^0)$, whereas the elastic force F_{e0} is a function of the coordinate position of all the nodes $F_{e0} = F_{e0}(Y^0)$, so that this equation can be written in the form

$$
F_{in0}(\ddot{Y}_0^0, \dot{Y}_0^0, Y_0^0) + G_0 = F_{e0}(Y^0) = F_{e0}^s + \Delta F_{e0}(Y^0),
$$

$$
\Rightarrow F_{in0}(\ddot{Y}_0^0, \dot{Y}_0^0, Y_0^0) = \Delta F_{e0}(Y^0) = \Delta F_{e0}(Y_0^0, Y_1^0, \ldots, Y_m^0).
\tag{240}
$$

As $Y_0^0(t)$ is a given function, the quantities $\ddot{Y}_0^0(t)$, $\dot{Y}_0^0(t)$ are also known functions so that the last relation can be written as

$$
\varphi_h(Y_0^0(t), \dot{Y}_0^0(t), \ddot{Y}_0^0(t), Y_1^0, \ldots, Y_m^0) = \varphi_h(t, Y_1^0, \ldots, Y_m^0) = 0.
\tag{241}
$$

This algebraic equation is non-linear by its arguments and it defines a hyper-surface in the subspace $\{Y_1^0, \ldots, Y_m^0\}$, and for the rest m differential equations (239) represents holonomic constraints. If the damping forces $F_t = F_t(Y^0, \dot{Y}^0)$, due to the spatial motion resistance, were also taken into account, then they had to be balanced by the elastic forces

$$
F_{in0}(Y_0^0, \dot{Y}_0^0, \ddot{Y}_0^0) + F_t(Y^0, \dot{Y}^0) = \Delta F_{e0}(Y_0^0, Y_1^0, \ldots, Y_m^0)
\tag{242}
$$

or, in a more compact form,

$$
\varphi_{nh}(t, Y_1^0, \ldots, Y_m^0, \dot{Y}_1^0, \ldots, \dot{Y}_m^0) = 0,
\tag{243}
$$

which for the rest m equations (239) represents non-holonomic constraints. Solving the nominal motion assumes the explicit calculation of the necessary contact forces and kinematic quantities of the contact points. If the trajectory $Y_0^0(t)$ is given, then there is obviously an infinite number of different ways in which this can be done. In each of these ways, the conditions attained at the end of gripping are unavoidably violated.

Approximate values of the contact forces are determined on the basis of attaining the elastic system's coordinated motion. If the motion were only static, the position Y_{0s} and elastic forces F_e^{0s} of the elastic system, calculated from (237) and (238) and corrected for ΔG, would ensure full coordination of the motion. To determine the nominal coordinates Y^0 in the presence of dynamic forces too, it is necessary to determine first the unloaded state position to which correspond the vectors Y_{0s} and F_e^{0s}. It will be allowed that the sum of the static and dynamic forces by their action displace, in a purely static way, the elastic system from the equilibrium state. The attained state is saved as a function of time.

As the gripping forces F_c^s (238) produce the displacement y^s, and the disposition of the forces during the static transfer with respect to the elastic system is unchanged, then the same displacement measured in the same coordinate system has to be produced in the new position (Figure 24). In the new static position, the elastic system is rotated relative to the state attained at the end of gripping by the orientation $\mathcal{A}_0^0 - \mathcal{A}_0^s$. The coordinates of displacement from the unloaded state are $y^{0s} = A_r(\mathcal{A}_0^0 - \mathcal{A}_0^s)y^s$. The absolute coordinates of that unloaded state are $Y^u = Y^{0s} - y^{0s}$, whereas the acting forces are equal to zero. In that state let the action of the resultant of dynamic and static forces begin (Figure 24).

$$\bar{F}_e = G + A_r^T(\mathcal{A}_0^0 - \mathcal{A}_0^s)\mathrm{col}(F_c^s, 0) + F_d^{0s}, \qquad (244)$$

where

$$F_d^{0s} = \mathrm{col}(F_{dc}^{0s}, F_{d0}^{0s}),$$

$$F_{dc}^{0s} = -(W_c(Y_c^{0s})\ddot{Y}_c^{0s} + F_{bc}(Y_c^{0s}, \dot{Y}_c^{0s}) + D_A(Y^{0s})\dot{Y}_c^{0s} + D_b(Y^{0s})\dot{Y}_0^{0s}),$$

$$F_{d0}^{0s} = -(W_0(Y_0^{0s})\ddot{Y}_0^{0s} + F_{b0}(Y_0^{0s}, \dot{Y}_0^{0s}) + D_c(Y^{0s})\dot{Y}_c^{0s} + D_d(Y^{0s})\dot{Y}_0^{0s})$$

are the dynamic forces; Y^{0s}, $\dot{Y}^{0s}$ and $\ddot{Y}^{0s}$ are the coordinates of the nodes and their derivatives, calculated on the basis of the prescribed trajectory of the manipulated object MC as if they belonged to the rigid body, with the relative distances attained at the end of gripping phase. Under purely static conditions, the action of the previous forces will produce the displacement y^0, which is calculated from

$$\bar{F}_e = A_r^T(\mathcal{A}_0^0 - \mathcal{A}_0^s)KA_r(\mathcal{A}_0^0 - \mathcal{A}_0^s)y^0, \qquad (245)$$

yielding the loaded state absolute coordinates

$$Y^0 = Y^{0s} - y^{0s} + y^0 \in R^{6m \times 6}. \qquad (246)$$

Due to the different disposition of the gravitational forces with respect to the loaded elastic system at the end of gripping and at the current position on the trajectory, the calculated values of the position of the manipulated object MC $Y_0^0 \in R^6$ will differ from the initial ones by $-y_0^{0s} + y_0^0 \in R^6$. This can be avoided by assuming that $y_0^{0s} = y_0^0$ when solving Equation (245) ($A_r^T K A_r \in R^{(6m+6) \times (6m+6)}$, rank $A_r^T K A_r = 6m$). Since the node coordinates and their derivatives are calculated as if the elastic system were a rigid body, the damping properties of the elastic system micro-motion were not taken into account. The repetition of the procedure for the series of positions on the trajectory (in the series of instants t_i), yields the discrete function $Y^0(t_i)$, which is adopted as the temporary nominal trajectory of the elastic system's nodes. Differentiating gives their derivatives. Substitution of the values of the node absolute coordinates and their derivatives into the equations of motion (217) yields the approximate values of the necessary contact forces as discrete functions. Such a procedure maximally reduces the error for the values of displacements and contact forces due to the action of the dynamic forces in the spatial macro-motion, but still the error remains due to the elastic system's micro-motion when the distances between contact points change. Hence, during the nominal motion of the elastic system, the elastic displacements have to be allowed, and the coordinated nominal trajectories determined by solving (115).

Let us note that if a simultaneous change of the elastic force F_e^{0s} (gripping conditions) during the motion along the nominal trajectory is also required, then this force must be included into calculation instead of the force F_e^{0s} determined from (238).

The previous procedure determines the discrete values of the contact forces in the process of the manipulated object MC passing through a series of selected points on the prescribed trajectory Y_0^0. It is proposed that the change of contact forces between the calculated values is a smooth monotonous function, which also retains such a character when passing from one trajectory segment to the other, determined by the selected points. Let us adopt the contact force thus determined as the nominal contact force and denote it by F_c^0. Further, it remains to determine for this nominal contact force the nominal trajectories that satisfy the elastic system equations of motion (115).

The elastic system properties are invariant under static and dynamic conditions. Only the origin of the forces acting on the elastic system varies. Under static conditions, the load is a result of the gravitational and contact forces, whereas for the dynamic conditions dynamic force is added. Therefore, the procedure for determining nominal trajectories possesses properties similar to the procedure for solving nominals under static condition. Under dynamic conditions, instead of the algebraic equations (217) and (218), differential equations should be solved.

Using the indexing system defined in (206), (207) and (203) for the structure of matrices and vectors, with the subscript v for the leader ($y_v = y_1$) and subscript s for the followers, to calculate the nominal motion conditions prescribed via the manipulated object MC equations (115), should be written in the form

$$W_v(Y_v)\ddot{Y}_v + F_{bv}(Y_v, \dot{Y}_v) + D_{uvs}(Y)\dot{Y}_c + D_{u0}(Y)\dot{Y}_0$$

$$+ u_{vs}(Y)Y_c + u_0(Y)Y_0 = G_v + F_v(Y),$$

$$W_s(Y_s)\ddot{Y}_s + F_{bs}(Y_s, \dot{Y}_s) + D_{Avs}(Y)\dot{Y}_c + D_{A0}(Y)\dot{Y}_0$$

$$+ A_{vs}(Y)Y_c + A_0(Y)Y_0 = G_s + F_s(Y),$$

$$W_0(Y_0)\ddot{Y}_0 + F_{b0}(Y_0, \dot{Y}_0) + D_{cv}(Y)\dot{Y}_v + D_{cs}(Y)\dot{Y}_s + D_d(Y)\dot{Y}_0$$

$$+ c(Y)Y_c + d(Y)Y_0 = G_0, \tag{247}$$

where

$$W_v(Y_v) = W_1(Y_1) \in R^{6\times 6},$$

$$F_{bv}(Y_v, \dot{Y}_v) = F_{b1}(Y_1, \dot{Y}_1) = \dot{W}_1(Y_1)\dot{Y}_1 - \partial T_1(Y_1, \dot{Y}_1)/\partial Y_1 \in R^{6\times 1},$$

$$D_{uvs}(Y) = (D_{uv}(Y) \mid D_{us}(Y)) = [D_{1i}(Y)]_{i=1...m} \in R^{6\times 6m},$$

$$u_{vs}(Y) = (u_v(Y) \mid u_s(Y)) = [A_{1i}(Y)]_{i=1...m} \in R^{6\times 6m},$$

$$G_v = G_1 \in R^{6\times 1}, \qquad F_v(Y) = F_{c1}(Y) \in R^{6\times 1},$$

$$W_s(Y_s) = \mathrm{diag}(W_2(Y_2), \dots, W_m(Y_m)) \in R^{(6m-6)\times(6m-6)},$$

$$F_{bs}(Y_s, \dot{Y}_s) = \mathrm{col}(F_{b2}(Y_2, \dot{Y}_2), \dots, F_{bm}(Y_m, \dot{Y}_m))$$

$$= \mathrm{col}\left(\dot{W}_2(Y_2)\dot{Y}_2 - \frac{\partial T_2(Y_2, \dot{Y}_2)}{\partial Y_2}, \dots, \dot{W}_m(Y_m)\dot{Y}_m - \frac{\partial T_m(Y_m, \dot{Y}_m)}{\partial Y_m}\right)$$

$$\in R^{(6m-6)\times 1},$$

$$D_{Avs}(Y) = (D_{Av}(Y) \mid D_{As}(Y)) = [D_{ij}(Y)]_{i=2...m, j=1...m} \in R^{6m-6},$$

$$A_{vs}(Y) = (A_v(Y) \mid A_s(Y)) = [A_{ij}(Y)]_{i=2...m, j=1...m} \in R^{(6m-6)\times 6m},$$

$$D_{cv}(Y) = D_{c1}(Y) \in R^{6\times 6},$$

$$D_{cs}(Y) = D_{c2}(Y) \dots D_{cm}(Y)) \in R^{6\times(6m-6)},$$

$$F_{b0} = \dot{W}_0(Y_0)\dot{Y}_0 - \frac{\partial T_0(Y_0, \dot{Y}_0)}{\partial Y_0} \in R^6. \tag{248}$$

For the given nominal conditions of the object MC, the variables $Y_0(t) = Y_0^0(t)$, $\dot{Y}_0(t) = \dot{Y}_0^0(t)$, $\ddot{Y}_0(t) = \ddot{Y}_0^0(t)$ are known, so that the last six equations in (247), describing the dynamics of the manipulated object, represent a differential constraint for the rest of the elastic system. In the case where the damping properties of the elastic contacts are not taken into account, these constraints become algebraic. The differentiation of the equation of constraint yields the possible leader's acceleration as a function of the accelerations of the followers. The leader can be only that manipulator whose contact point velocity is characterized by the non-singular matrix $D_{cv}(Y)$ (det $D_{cv}(Y) \neq 0$). By introducing the determined potential acceleration into the first equation, one obtains the contact force of the leader, so that all the quantities sought can be expressed as a function of the accelerations of the followers, i.e.

$$\ddot{Y}_v = -D_{cv}^{-1}(Y)D_{cs}(Y)\ddot{Y}_s - D_{cv}^{-1}(Y)\Omega,$$

$$F_v(Y) = -W_v(Y_v)D_{cv}^{-1}(Y)D_{cs}(Y)\ddot{Y}_s,$$

$$-W_v(Y_v)D_{cv}^{-1}(Y)\Omega + F_{bv}(Y_v, \dot{Y}_v)$$

$$+D_{uvs}(Y)\dot{Y}_c + D_{u0}(Y)\dot{Y}_0^0 + u_{vs}(Y)Y_c + u_0(Y)Y_0^0 - G_v,$$

$$F_s(Y) = W_s(Y_s)\ddot{Y}_s + F_{bs}(Y_s, \dot{Y}_s) + D_{Avs}(Y)\dot{Y}_c + D_{A0}(Y)\dot{Y}_0^0$$

$$+A_{vs}(Y)Y_c + A_0(Y)Y_0^0 - G_s, \tag{249}$$

where

$$\Omega = \Omega(Y_0^0, \dot{Y}_0^0, \ddot{Y}_0^0, \dddot{Y}_0^0, Y_v, \dot{Y}_v, Y_s, \dot{Y}_s)$$

$$= \dot{W}_0(Y_0^0)\ddot{Y}_0^0 + W_0(Y_0^0)\,\dddot{Y}_0^0 + \dot{F}_{b0}(Y_0^0, \dot{Y}_0^0) + \dot{D}_{cv}(Y)\dot{Y}_v$$

$$+ \dot{D}_{cs}(Y)\dot{Y}_s + \dot{D}_d(Y)\dot{Y}_0^0 + D_d(Y)\ddot{Y}_0^0$$

$$+ \dot{c}(Y)Y_c + c(Y)\dot{Y}_c + \dot{d}(Y)Y_0^0 + d(Y)\dot{Y}_0^0. \tag{250}$$

The matrix of inertia $W_s(Y_s)$ is always non-singular so that the followers' accelerations $\ddot{Y}_s$ are uniquely calculated as a function of the followers' contact forces, whose change is prescribed as the nominal $F_s = F_s^0 = F_s^0(Y_0^0)$. One obtains

$$\ddot{Y}_v = -D_{cv}^{-1}(Y)D_{cs}(Y)W_s^{-1}(Y_s)F_s^0(Y_0^0)$$

$$+ D_{cv}^{-1}(Y)D_{cs}(Y)W_s^{-1}(Y_s)(F_{bs}(Y_s, \dot{Y}_s) + D_{Avs}(Y)\dot{Y}_c + D_{A0}(Y)\dot{Y}_0^0$$

$$+ A_{vs}(Y)Y_c + A_0(Y)Y_0^0 - G_s) - D_{cv}^{-1}(Y)\Omega,$$

$$\ddot{Y}_s = W_s^{-1}(Y_s)F_s^0(Y_0^0) - W_s^{-1}(Y_s)(F_{bs}(Y_s, \dot{Y}_s) + D_{Avs}(Y)\dot{Y}_c + D_{A0}(Y)\dot{Y}_0^0$$

$$+ A_{vs}(Y)Y_c + A_0(Y)Y_0^0 - G_s),$$

$$F_v = -W_v(Y_v)D_{cv}^{-1}D_{cs}W_s^{-1}(Y_s)F_s^0(Y_0^0)$$
$$+ W_v(Y_v)D_{cv}^{-1}D_{cs}W_s^{-1}(Y_s)(F_{bs}(Y_s, \dot{Y}_s) + D_{Avs}(Y)\dot{Y}_c + D_{A0}(Y)\dot{Y}_0^0)$$
$$+ A_{vs}(Y)Y_c + A_0(Y)Y_0^0 - G_s),$$
$$F_s = F_s^0(Y_0^0). \tag{251}$$

The expression $\ddot{Y}_s$ for the followers' accelerations defines the complete set of $6m - 6$ second-order differential equations, whose solving gives the nominal trajectories $Y_s^0(t)$ of the followers' contact points in the nominal motion. Solving six second-order equations of (251) that describe the leader's acceleration $\ddot{Y}_v$, or the last six first-order equations of (247) for the nominal trajectories of the leader's velocity $\dot{Y}_v^0(t)$, with the preset parameters of the object MC trajectory included, gives the nominal trajectory of the leader's contact points $Y_v^0(t)$. The solution is most easily obtained by using numerical methods for the input $F_s = F_s^0(Y_0^0)$ being previously determined by the condition of attaining the coordinated motion. By substituting the obtained values into the expression for the leader's contact force F_v in (251), the nominal value of the leader's contact force, $F_v^0(t)$, is obtained so that all nominal values of the elastic system are determined.

Let us analyze the setting of the nominal motion. Namely, the question to be answered is: is it most convenient to prescribe the trajectory of the manipulated object MC? In view of the previous discussion, the choice of the nominal trajectory as the trajectory of the manipulated object MC will inevitably cause that the displacements and forces, at one contact point, become a function of the displacements and forces at the other nodes. Because of that, the analysis will also be carried out for differently defined nominal conditions of motion.

Let the trajectory of one contact point be prescribed as the nominal trajectory. Let us assume that the force and displacement of the object MC adapt to the arising conditions of motion. Further, the conditions of nominal motion correspond to such motion of the elastic system in which the geometrical figure formed after the gripping remains rigid. The positions of the instantaneous pole of rotation of that figure during the motion can be prescribed separately (for example, as a line formed by the object MC attached to that figure as a rigid body). To convey the idea in a simpler manner, it will be assumed that the instantaneous rotation pole is just at that contact point for which a nominal trajectory is prescribed.

Let the nominal trajectory be given for the contact point $i = 1 (= v)$ (of the leader) with the mass center CM_v. Similar to the previous case, holding for purely static conditions, for the transfer from the point CM_v^s to the point CM_v^0, it is necessary to do one translation using the vector $r_v^0 - r_v^s$ and one change of orientation about the instantaneous rotation pole, here adopted CM_v^0, by $\mathcal{A}_v^0 - \mathcal{A}_v^s$ of the gripped object and elastic system after gripping is completed in the whole

(Figure 25). Then the absolute coordinates of the nodes are (see expression (237))

$$Y^{0s} = \eta + A_r(\mathcal{A}_v^0 - \mathcal{A}_v^s)\rho_{vj}^s + a_r(\mathcal{A}_v^0 - \mathcal{A}_v^s), \tag{252}$$

with the forces acting at them

$$F_e^{0s} = A_r^T(\mathcal{A}_v^0 - \mathcal{A}_v^s)F_e^s = A_r^T(\mathcal{A}_v^0 - \mathcal{A}_v^s)(G + F^s) \tag{253}$$

where A_r and a_r are defined by (150) in which, instead of a, one should put $\mathcal{A}_v^0 - \mathcal{A}_v^s$; F_e^s is the elasticity force attained at the end of gripping; $\rho_{vj}^s = \text{col}(\rho_{10}^s, \rho_{11}^s, \rho_{12}^s, \ldots, \rho_{1m}^s)$, $\rho_{11}^s = 0$, is the vector of the distance of the nodes from the node CM_v at the end of the gripping phase, and $\eta = \text{col}(r_v^0 - r_v^s \ 0 \ r_v^0 - r_v^s \ 0 \ldots r_v^0 - r_v^s \ 0)$ is the expanded vector of absolute coordinates that defines the translation of the elastic system nodes at the end of the gripping phase, as if these were points of a rigid body. In such a displacement, at the object MC and at contact points, the gravitational and contact forces

$$G_c + F_c^{0s} \quad = F_{ec}^{0s},$$

$$G_0 + \Delta G_0 \quad = F_{e0}^{0s} \tag{254}$$

act, where F_c^{0s} and ΔG_0 are the forces that should act together with the gravitational forces at the nodes, to balance the elastic forces F_e^{0s}.

Let us assume that in the considered position of the CM_i^0 trajectory, the elastic system transferred to this position is also under the action of dynamic forces, in a purely static manner. Since the contact points in the assumed case of nominal motion are not relatively displaced and if displacement of the manipulated object CM_0 is not allowed, the acting dynamic forces will be equal to those dynamic forces that would be produced if the elastic system were rigid. These forces can be explicitly calculated from (115), whereby it is necessary to introduce the nodes coordinates and their derivatives, determined on the basis of rigid-body kinematics for the known kinematic quantities at one of its points (of the trajectory $Y_v^0(t)$). By introducing only the dynamic forces produced at contact points into the previous equations, only the contact forces will be altered, whereas the relative positions of the elastic system's nodes will not change, and neither will the elastic forces

$$F_{dc}^0 + \ G_c + F_c^{0d} \quad = F_{ec}^{0s}, \qquad F_c^{0d} = F_c^{0s} - F_{dc}^0$$

$$G_0 + \Delta G_0 \quad = F_{e0}^{0s}. \tag{255}$$

Let the static force G_0 and dynamic force F_{d0}^0 displace the object mass center CM_0 to an equilibrium state. Since it is assumed that the positions of contact points

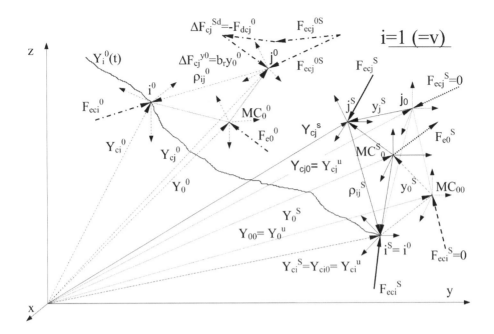

Figure 25. Nominal trajectory of a contact point

do not change, to satisfy the equilibrium state, the position of the object MC must change. This will result in a change of elastic forces. If the change is considered with respect to the state prior to the action of F_{d0}^0 and G_0, all the displacements will be small so that the relation (95) with the constant stiffness matrix will hold, whereby only $y_0^0 \neq 0$ changes, all other displacements being zero. By changing y_0^0 by the value Δy_0^0 elastic forces at all nodes change, but only as a function of Δy_0^0, so that the equilibrium conditions (255), once when the action of F_{d0}^0 and G_0 is over, will be given by the expression (Figure 25)

$$F_{dc}^0 + G_c + F_c^0 = F_{ec}^{0s} + b_r \Delta y_0^0,$$

$$F_{d0}^0 + G_0 = F_{e0}^{0s} + d_r \Delta y_0^0, \tag{256}$$

where F_c^0 is the force at contact points that should be realized to preserve the distance between the nodes attained at the end of the gripping phase for such a choice of the nominal general motion.

Therefore, the contact forces F_c^{0s} needed to maintain the elastic system in the gripped state after the static displacement must be changed in the motion along the prescribed path of the contact point by

$$\Delta F_c = F_c^0 - F_c^{0s} = \Delta F_c^{sd} + \Delta F_c^{y0} = -F_{dc}^0 + b_r \Delta y_0^0, \tag{257}$$

where $\Delta F_c^{sd} = F_c^{0d} - F_c^{0s} = -F_{dc}^0$ is the increment of contact forces due to the action of dynamic forces at contact points; $\Delta F_c^{y0} = b_r \Delta y_0^0$ is the increment of contact forces due to the action of dynamic forces at the manipulated object MC, at the contact points manifested as the support reaction; Δy_0^0 is the displacement vector of the object MC calculated from the second equation of (256) and measured from the state in which the elastic system would have found if it were statically transferred along trajectory; and b_r, d_r are the submatrices of the stiffness matrix $K_r = A_r^T (\mathcal{A}_v^0 - \mathcal{A}_v^s) K A_r (\mathcal{A}_v^0 - \mathcal{A}_v^s)$ associated to the vector Δy_0^0. At that, all the coordinates of contact points are determined by (252), the only exception being the coordinates of the manipulated object MC, to which the vector Δy_0^0 should be added.

It can be finally concluded that such a choice of nominal motion ensures the preservation of the geometric conditions of the contact at the end of the gripping phase, whereas the contact forces, elastic forces, and coordinates of the object's MC changed in proportion to the dynamics, are directly dictated by the choice of nominal trajectory. Despite the existence of the mentioned changes in forces and positions during the motion, this choice of nominal motion has its essential advantages:

- All the quantities are relatively easily and exactly calculated.

- For each phase of motion, it suffices to use only the existing theory of mechanics. The influence of elastic properties can be fully covered by the theory of elasticity with one constant stiffness matrix, for both the gripping phase and motion phase.

- The assessment of nominal motion stability is reduced to the examination of the MC motion of the manipulated object elastically connected to the rigid geometric figure formed by the contact points (as a cage). Since the stiffness matrix is constant and the displacements are small, the nominal motion stability can be also examined, based on the existing theory of the stability of linear systems.

- All that was said above, along with the analysis of human motion in the process of transferring, indicates that such a setting of the nominal motion in cooperative manipulation is appropriate. This statement is based on the fact that it suffices to know the position of one contact point and force at all the other points to derive a conclusion about the displacements and forces at the manipulated object MC, i.e. to control its motion.

Irrespective of the algorithm chosen, the result of the synthesis of nominal conditions for the elastic system is the absolute positions Y_c^0, Y_0^0, velocities $\dot{Y}_c^0$,

$\dot{Y}_0^0$, accelerations $\ddot{Y}_c^0$, $\ddot{Y}_0^0$, and the forces F_c^0 at contact points.

5.3   Nominal Driving Torques

As the manipulators are non-redundant, the relation $Y_c = \Theta(q) \in R^{6m \times 1}$ between the internal and external coordinates is unique. Without dealing with the complexity of obtaining the inverse transformation of coordinates for the known external nominal coordinates Y_c^0, the manipulator's internal nominal coordinates q^0 and their derivatives are uniquely determined from (172) by the expressions

$$q^0 = \Theta^{-1}(Y_c^0) \in R^{6m \times 1},$$

$$\dot{q}^0 = J^{-1}(\Theta^{-1}(Y_c^0))\dot{Y}_c^0 \in R^{6m \times 1},$$

$$\ddot{q}^0 = J^{-1}(\ddot{Y}_c^0 - \dot{J}(\Theta^{-1}(Y_c^0))J^{-1}(\Theta^{-1}(Y_c^0))\dot{Y}_c^0) \in R^{6m \times 1}. \tag{258}$$

For the known kinematic quantities q^0, $\dot{q}^0$, $\ddot{q}^0$ and the contact forces $F_c^0 = -f_c^0$, the equations (167), defining the model of motion of non-elastic manipulators with six DOFs with non-compliant joints and with the force at the gripper tip in the space of internal coordinates, allow us to uniquely calculate the nominal driving torques τ^0

$$\tau^0 = H(q^0)\ddot{q}^0 + h(q^0, \dot{q}^0) - J^T(q^0)f_c^0. \tag{259}$$

When the manipulator driving torques are being calculated, all nominal quantities of the cooperative system's coordinated motion are determined on the basis of the known trajectory of the manipulated object MC or trajectory of the selected contact point and contact forces defined by the condition of maintaining the gripped state at the beginning of the motion.

The proposed algorithm assumes that the gripping phase ends in a stationary state of the cooperative system. However, in the applications the system of differential equations that is numerically solved may be unstable or simplified by neglecting damping forces. Then, the time of transient process relaxation in the action of any input quantity on the system will be infinite. This means that, in the time interval predicted for the gripping, the elastic system will not attain a stationary state. Or, more precisely, the sought positions of contact points and required values of their velocities and accelerations will not be realized simultaneously. The same will also happen if a small interval is adopted for the motion between the particular points on the trajectory (e.g. for the duration of the gripping process). When the time predicted for gripping has ended, the general motion from the initial state determined by the moment of termination of the gripping process will continue irrespective of the realized values of the state quantities. To avoid the occurrence of discontinuity in the changes of nominal quantities, using the previously presented procedure, it is possible to determine the driving torques at the beginning and at the end of the gripping phase, as well as during the quasi-static displacement. To the driving torques thus determined, time is associated as an independent variable, t_0 and t^s for the beginning and end of gripping, and t^0 for the position on the trajectory. In these moments, the driving torques have accurately determined values, whereas between these moments the function of the change of driving torques may be an arbitrary monotonously continuous function. As the systems of differential equations (175) for the description of the cooperative manipulation for the immobile unloaded state and (181) for the description of the cooperative manipulation dynamics for a mobile unloaded state are unstable, the solutions of these systems will diverge. If these systems are locally stabilized first, and if the above procedures are used to determine the nominal inputs for the thus stabilized systems, then the numerical solving of the locally stabilized system will give the nominal trajectories without any discontinuity. Such an approach is very convenient when the precise position of the manipulated object MC is required.

5.4   Algorithms to Calculate the Nominal Motion in Cooperative Manipulation

The nominal motion is given for the phases of gripping and of general motion. For both phases, the nominal conditions are prescribed either for the manipulated object MC or a selected contact point. The initial conditions of motion are attained at the end of the gripping phase. The nominal conditions for the gripping and the motion phases are prescribed independently because it is not important whether the conditions at the end of gripping are achieved by prescribing the conditions for the object MC or for the contact point. For example, it is possible to perform gripping in accordance with the conditions given for the object MC and nominal general motion according to the conditions for the contact point. Hence the algorithm to calculate the nominal motion in gripping and nominal general motion is given for four possible cases. The initial step in calculating the nominal motion for all the algorithms is defined by the system model.

The independent input parameters to the algorithm to calculate the gripping nominals are the displacements of the nodes or the gripping force. These parameters are determined based on the requirement for the magnitude of the stress state for each concrete manipulated object. Calculation of the input parameters is carried out by conventional procedures of the theory of elasticity, and will not be considered here.

The initial state from which we consider the process of gripping the manipulated object by the manipulators, is determined by the immobile unloaded state 0 of the elastic system, for which all displacements of nodes are equal to zero.

5.4.1 Algorithm to calculate the nominal motion in gripping for the conditions given for the manipulated object MC

Step 1.
Equations (217) are formed for the static conditions of the elastic system equilibrium. Displacement of the manipulated object MC is known, $y_0^s = 0$, and if the necessary displacements of contact points y_c^s are known, the forces at all nodes of the elastic system at the end of gripping F_{ec}^s and F_{eo}^s are calculated from (217).

Step 2.
If the displacements of contact points at the end of the gripping phase are not known for the condition of the immobile MC of the manipulated object, $y_0^s = 0$, displacements of contact points at the end of gripping y_c^s are determined from (218) as a function of the given forces $F_{ec}^s = G_c + F_c^s$ as independent variables.

Step 3.
This step exists if the exactly determined force at the manipulated object MC at the end of the gripping phase is required. Then, it is necessary to do the following:

- To request the force F_{e0}^s at the manipulated object MC at the end of the gripping phase (e.g., $F_{e0}^s = G_0$).

- To determine displacement of the leader's contact point y_v^s from (219) as a function of the displacements of the contact points of the manipulators followers y_s^s and forces at the manipulated object MC, F_{e0}^s.

- To determine the forces at the contact points of the leader and followers according to (221) as a function of the displacements of the contact points of the followers y_s^s and required forces at the MC of the manipulated object, F_{e0}^s, whereby the quantities y_s^s and F_{e0}^s must be given as independent variables.

- If, instead of the displacement y_s^s, the forces at the contact points of the followers, F_{es}^s, are given as independent variables, then all the displacements of contact points y_v^s, y_s^s and force at the leader's contact point, F_{ev}^s, are calculated from (222) as a function of the forces F_{es}^s and F_{e0}^s, given as independent variables.

In the first three steps, all the quantities characterizing static conditions at the end of the gripping phase are determined.

Step 4.
Equation (217) is used to calculate the contact force at the end of the gripping phase $F_c^s = F_{ec}^s - G_c = \mathrm{col}(F_v^s,\ F_s^s)$. It is necessary to select a monotonous function for the change of the contact forces of the followers with time, $F_s^s(t)$, from the value at the beginning of gripping to its end.

Step 5.
Numerical methods are used to solve the system of differential equations (236) for the forces $F_s^s(t)$ and the nominal trajectories of contact points $y_s(t)$ and $y_v(t)$ are determined, as well as their derivatives $\dot{y}_s(t)$, $\dot{y}_v(t)$ and $\ddot{y}_s(t)$, $\ddot{y}_v(t)$ and contact force at the leader's contact point $F_v^s(t)$ during the gripping phase.

Step 6.
Starting from the assumption that the absolute coordinates of the contact points of the immobile unloaded state 0 are known, and that they are determined by the vector $Y_{c0} = \mathrm{const}$, the absolute coordinates of the contact points during the gripping are calculated, $Y_c^0(t) = Y_{c0} + y_c(t)$, whereby the trajectories $y_c(t)$ were determined in the preceding step. By introducing the absolute coordinates of the contact points and their derivatives into (258) the internal coordinates and derivatives of those internal coordinates to be realized in the nominal gripping are calculated.

Step 7.
By introducing the calculated internal coordinates and their derivatives into (259) the nominal driving torques to realize the nominal gripping are determined.

5.4.2   Algorithm to calculate the nominal motion in gripping for the conditions of a selected contact point

Step 1.
Equations (223) are formed for the static equilibrium conditions of the elastic system. The displacement of the leader's contact point y_v^s is known, and if the displacements of the other nodes y_{s0}^s are also known, the forces at all contact points of the elastic system at the end of the gripping phase, F_{ec}^s and F_{eo}^s, are calculated from (223).

Step 2.
If the displacements of the nodes at the end of the gripping phase are not known, but the forces at contact points, F_{ec}^s, are, then the nodes displacements $y_{s0}^s = \mathrm{col}(y_s^s,\ y_0^s)$ and the force at the MC of the manipulated object at the end of grip-

ping F_{e0}^s is determined from (224) and (225) as a function of the displacements, y_v^s, and prescribed forces $F_{ec}^s = G_c + F_c^s$ as independent variables.

Step 3.
This step exists if the exactly determined force at the manipulated object MC at the end of the gripping phase is required. Then, it is necessary:

- To request the force F_{e0}^s at the manipulated object MC at the end of the gripping phase (e.g., $F_{e0}^s = G_0$).

- To determine the displacement of the manipulated object MC, y_0^s, from (227) as a function of the displacements of the contact points of the leader y_v^s and of followers y_s^s, and of the force at the manipulated object MC, F_{e0}^s.

- To determine the forces at the contact points of the leader F_{ev}^s and followers F_{es}^s from (228) depending on the contact point displacements $y_c^s = \mathrm{col}(y_v^s, \ y_s^s)$ and required force at the MC of the manipulated object F_{e0}^s, the quantities y_v^s, y_s^s and F_{e0}^s must be given as independent variables.

- If, instead of the displacements y_s^s, the forces at the contact points of the followers F_{es}^s are prescribed as independent variables, then the displacements of the contact points of followers y_s^s, displacement of the object MC y_0^s, and the force at the leader's contact point F_{ev}^s are calculated from (229) and (230) as a function of the displacements y_v^s and forces F_{es}^s and F_{e0}^s, given as independent variables.

For any variant, all independent variables characterizing static conditions at the end of the gripping phase are prescribed in the first three steps.

Step 4.
Using (217), the contact forces at the end of the gripping phase are calculated, $F_c^s = F_{ec}^s - G_c = \mathrm{col}(F_v^s, \ F_s^s)$. It is necessary to choose a monotonous function of the change of contact forces in time, $F_c^s(t)$, from the value at the beginning of gripping to its end.

Step 5.
Numerical methods are used to solve the system of differential equations (102) for the force $F_c^s(t)$, to determine the nominal trajectories of contact points $y_c(t)$ and of the manipulated object MC $y_0(t)$, as well as the derivatives $\dot{y}_c(t)$, $\dot{y}_0(t)$ and $\ddot{y}_c(t)$, $\ddot{y}_0(t)$ during the gripping phase.

Steps 6 and 7.
These steps are identical to Steps 6 and 7 of the algorithm in Section 5.4.1, to calculate the nominal motion during the gripping when the conditions for the manipulated object MC are prescribed.

All the above calculations are carried out on the basis of the unstabilized model of cooperative manipulation. If the nominal trajectories are to be determined by numerically solving the system of differential equations (175) for the known driving torques, then it is convenient to first carry out local stabilization of the system and replace Steps 4, 5, 6 and 7 by Steps 4a, 5a, 6a and 7a.

Step 4a.
Using (217), the contact forces at the end of the gripping phase are calculated, $F_c^s = F_{ec}^s - G_c = \mathrm{col}(F_v^s, F_s^s)$. Starting from the assumption that the absolute coordinates of the contact points of the immobile unloaded state 0 are known and that they are determined by the vector $Y_{c0} = const$, the absolute coordinates at the end of the gripping process are calculated, $Y_c^s(t) = Y_{c0} + y_c^s$, whereby the displacements of contact points y_c^s are determined in Step 3. Using (172), i.e. (258), the internal coordinates at the beginning (q_0^s) and in the end (q^s) of the gripping process are calculated.

Step 5a.
Local stabilization of the system (175) is carried out according to the specially preset requirement. As it has been assumed that the elastic system is immobile at the beginning and at the end of gripping, the derivatives of internal coordinates at the beginning and end of gripping are zero. At the end of the gripping process, it can be realized that the internal coordinate derivatives are not zero, but their exact and matched values have to be known. By introducing the internal coordinates q_0^s determined in the preceding step, the values of the derivatives of internal coordinates and contact forces in the system of equations describing the locally stabilized system, the driving torques in the beginning of gripping τ_0^s are calculated. By introducing the internal coordinates q_s determined in the preceding step and the values of the derivatives of the internal coordinates and contact forces F_c^s calculated in Step 4a into the system describing the locally stabilized system, the driving torques at the end of the gripping phase τ^s are calculated.

Step 6a.
The duration of the gripping process, determined by the beginning t_0 and the end t^s of the process, is selected. Also, the function of the change of driving moments with time is selected. For a linear change, the nominal driving torques are calcu-

lated from the expression

$$\tau(t) = \frac{\tau_s - \tau_0^s}{t_s - t_0}(t - t_0) + \tau_0^s.$$

Step 7a.
By numerically solving the locally stabilized system of differential equations (175) for the input driving torque $\tau(t)$, the nominal trajectories $q(t)$ of the leading links and the nominal values of any quantity existing in the description of the cooperative system, are determined.

By ending the calculation from Step 7 (7a) in any of the above algorithms, all the calculations concerning the gripping phase are finished. The calculated displacements, absolute coordinates, and forces at the nodes describe in full the coordinated gripping of the manipulated object in all phases of the gripping process. The state of the absolute coordinates Y^s, their derivatives $\dot{Y}^s$ and $\ddot{Y}^s$ and forces at the elastic system nodes F_c^s and F_0^s attained at the end of the gripping process determine the initial state of the nominal general motion. The known vector of absolute coordinates Y^s at the end of the gripping phase serves as the basis to determine the vector of distance of the nodes from the manipulated object MC $\rho_0^s = \mathrm{col}(\rho_{00}^s, \rho_{01}^s, \dots, \rho_{0m}^s)$, $\rho_{00}^s = 0$, and distance vector for the nodes with respect to the leader's contact point CM_v $\rho_{vj}^s = \mathrm{col}(\rho_{10}^s, \rho_{11}^s, \rho_{12}^s, \dots, \rho_{1m}^s)$, $\rho_{11}^s = 0$.

5.4.3 Algorithm to calculate the nominal general motion for the conditions given for the manipulated object MC

Step 1.
The nominal trajectory of the manipulated object MC $Y_0^0 = \mathrm{col}(r_0^0, \mathcal{A}_0^0) \in R^{6\times 1}$ is prescribed as a line in space. On this trajectory, the manipulated object MC is found at the end of the gripping phase $Y_0^s = Y_0^0(t_0) = \mathrm{col}(r_0^s, \mathcal{A}_0^s)$.

Step 2.
The trajectory time profile $Y_0^0(t)$ is selected and its derivatives $\dot{Y}_0^0(t)$ and $\ddot{Y}_0^0(t)$ are determined.

Step 3.
The trajectory is divided into a finite number of segments. The number of divisions depends on the form of the trajectory in space and time. For the linear parts of the trajectory, it suffices to select two points at the beginning and end of the linear interval. The circular and oscillatory parts of the trajectory should be divided so that full circumference or oscillation is approximated by not less than 32 points.

Let $Y_0^0(t)$ be the point representing the trajectory at the instant t.

Step 4.

The translatory, $r_0^0(t) - r_0^s$, and angular, $\mathcal{A}_0^0(t) - \mathcal{A}_0^s$, static displacements of the manipulated object MC and of the overall elastic system from the initial to the current state on the trajectory at the time t is determined. In this algorithm, the instantaneous rotation pole coincides with the instantaneous position of the object MC on the given nominal trajectory. The relation (150) serves to determine the transformation matrix $A_r(\mathcal{A}_0^0(t) - \mathcal{A}_0^s) = A_r(t)$ and vector $a_r(\mathcal{A}_0^0(t) - \mathcal{A}_0^s) = a_r(t)$. Using (237), the absolute coordinates of the elastic system nodes $Y^{0s}(t)$ after the static transfer from the initial to the current position on the trajectory are determined.

Step 5.

The absolute coordinates of the fictitious unloaded state 0 of the elastic system for the current position on the trajectory are determined by mapping the unloaded state 0 at the beginning of the gripping phase. Namely, the vector of the node displacements in gripping y^s is mapped into the vector of the fictitious node displacements $y_{00}^s(t) = A_r(t)y^s$, and the absolute coordinates of the nodes of the fictitious unloaded state 0 of the elastic system at the current position on the trajectory is determined by the expression $Y_{00}^s(t) = Y^{0s} - y_{00}^s(t)$.

Step 6.

The derivatives of the absolute coordinates $Y^{0s}(t)$, calculated on the basis of the given nominal trajectory of the manipulated object MC are determined. By introducing the current coordinates of nodes $Y^{0s}(t)$ and their derivatives $\dot{Y}^{0s}(t)$, $\ddot{Y}^{0s}(t)$ into (244), we obtain the approximate values of the forces $\bar{F}_{ec}$ and $\bar{F}_{e0}$ that would act at the nodes in the current position on the given trajectory if the elastic system moved as a rigid body.

Step 7.

Assuming that $(y_e^0)_0 = (y_{00}^s)_0$, and using (245), the displacements y_e^0 from the current fictitious unloaded state 0 are determined.

Step 8.

From (246), it is necessary to determine the absolute coordinates of elastic system nodes $Y^0(t)$ after the action of the forces determined in Step 6. The differentiation gives the derivatives $\dot{Y}^0(t)$ and $\ddot{Y}^0(t)$.

Step 9.
By introducing the absolute coordinates and their derivatives determined in the preceding step into the equations of behavior (115), the contact forces are calculated. The calculated contact forces at the nodes of the manipulators-followers can be adopted as the nominal forces $F_s^0(Y_0^0(t)) = F_s^0(t)$. Such a choice ensures the realization of the coordinated nominal motion of the manipulated object MC without additional requirements concerning the accompanying changes in the gripping. If a simultaneous change in gripping is also required during the motion, then these forces can be prescribed as independent variables.

Step 10.
For the known nominal trajectory of the manipulated object MC, $Y_0^0(t)$, and its derivatives $\dot{Y}_0^0(t)$, $\ddot{Y}_0^0(t)$ and the nominal input force $F_0^0(t)$ from Step 9, the numerical solving of the system of differential equations (251) gives the nominal trajectories of all the contact points $Y_c^0 = \text{col}(Y_v^0, \ Y_s^0)$ and the nominal force F_v^0 at the leader's contact point.

Step 11.
By replacing the absolute coordinates of the nominal trajectories of the contact points and their derivatives in (258), the internal coordinates and their derivatives that are to be realized during the nominal general motion are calculated.

Step 12.
By introducing the calculated internal coordinates and their derivatives into (259), the nominal driving torques to be introduced at the manipulator joints in order to realize the nominal general motion are determined.

5.4.4 Algorithm to calculate the nominal general motion for the conditions given for one contact point

Step 1.
The nominal trajectory of one (leader's) contact point $Y_v^0 = \text{col}(r_v^0, \ \mathcal{A}_v^0) \in R^{6\times1}$, is prescribed as a line in space. On that line there is a selected contact point corresponding to the end of the gripping phase $Y_v^s = Y_v^0(t_0) = \text{col}(r_v^s, \ \mathcal{A}_v^s)$.

Step 2.
The trajectory time profile $Y_v^0(t)$ and its derivatives $\dot{Y}_v^0(t)$ and $\ddot{Y}_v^0(t)$ are determined.

Step 3.
The trajectory is divided in the same way as in Step 3 of the algorithm in

Section 5.4.3 to calculate the nominal general motion for the conditions given for the manipulated object MC. Let $Y_v^0(t)$ be the point that represents the leader's contact point at the moment t.

Step 4.
The translatory, $r_v^0(t) - r_v^s$, and rotational, $\mathcal{A}_v^0(t) - \mathcal{A}_v^s$, static displacements of the elastic system from the initial state to the current state on the trajectory at time t is determined. The instantaneous rotation pole is at the instantaneous position of the leader's contact point on the trajectory. Relation (150) is used to determine the transformation matrix $A_r(\mathcal{A}_v^0(t) - \mathcal{A}_v^s) = A_r(t)$ and the vector $a_r(\mathcal{A}_v^0(t) - \mathcal{A}_v^s) = a_r(t)$. Using (252), the absolute coordinates of nodes $Y^{0s}(t)$ after the static displacement of the elastic system as a rigid body from the initial to the current position, are determined. Differentiating gives the derivatives $\dot{Y}^{0s}(t)$ and $\ddot{Y}^{0s}(t)$.

Step 5.
Now, from (253) and (254) it is necessary to determine the elastic $F_e^{0s} = \mathrm{col}(F_{ec}^{0s}, F_{e0}^{0s})$ and contact forces F_c^{0s} that should act at the elastic system's nodes in the current position on the trajectory in order that the distances between the nodes remain unchanged with respect to the distances attained at the end of the gripping process.

Step 6.
The introduction of the absolute coordinates of nodes $Y^{0s}(t)$ and their derivatives $\dot{Y}^{0s}(t)$ and $\ddot{Y}^{0s}(t)$ into (115) allows the determination of the dynamic forces F_{dc}^0 and F_{d0} that would act at the elastic system's nodes so that it moved along the prescribed trajectory as a rigid body.

Step 7.
The stiffness matrix $K_r = A_r^T(\mathcal{A}_v^0 - \mathcal{A}_v^s)KA_r(\mathcal{A}_v^0 - \mathcal{A}_v^s)$ is determined and the submatrices b_r and d_r are separated.

Step 8.
The second equation of (256) is used to determine Δy_0^0.

Step 9.
After introducing Δy_0^0, determined in the previous step, F_{dc}^0 determined in Step 6, and F_{ec}^{0s} determined in Step 5 into the first equation of (256), it is necessary to calculate the contact forces F_c^0 that ensure a coordinated motion and can be adopted as the nominal forces. As in the previous algorithm, if the simultaneous change in gripping during the motion is required, the contact forces can be given

as independent variables.

Step 10.
By solving the stabilized system of differential equations (115) for the input force F_c^0 calculated in Step 8, the trajectory coordinates $Y^0 = \text{col}(Y_v^0, Y_s^0, Y_0^0)$ of all the nodes of the elastic system are determined. At the same time, the derivatives $\dot{Y}^0$ and $\ddot{Y}^0$ are also determined. The trajectories thus determined are adopted as the nominal trajectories.

Steps 11 and 12.
These steps are identical to Steps 11 and 12 in the algorithm in Section 5.4.3 to calculate the nominal general motion for the conditions given for the manipulated object MC.

The above calculations were done on the basis of the unstabilized model (181) for the description of the dynamics of cooperative manipulation for the mobile unloaded state. Like in the algorithm to calculate the nominal motion in gripping for the conditions of a selected contact point (Section 5.4.2), whereby the nominal trajectories are determined by numerically solving the system of differential equations (181) for the known driving torques, the system can be stabilized first and then Steps 10, 11 and 12 replaced by Steps 10a, 11a and 12a.

Step 10a.
By introducing the coordinates, velocities, and accelerations of the nodes, determined in Step 4, and the coordinates of the manipulated object MC $Y_0^0 + \Delta y_0^0$ into (258), the internal coordinates and their derivatives are calculated.

Step 11a.
Local stabilization of the system (181) is carried out according to a specially given requirement. The introduction of the coordinates and their derivatives, calculated in the preceding step, and the contact forces F_c^0, calculated in Step 9, into the system of equations describing the locally stabilized system, serves to determine the driving torques τ_0 at the selected points on the trajectory. The obtained discrete time functions of driving torques are approximated by a smooth time function $\tau(t)$.

Step 12a.
By numerically solving the locally stabilized system of differential equation for the input driving torque $\tau(t)$, the nominal trajectories $q(t)$ of the leading links and nominal values of every quantity present in the description of the cooperative system are determined.

5.4.5   Example of the algorithm for determining the nominal motion

The algorithms for the synthesis of nominals in the gripping phase and nominal motion of the cooperative system will be illustrated on the 'linear' cooperative system (Figure 26) considered in Chapter 3 (Figures 8 and 9). It is assumed that the masses of the object-manipulators' elastic interconnections are much smaller than the mass of the manipulated object, so that they are neglected.

The basis for the synthesis of the nominals is the mathematical model of the cooperative system that describes faithfully enough the statics and dynamics of the cooperative system.

The motion in the gripping phase can be described using the elastic system model given with the aid of the coordinates of deviation y from the immobile unloaded state 0 given by (42), in which it is necessary to put $\ddot{Y}_{10} = 0$ and add the damping forces of elastic interconnections, thus yielding the model

$$\ddot{y}_2 + \frac{(d_p + d_k)}{m}\dot{y}_2 + \frac{(c_p + c_k)}{m}y_2 = \frac{d_p}{m}\dot{y}_1 + \frac{d_k}{m}\dot{y}_3 + \frac{c_p}{m}y_1 + \frac{c_k}{m}y_3 - g,$$

$$F_{e1} = c_p y_1 - c_p y_2,$$

$$F_{e3} = -c_k y_2 + c_k y_3,$$

$$F_{c1} = d_p \dot{y}_1 - d_p \dot{y}_2 + c_p y_1 - c_p y_2,$$

$$F_{c2} = -d_k \dot{y}_2 + d_k \dot{y}_3 - c_k y_2 + c_k y_3, \tag{260}$$

where d_p and d_k are the coefficients of damping of elastic interconnections; F_{ei}, $i = 1, 2, 3$ are the elasticity forces produced at the nodes, and F_{cj}, $j = 1, 2$ are the contact forces. Equations (260) represent the developed form of Equations (102) of the model of elastic system dynamics for the immobile unloaded state, given in Section 4.5. In this example, the masses of elastic interconnections are neglected, so that $W_c(y_c) = 0_{3\times3}$, $w_{c1}(y, \dot{y}) = d_p \dot{y}_1 - d_p \dot{y}_2 + c_p y_1 - c_p y_2$, $w_{c2}(y, \dot{y}) = -d_k \dot{y}_2 + d_k \dot{y}_3 - c_k y_2 - c_k y_3$, $W_0(y_0) = m$, $w_0(y, \dot{y}) = -d_p \dot{y}_1 + (d_p + d_k)\dot{y}_2 - d_k \dot{y}_3 - c_p y_1 + (c_p + c_k)y_2 - c_k y_3 + mg$ and $F_c = (F_{c1}, F_{c2})^T$.

The general motion is described using the elastic system defined by the absolute coordinates Y and given by the expressions (43), which have to be supplemented by the damping of elastic interconnections, to obtain the model

$$\ddot{Y}_2 + \frac{(d_p + d_k)}{m}\dot{Y}_2 + \frac{(c_p + c_k)}{m}Y_2 = \frac{d_p}{m}\dot{Y}_1 + \frac{d_k}{m}\dot{Y}_3 + \frac{c_p}{m}Y_1 + \frac{c_k}{m}Y_3 - g + \frac{c_p}{m}s_1 - \frac{c_k}{m}s_3,$$

$$F_{e1} = c_p Y_1 - c_p Y_2 + c_p s_1,$$

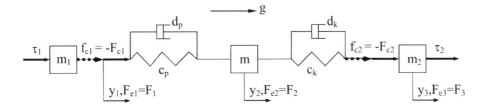

Figure 26. 'Linear' cooperative system

$$F_{e3} = -c_k Y_2 + c_k Y_3 - c_k s_3,$$

$$F_{c1} = d_p \dot{Y}_1 - d_p \dot{Y}_2 + c_p Y_1 - c_p Y_2 + c_p s_1,$$

$$F_{c2} = -d_k \dot{Y}_2 + d_k \dot{Y}_3 - c_k Y_2 + c_k Y_3 - c_k s_3. \quad (261)$$

Although the model has been taken over from (43), it represents a developed form of Equations (115) for the model of elastic system dynamics for the mobile unloaded state given in Section 4.6. The obtained model is relatively simple because there is no rotation, so that (82) is reduced to $y_{ij}^D = \rho_{ija} - \rho_{ij0}(\rho_{ija}/\|\rho_{ija}\|) = \rho_{ija} - \rho_{ij0}$, $\rho_{ij0} = $ const. By comparing with (115), it can be concluded that $W_{ca}(Y_c) = 0_{3\times3}$, $w_{ca1}(Y, \dot{Y}) = d_p \dot{Y}_1 - d_p \dot{Y}_2 + c_p Y_1 - c_p Y_2 + c_p s_1$, $w_{ca2}(Y, \dot{Y}) = -d_k \dot{Y}_2 + d_k \dot{Y}_3 - c_k Y_2 - c_k Y_3 - c_k s_3$, $W_{0a}(Y_0) = m$, $w_{0a}(Y, \dot{Y}) = -d_p \dot{Y}_1 + (d_p + d_k)\dot{Y}_2 - d_k \dot{Y}_3 - c_p Y_1 + (c_p + c_k)Y_2 - c_k Y_3 + mg$ and $F_c = (F_{c1}, F_{c2})^T$. Models of the manipulators are taken in the form

$$m_1 \ddot{q}_1 + m_1 g = \tau_1 + f_{c1}, \quad f_{c1} = -F_{c1},$$

$$m_2 \ddot{q}_2 + m_2 g = \tau_2 + f_{c2}, \quad f_{c2} = -F_{c2}. \quad (262)$$

Kinematic relations between the external and internal coordinates are given by the expressions

$$q_1 = Y_1 = Y_{10} + y_1, \qquad\qquad q_2 = Y_3 = Y_{30} + y_3,$$

$$\dot{q}_1 = \dot{Y}_1 = \dot{Y}_{10} + \dot{y}_1 = \dot{y}_1|_{Y_{10}=\text{const}}, \qquad \dot{q}_2 = \dot{Y}_3 = \dot{Y}_{30} + \dot{y}_3 = \dot{y}_3|_{Y_{30}=\text{const}},$$

$$\ddot{q}_1 = \ddot{Y}_1 = \ddot{Y}_{10} + \ddot{y}_1 = \ddot{y}_1|_{Y_{10}=\text{const}}, \qquad \ddot{q}_2 = \ddot{Y}_3 = \ddot{Y}_{30} + \ddot{y}_3 = \ddot{y}_3|_{Y_{30}=\text{const}}.$$

$$(263)$$

Numerical values of the parameters of the elastic system model (Figure 26) are $s_1 = s_2 = 0.05$ [m], $m = 25$ [kg], $c_p = 20 \cdot 10^3$ [N/m], $c_k = 10 \cdot 10^3$ [N/m], $d_p = 500$ [N/(m/s)] and $d_k = 1000$ [N/(m/s)]. Numerical values of the model parameters of the manipulators are $m_1 = 12.5$ [kg] and $m_2 = 12.5$ [kg]. The initial position of the cooperative system before the beginning of gripping is determined by the node coordinates $Y_{10} = 0.150$ [m], $Y_{20} = 0.200$ [m] and $Y_{30} = 0.250$ [m].

In the next example, we give the algorithm to calculate the general nominal motion for the conditions given for the manipulated object MC on the basis of the unstabilized model of the cooperative system dynamics described in Section 5.4.3. This algorithm can also be used for the gripping phase if the required trajectories of the nominal quantities are also given for the gripping phase.

Step 1.
The motion of the manipulated object takes place along a vertical straight line that is adopted as the Y axis, defining the line in space along which the motion takes place.

Step 2.
The motion of the manipulated object MC from the beginning of the gripping process to the moment t is described by the function

$$Y_2^0 = \begin{cases} Y_2^s, & 0 \le t \le T_{ks}, \\ Y_2^s + c(t - T_{ks})^2, & T_{ks} < t < T_d, \quad c = \dfrac{Y_2^d - Y_2^0}{T_d - Tks}, \\ Y_2^d + A_y \sin\left(\dfrac{2\pi}{T_y}(t - T_d)\right), & T_d \le t \le T_{kraj}, \end{cases} \quad (264)$$

where Y_2^s [m] is the unchanged position of the manipulated object MC during the gripping; Y_2^d [m] is the position to which the object is lifted upon gripping, and about which proceeds the oscillatory motion of the MC; A_y and T_y are the amplitude and period of oscillation; T_{ks}, T_d and T_{kraj} are the respective moments at which gripping, lifting, and motion are terminated. In parameter selecting, it is requested that the conditions $Y_2^s \cdot Y_2^d > 0$ and $\|Y_2^s\| < \|Y_2^d\|$ are fulfilled. By combining the parameters, it is possible to obtain different functions. For example, if $Y_2^s = Y_2^d$ and $A_y = 0$ are selected, the manipulated object MC remains immobile in the initial position.

The derivatives of this function are

$$\dot{Y}_2^0 = \begin{cases} 0, & 0 \le t \le T_{ks}, \\ 2c(t - T_{ks}), & T_{ks} < t < T_d, \\ A_y \dfrac{2\pi}{T_y} \cos\left(\dfrac{2\pi}{T_y}(t - T_d)\right), & T_d \le t \le T_{kraj}, \end{cases} \quad (265)$$

$$\ddot{Y}_2^0 = \begin{cases} 0, & 0 \le t \le T_{ks}, \\ 2c, & T_{ks} < t < T_d, \\ -A_y \left(\dfrac{2\pi}{T_y}\right)^2 \sin\left(\dfrac{2\pi}{T_y}(t - T_d)\right), & T_d \le t \le T_{kraj}, \end{cases} \tag{266}$$

$$\overset{...}{Y}_2^0 = \begin{cases} 0, & 0 \le t \le T_{ks}, \\ 0, & T_{ks} < t < T_d, \\ -A_y \left(\dfrac{2\pi}{T_y}\right)^3 \cos\left(\dfrac{2\pi}{T_y}(t - T_d)\right), & T_d \le t \le T_{kraj}. \end{cases} \tag{267}$$

Step 3.
Since the results of the calculation of nominal quantities are input quantities to the closed control system, the nominal trajectories are calculated for each integration step by which the closed control system is simulated. In this example, the integration step is 0.0005 (s).

Steps 4, 5, 6, 7 and 8.
Masses of elastic interconnections are neglected, so that all further calculations are algebraic. If the elastic system moves as a rigid body, the contact point coordinates and their derivatives will be $Y_1^0 = Y_2^0 - s_1$, $Y_3^0 = Y_2^0 + s_3$ and $\overset{(k)^0}{Y_1} = \overset{(k)^0}{Y_2} = \overset{(k)^0}{Y_3}$, $k = 1, 2, \ldots$. In the course of static transfer along a vertical, the gravitation and contact forces do not change either their direction or orientation. The general motion produces only the inertial force $-m\ddot{Y}_2^0$ at the manipulated object MC. This force will displace the manipulated object MC. However, in view of the fact that the nominal trajectories were determined while neglecting the friction of connections and that the connections are massless, the same displacement will experience the contact points too.

Step 9.
It is adopted that the contact force F_{c2} at node 3 is an independent variable, given by the function

$$F_{c2} = \tag{268}$$

$$\begin{cases} F_{c2}^0(1 + e^{c_f t}), & 0 \le t \le T_{ks}, \quad c_f = \dfrac{1}{T_{ks}} \ln\left(\dfrac{F_{c2}^s}{F_{c2}^0} - 1\right) > 0, \\ F_{c2}^s, & T_{ks} < t < T_d, \\ F_{c2}^s + A_f \sin\left(\dfrac{2\pi}{T_f}(t - T_d)\right), & T_d \le t \le T_{kraj}, \end{cases}$$

whose derivative is

$$\dot{F}_{c2} = \begin{cases} c_f F_{c2}^0 e^{c_f t}, & 0 \le t \le T_{ks}, \\ 0, & T_{ks} < t < T_d, \\ A_f \dfrac{2\pi}{T_f} \cos\left(\dfrac{2\pi}{T_f}(t - T_d)\right), & T_d \le t \le T_{kraj}, \end{cases} \qquad (269)$$

where F_{c2}^0, F_{c2}^s are the respective contact forces at the moment of observation and at the end of gripping ($F_{c2}^0 \cdot F_{c2}^s > 0$, $\|F_{c2}^0\| < \|F_{c2}^s\|$); A_f and T_f are the amplitude and period of oscillation of the contact force.

Step 10.
Algebraic solving of (261) for the known Y_2^0 and F_{c2}^0 gives the nominal coordinates of contact points

$$Y_3^N = Y_2^0 + \frac{1}{c_k} F_{c2}^0 + s_3,$$

$$Y_1^N = -\frac{c_k}{c_p} Y_3^N + \frac{1}{c_p} f^N(t), \qquad (270)$$

where $f^N(t) = \ddot{Y}_2^0 + (c_p + c_k)Y_2^0 - c_p s_1 + c_k s_3 + mg$.
    The nominal contact force at the first node is given by the expression

$$F_{c1}^N = c_p(Y_1^N - Y_2^0 + s_1). \qquad (271)$$

Step 11.
The relation between the internal and external coordinates is given by the expression (263).

Step 12.
By introducing the calculated values into (262), the driving torques are determined as

$$\tau_1 = m_1 \ddot{Y}_1^N + m_1 g + F_{c1}^N,$$

$$\tau_2 = m_2 \ddot{Y}_3^N + m_2 g + F_{c2}^0. \qquad (272)$$

The results of the calculation of nominal quantities are given in Figures 27 to 32. All nominal quantities for the gripping phase are presented in Figure 27, while those for the nominal motion are shown in Figure 30. In Figures 28 and 31, the nominal quantities are given as the desired input values to be tracked by the control system of cooperative manipulation in the phase of gripping and general motion, respectively. As the damping and masses are neglected, identical results are also obtained for the conditions of the manipulated object MC and for the conditions

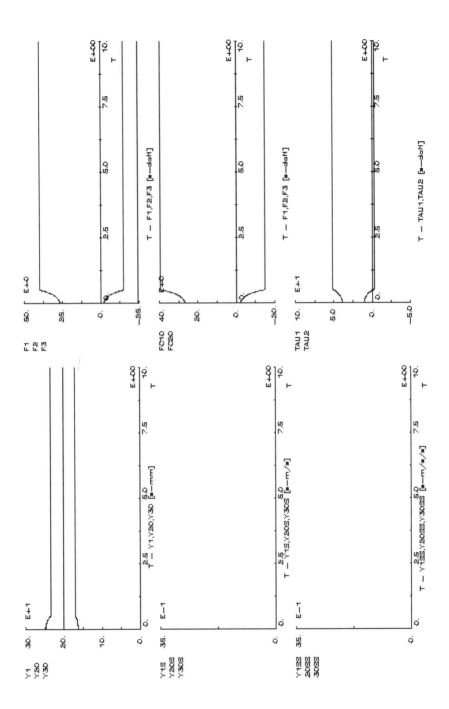

Figure 27. Nominals for gripping a manipulated object

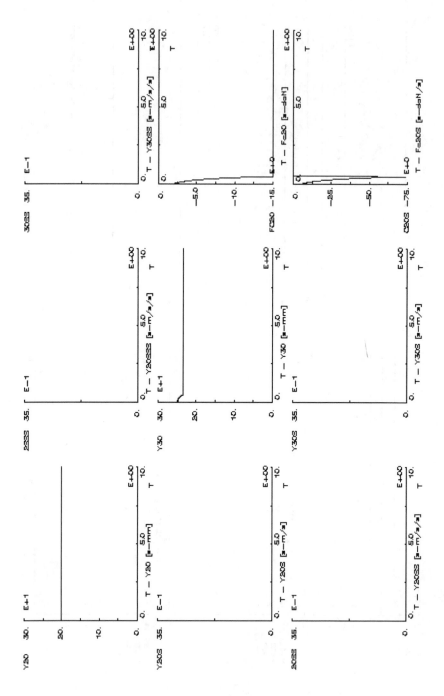

Figure 28. Nominal input to a closed-loop cooperative system for gripping

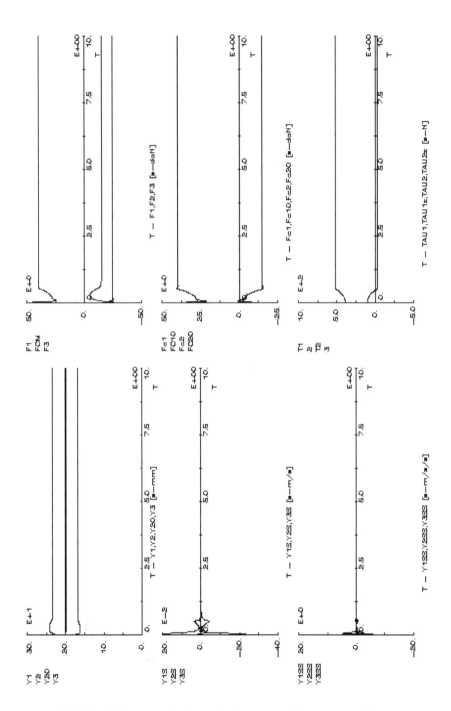

Figure 29. Simulation results for gripping (open-loop cooperative system)

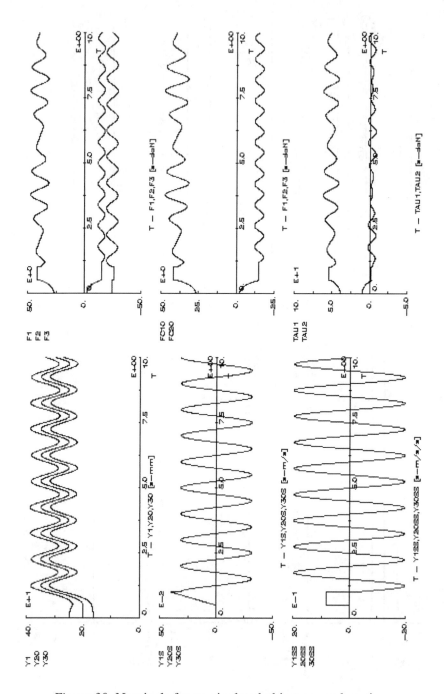

Figure 30. Nominals for manipulated object general motion

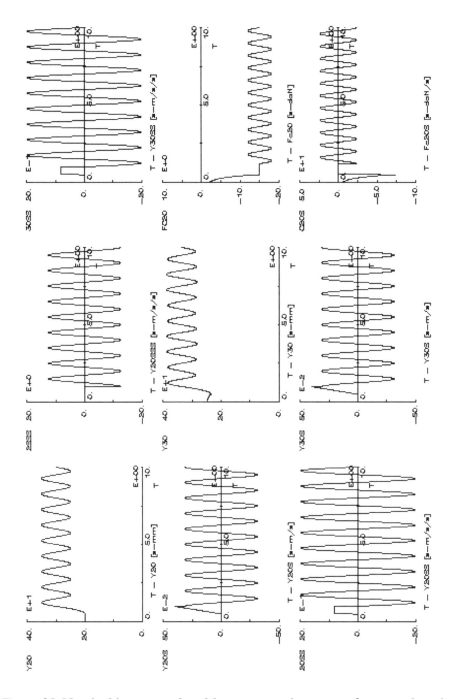

Figure 31. Nominal input to a closed-loop cooperative system for general motion

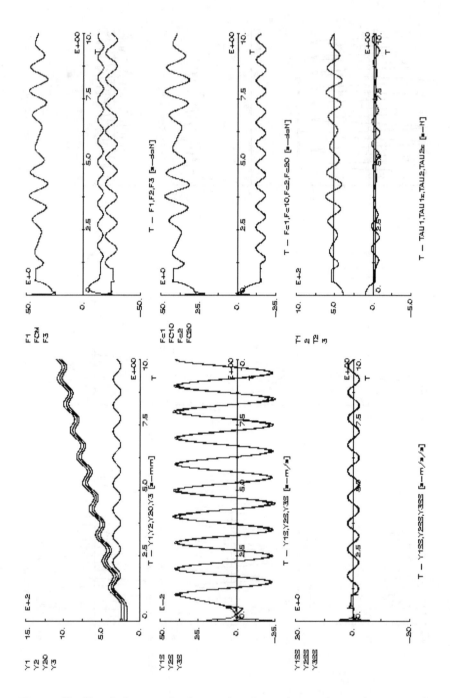

Figure 32. Simulation results for motion (open-loop cooperative system)

given for one contact point. The simulation responses of the non-controlled un-stabilized cooperative system under the action of the calculated nominal driving torques, given for the phases of gripping and nominal motion, are presented in Figures 29 and 32, respectively. In the general motion of the non-controlled un-stabilized cooperative system, the action of nominal driving torques produces the nominal contact forces, but the absolute position of contact points diverge, retaining the prescribed relative distances (Figure 32).

6 COOPERATIVE SYSTEM CONTROL

In this chapter, the problem of cooperative manipulation of an object by several non-redundant manipulators with six DOFs is solved as the problem of controlling a mobile elastic structure while taking into account all specific features of cooperative manipulation. We give a classification of control tasks and propose a procedure to calculate the driving torques to be introduced at the joints of the manipulators in order to ensure tracking of the nominal trajectory of the manipulated object MC and nominals of the followers' contacts. A theoretical analysis of the behavior of the closed-loop cooperative system is given, with a special reference to the behavior of non-controlled quantities. The procedure for calculating driving torques and the behavior of the closed-loop cooperative system are illustrated in a simple cooperative system consisting of the manipulated object and two one-DOF manipulators.

6.1 Introduction to the Problem of Cooperative System Control

Generally, the task of control is to provide a set of drives (inputs) that will produce such a state of the object to satisfy its desired outputs. The control can neither change nor improve the physical characteristics of the object. Through the control, on the basis of the instantaneous requirement (desired input) for the object's behavior and instantaneous state of the object, such drive (input) is synthesized that will force the object to behave in the desired way. At that, it is assumed that the required object's behavior is realizable (within its working envelope), i.e. the states of the object excited by the synthesized drives should be all the time within the allowed limits. The synthesized drives establish a functional relationship between the requirements for the object behavior and object state, and they are called *control laws*. In the rest of this chapter, the quantities used to guide the system are called controlled (directly tracked) outputs, and the quantities that are not involved in the system guiding bear the attribute 'non-controlled'. Similarly, a cooperative system without feedback loops is 'non-controlled', whereas the one involving feedback loops is a 'controlled' cooperative system.

Control laws for a cooperative system are selected on the basis of the model of its dynamics and they will have sense only if the model describes sufficiently well the system's statics and dynamics. The main reason for not finding an adequate solution to the cooperative system control is the presence of force uncertainty in the description of its dynamics. A unique solution of this problem was given first in [8]. It was shown that the problem of force uncertainty, as described in the available literature, is a consequence of the assumption about the non-elastic properties of the cooperative system in its part where the force at the manipulated object MC is decomposed into contact forces.

Numerous propositions of cooperative manipulation control laws based on the models involving force uncertainty that can be found in the available literature, cannot be accepted as an appropriate solution to cooperative manipulation control. There are only a few solutions proposed for the model and control of cooperative manipulation of elastic objects [1, 3–5]. The model given in [1, 3] correctly describes the motion about the immobile unloaded state, and was used to derive a conclusion about the cooperative system general motion. The model presented in [4, 5] starts from the erroneous implicit assumption that the position of the unloaded elastic system during the motion is known. Irrespective of the validity of the model, the control laws proposed by all these authors rely upon the prescribed behavior of deviations from the nominal trajectories or nominal forces. Stability of the closed-loop cooperative system has been proved by simulations or by experiment, but not analytically.

The basic task of cooperative manipulation is the controlled transfer of the working object in space and time. From the point of view of control theory, the task is reduced to tracking the nominal trajectory. The nominal trajectory expresses the explicit or implicit requirement for an ideal motion of the manipulated object MC. This requirement represents input to the control system. It is given as the hodograph of a time-variable six-dimensional position vector, determining the position and orientation of the manipulated object. In order to be given, the input has to be synthesized first. Hence, the first task to be solved is the synthesis of the nominal motion (nominals). The nominals are synthesized analytically on the basis of the mathematical model of the controlled object dynamics. The solution of the task of the synthesis of nominals gives a set of nominal quantities ($6m$ inputs and $6m + 6$ states) of the non-controlled cooperative system (Chapter 5 and [10]).

The model of cooperative system dynamics has more equations of motion than physical inputs (Chapter 4 and [8]). A consequence of this is the number of nominal quantities that exceeds the number of real inputs (driving torques), so that a prerequisite to control is to select the quantities by which the system will be guided. Hence, the control in cooperative manipulation must be hierarchical. The algorithms defined at a higher hierarchical level select for certain classes of tasks, the form of nominal motion and nominal quantities as controlled outputs. These algorithms also define the transient states in the change of guidance and nominals during the manipulation. However, the higher control level is not of concern to us. At the lower control level, control laws are defined for the selected class of controlled outputs.

To answer the question of what can one require from a cooperative system, i.e. what classes of controlled outputs can be selected, this section offers a special analysis. Namely, if only six driving torques (inputs) are used to control the motion along a prescribed trajectory, the question arises as to the remaining $6m - 6$ driving torques. In other words, apart from the prescribed trajectory, it is necessary to know which and how many of the $6m + 6m$ remaining nominal quantities can be adopted as controlled output quantities.

In this chapter, the synthesis of control laws is performed by the method of calculating inputs, i.e. driving torques. Driving torques are calculated using the model of cooperative manipulation and the law of control error, given in advance. The calculated driving torques ensure that the error of controlled outputs has the prescribed properties. The quality of the synthesized driving torques is determined by the quality of the mathematical model (model order and accuracy of the model parameters). A shortcoming of the obtained control laws is that they involve all the state quantities and their derivatives. Their advantage is that the driving torques are exactly determined on the basis of the non-linear model of the cooperative system dynamics. Also, it is relatively easy to perform theoretical analysis of the behavior of the controlled cooperative system with the possibility of using the physical laws that determine its statics and dynamics. This advantage enables us to carry out an exact theoretical analysis of the behavior of non-controlled quantities and define the behavior of all the quantities (not only the controlled ones) of the controlled cooperative system, and derive correct conclusions about the stability of the overall system.

6.2 Classification of Control Tasks

6.2.1 Basic assumptions

A problem arises as to the determination of the number and properties of the requirements concerning the functioning of the cooperative system. To this end, we will consider the properties of controllability and observability of the states and of the system on the basis of which the characteristics and number of possible requirements will be determined.

For a linear system of n_x ordinary first-order differential equations with the matrices $\bar{A}, \bar{B}, \bar{C}, \bar{D}$, states $x \in R^{n_x \times 1}$, inputs $\upsilon \in R^{n_\upsilon \times 1}$, and outputs $y \in R^{n_y \times 1}$,

$$\dot{x} = \bar{A}x + \bar{B}\upsilon,$$

$$y = \bar{C}x + \bar{D}\upsilon, \tag{273}$$

the condition [48]

$$\text{rank}\,(\bar{C}^T \bar{B}, \bar{C}^T \bar{A}\bar{B}, \ldots, \bar{C}^T \bar{A}^{n_x-1} \bar{B}, \bar{D}) = n_x, \tag{274}$$

according to the Caley–Hamilton theorem, is a necessary and sufficient condition that on the basis of the solution

$$x(t) = e^{\bar{A}t}x(0) + \int_0^t e^{\bar{A}(t-\tau)} \bar{B}\upsilon(\tau)\,d\tau$$

$$y(t) = \bar{C}\,e^{\bar{A}t}x(0) + \int_0^t \bar{C}\,e^{\bar{A}(t-\tau)} \bar{B}\upsilon(\tau)\,d\tau + \bar{D}\upsilon(t), \tag{275}$$

for $x(t) = 0$ and for some $t \neq 0$, from the obtained dependence for an arbitrary initial state

$$x(0) = -\int_0^t e^{-\bar{A}\tau} \bar{B}\upsilon(\tau)\,d\tau, \tag{276}$$

we can uniquely determine the control that will bring that initial state to the state $x(t) = 0$. If the rank of the above matrix is lower than n_x, then it is not possible to find the input that would bring all the states to the state $x(t) = 0$. This means that there exist some other inputs (drives) that produce states that are not due to the inputs υ. Also, the initial state $x(0)$ can be uniquely determined as a function of the known expression $y(t)$, for $\upsilon(t) = 0$, if and only if the columns of the matrix $\bar{C} \exp(\bar{A}\tau)$ are linearly independent. This will be fulfilled if the matrix rank is equal to the order of the system

$$\text{rank}\,(\bar{C}^T, \bar{A}^T\bar{C}^T, (\bar{A}^T)^2\bar{C}^T, \ldots, (\bar{A}^T)^{n-1}\bar{C}^T) = n_x. \tag{277}$$

If, however, the rank of this matrix is lower than n_x, then it is not possible to determine all initial states of the system on the basis of the known output.

In accordance with the above, control theory defines the state controllability, output controllability, and state observability. The system state $x(0)$ is controllable if and only if there exists a defined control υ which brings the system from a state $x(0)$ to the zero state $x(t) = 0$ in a finite time t. The system's output quantity $\gamma(t_0)$, is controllable if and only if there is a control υ that will bring the system from the initial state $x(t_0)$, to which corresponds the initial value of the output $\gamma(t_0)$, to the state to which corresponds the output value $\gamma(t) = 0$. In order that the linear system with one input $(n_\upsilon = 1)$ and one output $(n_\gamma = 1)$ is output-controllable, it is necessary that

$$\text{rank } (\bar{C}^T \bar{B}, \bar{C}^T \bar{A}\bar{B}, \ldots, \bar{C}^T \bar{A}^{n_x-1}\bar{B}, \ Db) = 1. \tag{278}$$

The system state $x(t_0)$ is observable if only if it is uniquely determined by the output $\gamma(t)$ and control $\upsilon(t)$ on some limited time interval $t \in [0, T]$. If all the system states are controllable, the system is (completely) controllable, and if the output is completely controllable, the system is fully controllable. If all the system states are observable, the system is completely observable. Kalman [49] showed that the linear system with one input and one output is controllable (observable) if and only if its dual system is observable (controllable).

It has been shown that for a linear stationary time-continuous dynamic system, the positive solution of the controllability problem guarantees the existence of control in the closed system, which will guarantee stability of the overall control system. Applying intuitively the same logic to non-linear systems, it turns out that the solution of controllability is also of crucial importance for the existence of the solution of any task of theory of control such as, for example, the problem of ensuring the system's stability.

The criteria of linear systems cannot be directly applied to derive conclusions about the properties of non-linear systems, but it can be expected that from the part of the necessary conditions for controllability of the non-linear system, should at least come the conditions for the number and characteristics of requirements (in this case, the cooperative manipulation) that can be imposed on it.

Part of the necessary conditions of controllability of a non-linear system can be obtained on the basis of the following reasoning.

The general solution (275) of the linear system (273) and some non-linear system over the same sets of inputs $\mathcal{D}_\upsilon$, states $\mathcal{D}_x$ and outputs $\mathcal{D}_\gamma$ is of the same mathematical form

$$x(t) = x(x_0, t_0, t, \upsilon)),$$

$$\gamma(t) = \gamma(x, \upsilon) = \gamma(x(x_0, t_0, t, \upsilon), \upsilon) = \gamma(t, \upsilon), \tag{279}$$

whereby the time t_0, t and the initial state x_0 are parameters. By eliminating the

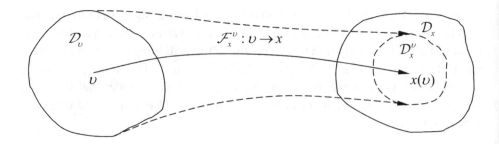

Figure 33. Mapping from the domain of inputs to the domain of states

parameter t, we obtain the functional relations

$$x = x(x_0, \upsilon),$$

$$\gamma = \gamma(x, \upsilon) = \gamma(x(x_0, \upsilon), \upsilon), \tag{280}$$

that define the mapping of the input domain to the state domain and both of them to the output domain. According to the assumption, the domains of input, state, and output are subsets of the n_υ-, n_x- and n_y-dimensional space, respectively. A physical system whose description contains control as an independent variable, is an open system. This means that there exists some other source system from which energy, matter, and/or information are introduced to that system. Part of the output space of the source system is the input space to the system under consideration. The system considered can 'see' the source system only through its projection into the input space, so that on the spaces of input and state there exists a 'picture' of an isolated system from the point of view of the system considered. Part of that space, or the whole space, can be called the *natural output space*, and it is equal to the product of the input space and the state space. Hence, the dimension of the space in which the overall system is 'seen' is $n_\upsilon + n_x$. At the same time, this is also the maximal dimension of the natural output space for the considered system $\max\{n_{yp}\} = n_\upsilon + n_x$. All the other output spaces represent the transformation or mapping of the natural output space. The dimension of the output space can be smaller than, equal to, or higher than the dimension of the natural output space.

Solutions of (279), (280) define the mappings from one domain to the other.

The function $x = x(x_0, \upsilon)$ determines the mapping $\mathcal{F}_x^\upsilon: \upsilon \to x$ by which the whole input domain is mapped into the whole/part of the state domain $\mathcal{D}_\upsilon \to \mathcal{D}_x^\upsilon \subseteq \mathcal{D}_x$ (Figure 33).

The function $y(t)$ of the outputs (279), (280) can be considered as the image of the pair $(\upsilon, x(\upsilon))$. In other words, the function of the controlled outputs $\gamma(t)$

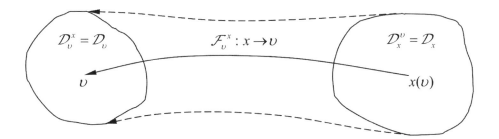

Figure 34. Mapping from the domain of states to the domain of inputs

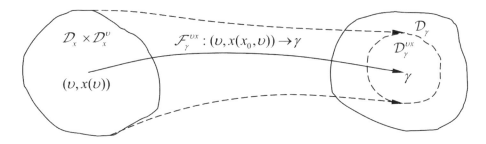

Figure 35. Mapping from the domain of inputs to the domain of outputs

defines the mapping $\mathcal{F}_\gamma^{ux} : (v, x(x_0, v)) \rightarrow \gamma$ of the whole product of the whole input domain and part of the state domain (Figure 34) (obtained by mapping from the input domain) to the domain of controlled outputs, which is part of the output domain $\mathcal{D}_v \times \mathcal{D}_x^v \rightarrow \mathcal{D}_\gamma^{ux} \subseteq \mathcal{D}_\gamma^v$ (Figure 35).

Definitions and theorems of controllability and observability specify the properties and conditions of mapping between the domains of inputs, states, and outputs.

The necessary condition of state controllability (274) defines the condition of the existence of the inverse mapping $\mathcal{F}_v^x: x_0 \rightarrow v$. As the initial states of mapping cover the whole state domain, (274) defines the condition of mapping of the whole set of states into part of the output set $\mathcal{D}_x \rightarrow \mathcal{D}_v^x \subseteq \mathcal{D}_v$ (Figure 34). However, the condition (274) is necessary and sufficient, which, from the point of view of mapping, means that it defines conditions of the existence of mapping of the whole input domain to the whole state domain.

Definition of observability specifies the mapping of the pair $(v, \gamma(v))$ into the state x, i.e. $\mathcal{F}_x^{v\gamma}: (v, \gamma(v)) \rightarrow x$. In terms of sets, this can be expressed in the following way.

Let the set $\mathcal{D}_\gamma^{ux}$ be obtained by mapping from the part of the state domain $\mathcal{D}_x^v$

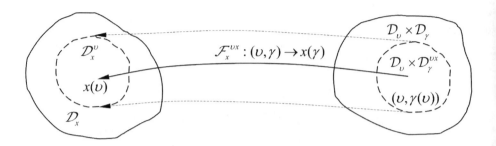

Figure 36. Mapping from the domain of outputs to the domain of inputs

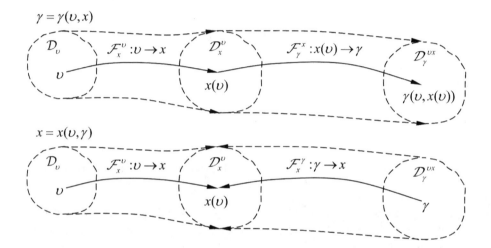

Figure 37. Mapping through the domain of states

to which is mapped the input domain $\mathcal{D}_v$. The definition of observability is related to the mapping of the direct product of the whole input domain and of the part of the output domain into the part of the state space $\mathcal{D}_v \times \mathcal{D}_{\gamma}^{vx} \subseteq \mathcal{D}_v \times \mathcal{D}_{\gamma} \rightarrow \mathcal{D}_x^v \subseteq \mathcal{D}_x$ (Figure 36).

The observability condition (277) specifies the conditions for which the subset $\mathcal{D}_x^v$ will be the whole state domain $\mathcal{D}_x^v = \mathcal{D}_x$ and the subset $\mathcal{D}_{\gamma}^{vx}$ will be equal to the whole output domain $\mathcal{D}_{\gamma}^{vx} = \mathcal{D}_{\gamma}$. The above discussion is based on considering the properties of the function composition $\gamma = \gamma(x(v), v)$. If the direct mapping from the input set to the output set (in (275), $\bar{D}v(t) = 0$) is not considered, the function composition acquires the form $\gamma = \gamma(x(v))$, which is graphically presented in Figure 37.

Still, it remains to consider the output controllability.

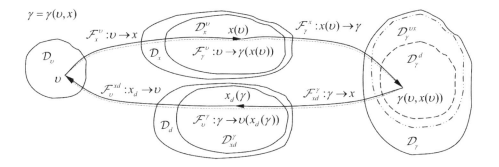

Figure 38. Mapping of the control system domain

The definition of the output controllability gives precisely the properties of mapping the input domain $\mathcal{D}_v$ and part of the state domain $\mathcal{D}_x^v$ to the domain of controlled outputs $\mathcal{D}_\gamma^v$. From the condition rank $(\bar{C}^T \bar{B}, \bar{C}^T \bar{A} \bar{B}, \ldots, \bar{C}^T \bar{A}^{n_x-1} \bar{B}, \bar{D}) = 1$ and Kalman's works [49] it comes out that the dimensions of the input space $\mathcal{D}_v$ and space of controllable outputs must be the same, $\dim\{\mathcal{D}_v\} = \dim\{\mathcal{D}_\gamma^v\}$, and that there must exist the inverse mapping

$$\gamma = \gamma(x, v) = \gamma(x(v), v) = \gamma(v)$$

$$\exists \gamma^{-1}: v = \gamma^{-1}(\gamma) = v(\gamma), \quad \gamma = \gamma(\gamma^{-1}(\gamma)) = \gamma$$

from the space of controlled inputs, in order that the system can be controllable. In other words, in order to have an output-controllable system, a prerequisite is the existence of a one-to-one correspondence between the whole space of inputs $\mathcal{D}_v$ and the whole space of controlled outputs $\mathcal{D}_\gamma^v$. The criteria of controllability/observability of the system states express the conditions of mapping of the whole space of inputs/outputs into the whole space of states.

With dynamic systems, mapping from the set $\mathcal{D}_v$ to the set $\mathcal{D}_\gamma^v$ must proceed indirectly via the set $\mathcal{D}_x$. The opposite mapping from the set $\mathcal{D}_\gamma^v$ to the set $\mathcal{D}_v$ may be either direct or via some other set $\mathcal{D}_d$, which, if it exists, represents for the control system, a set of states of the sensors $x_d(\gamma)$ (Figure 38). With dynamic systems, mapping from the set $\mathcal{D}_v$ to the set $\mathcal{D}_\gamma^v$ must proceed indirectly via the set $\mathcal{D}_x$. The opposite mapping from the set $\mathcal{D}_\gamma^v$ to the set $\mathcal{D}_v$ may be either direct or via some other set $\mathcal{D}_d$, which, if it exists, represents for the control system, a set of states of the sensors $x_d(\gamma)$ (Figure 38).

The consideration of the mapping from one domain to another is based on the functional dependence of the solution of the system of differential equations and controlled outputs. As these relationships are of the same form for both linear and non-linear systems, and no special properties of linear systems have been used in the inference, the conclusions hold for both systems.

It is known [31] that for the continuously differentiable functions $\gamma_i = \gamma_i(\upsilon_1, \ldots, \upsilon_{n_\upsilon})$, $i = 1, \ldots, n_\upsilon$ in an area of the n_υ-dimensional space, provided the Jacobian is different from zero,

$$\partial(\gamma_1, \gamma_2, \ldots, \gamma_{n_\upsilon})/\partial(\upsilon_1, \upsilon_2, \ldots, \upsilon_{n_\upsilon}) = \det(\partial\gamma_i/\partial\upsilon_j) \neq 0 \qquad (281)$$

the mapping $\gamma = \gamma(\upsilon)$ from a sufficiently small neighborhood of each selected point υ of the space $\mathcal{D}_\upsilon$ into the uniquely determined neighborhood of the point $\gamma(\upsilon)$ of the space $\mathcal{D}_\gamma^\upsilon$, and vice versa, takes place as a biunique mapping. Obviously, the necessary condition for the one-to-one mapping is that the dimensions of the space of inputs $\mathcal{D}_\upsilon$ and space of controlled outputs $\mathcal{D}_\gamma^\upsilon$ are the same in order that the previous Jacobian would exist at all. In this mapping can be set exactly $\dim\{\mathcal{D}_\upsilon\} = \dim\{\mathcal{D}_\gamma^\upsilon\}$ independent variables υ or γ and obtain the same number of dependent variables γ or υ. Each υ or γ selected as independent variable, may express one independent control requirement for the system. Independent variables may be either only the outputs γ, only the inputs υ, or even their combination. In other words, the control requirements can be set up to the system's output, to its input, or even to their combination. The preset requirements must be congruent, which means that for one independent variable or the variable dependent on it, it is possible to preset only one independent requirement. It follows that the number of controlled outputs should be equal to the number of control inputs.

Therefore, it has to be allowed that, on the basis of the known selected control inputs, it is possible to uniquely determine the controlled quantities. This is possible to realize if there is a mutually unique relation between the inputs and the outputs. With dynamic systems, this relationship is differential and it is indirectly established via the system dynamics, whereas with non-dynamic systems, this relation is algebraic.

In principle, the number of inputs n_υ and the number of outputs n_γ are not the same. If $n_\upsilon > n_\gamma$, the problem is easily solvable by canceling the excessive inputs. This can be achieved by establishing a functional dependence between $n_\upsilon - n_\gamma$ inputs or by applying the hierarchical control, so that for a concrete control task at higher control level n_γ suitable inputs are selected, whereas the rest of the inputs are kept constant. If the number of really possible outputs n_γ (mutually independent variables) is higher than the number of inputs n_υ, $n_\upsilon < n_\gamma$ then n_υ outputs can be controlled, while $n_\gamma - n_\upsilon$ outputs remain non-controlled, and their behavior is determined only by the dynamics of the controlled object. This ratio of the number of physical inputs and outputs also exists in cooperative manipulation. The task of the control is reduced to selecting the set with $n_\gamma = n_\upsilon$ outputs that will be controlled and such inputs that will result in the acceptable character of change of the non-controlled quantities. The control system should be hierarchical. The higher control level is to define a set of nominal motions of the system, select the control quantities, and define the mode of transition from one set of nominals to another. The control of selected quantities is realized at the lower control level by concretely selected control laws.

To determine the control on the basis of the requirements for input and/or output (e.g. trajectory tracking), means finding such a mapping that will map the domain of inputs exactly to the required subset of the set of outputs. Often, it happens that the explicit requirement concerning an output quantity has negative consequences to the other non-controlled output quantities. Correction of the characteristics of non-controlled quantities can be achieved in two ways (Figures 39 and 40). The simplest way is to change the preset requirements without changing the system's structure, and select the mapping to that part of the set of outputs in which all the inputs have satisfactory characteristics. The other way to solve this problem is to change the values of the independent variables as a function of the outputs. Namely, in the output function $\gamma = \gamma(x, \upsilon)$, independent variables are the state quantities $x \in \mathcal{D}_x$ and the inputs $\upsilon \in \mathcal{D}_\upsilon$. The input domain $\mathcal{D}_\upsilon$ cannot usually be essentially changed, so that it remains to change the part of the state space to which the input domain is mapped, and which represents a set of independent state variables x for the function of output as an independent quantity. This change is performed by changing the function $x = x(\upsilon)$ of mapping from the domain of inputs to the domain of states. As this function represents the solution of the system of differential equations in which the input υ is the drive, this means that it is necessary to modify the differential equations by which the system is described. This modification assumes the alteration of the constant parameters and functional relations in which states and their derivatives exist. In control theory, this procedure is called *synthesis of control laws* or *synthesis of the control system*. In other words, it is necessary to improve first the system characteristics and, on the improved system, impose the requirements for the controlled output quantities on the basis of which the necessary input quantities will be determined in a mutually unique way.

The synthesized control laws should ensure such object inputs υ^{ob} that will, in accordance with the object dynamics $\dot{x} = f(x, \upsilon^{ob})$ only, produce the states $x = x(x_0, \upsilon^{ob})$ of the control object such that the obtained outputs $\gamma = g(x)$ satisfy the preset requirements (Figure 39). For a control object, physical laws dictate its domains of inputs $\mathcal{D}_\upsilon$, states $\mathcal{D}_x$ and outputs $\mathcal{D}_y$, as the functions of mapping between them $x = x(x_0, \upsilon^{ob})$ and $\gamma = g(x)$ (Figure 40a, longer bold dotted line). These physical laws determine the object dynamics and cannot be changed by any control. The selected control laws can produce only such outputs

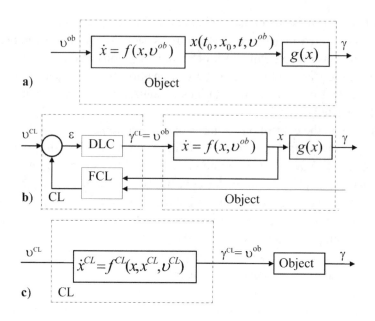

Figure 39. Structure of the control system

$\gamma^{CL} = \upsilon^{ob}$ of the control system that, as such, drives to the control object, and that the mapping $g(x)$ from the domain of states, realized by these drives, to the domain of outputs, is carried out only in that part of the output function $g(x)$ satisfying the preset requirements (Figures 39b, c and 40, short bold line).

The control law output γ^{CL} is a function of the preset required object input υ^{CL}; it may also be a function of the object state x and, if the control system is dynamic, of its state x^{CL} too, so that we finally have $\gamma^{CL} = \upsilon^{ob}(x, x^{CL}, \upsilon^{CL})$ (Figure 40c). Like the object's physical features determine the mapping functions, they also condition the requirements to be preset for the object, and which are expressed as the requirements υ^{CL} to the closed system. Also, the nominal regimes cannot be prescribed arbitrarily but in concordance with the physical laws that determine the controlled object dynamics. At that, the nominal motion should be determined only as a realizable object's motion in concordance with the physical laws, once the maximal number of preset requirements, equal to the number of the object physical inputs, is realized. Solution of the problem thus stated is given in Chapter 5 and in [10].

The analysis of the object's dynamic behavior represents consideration of the character of the solution of the system of differential equations describing the system's dynamic behavior for the excitations that are only functions of time for the non-controlled object and of time and state for the controlled object. For the con-

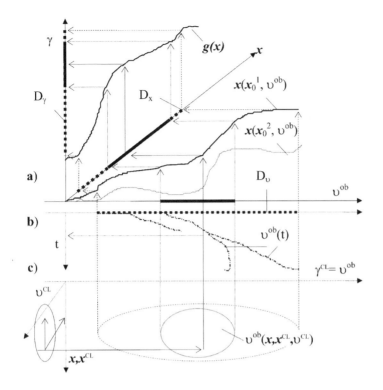

Figure 40. Mapping of the control object domain

trolled object, the analysis of dynamic behavior can be carried out based on the closed-loop model, open-loop model, and based on the object model only, if its input is fully determined. In this section, the analysis of the behavior of controlled quantities is carried out on the basis of the closed-loop model. The analysis of the output quantities that are not controlled is performed on the basis of the object model only.

It should be pointed out that the elastic part of the cooperative system has an infinite number of characteristic frequencies and modal forms corresponding to them. By choosing the cooperative manipulation model in the form (211), it is possible to encompass $6m$ characteristic frequencies and modal forms. At that, the first characteristic frequency is closest to the real one, provided the adopted approximation of the elastic system is valid. The character of the motion of a concrete elastic system is dictated by the character of its acting load. In other words, in the stage of selecting nominals, we can define the character of elastic system motion (i.e. of the cooperative system). The (quasi)static motion can be selected as nominal motion, but so can the motion that is far from or close to the characteristic modal forms of the elastic structure (resonance states). The analysis of dynamic behavior of the closed-loop cooperative system should show whether the system, using the selected vector of controlled outputs and control laws, can realize the required nominal motion in an asymptotically stable way.

6.2.2   Classification of the tasks

Before making a choice of the cases for which control laws will be sought, we will give a global classification of the conditions and tasks in cooperative manipulation.

In view of the elastic properties of the manipulated object, there are two possibilities:

- manipulation of a rigid object, and

- manipulation of an elastic object.

Depending on the required accuracy of the description of its physical characteristics, the object elasticity can be presented in different ways, from the description by partial differential equations to the description by algebraic equations.
In view of the space conditions in which the object manipulation takes place, it is possible to distinguish that:

- the motions of the object and manipulators proceed without constraints, and

- the motions of the object and manipulators are constrained in respect of the contact with the obstacles, limited ergonomic characteristics of the manipulators, etc.

In view of the spatial position of the manipulated object, there are three cases:

- The object is on support during the whole gripping and moving phases. In the state of rest, it is possible that the force at the object MC may be different from the weight $F_{e0} = G_0 + F_0$ and $\ddot{Y}_0 = \dot{Y}_0 = 0$.

- The object is held with no motion in space. In the state of rest, $F_{e0} = G_0$ and $\ddot{Y}_0 = \dot{Y}_0 = 0$.

- The object moves in space, and then $F_{e0} = F_{d0} + G_0$, $\ddot{Y} \neq 0$ and $\dot{Y}_0 \neq 0$.

The manipulators participating in cooperation may be

- in view of elastic properties

    - rigid, and

- elastic;

- in view of their redundancy

  - non-redundant, and

  - redundant;

- in view of joint compliance

  - with non-compliant joints, and

  - with compliant joints;

- in view of the DOFs of the object and manipulators

  - the manipulators (grippers) and manipulated object have the same number of DOFs, and

  - the manipulators and manipulated object have different numbers of DOFs;

- in view of the comparative characteristics of the manipulators

  - the manipulators in cooperation have the same characteristics (e.g. all manipulators are rigid with six DOFs), and

  - the manipulators in cooperation have different characteristics (e.g. the manipulators have a different number of DOFs).

In view of the type of contact in the sense of their elasticity, the classification is

- rigid contact, and

- elastic ('soft') contact.

In view of the contact in the sense of the mutual mobility of the contacting surfaces of the manipulator tip areas and the object envelope, the characteristic cases are

- Sliding contact. In this case the contact point of the manipulators' tips and object can move on the object envelope in the course of manipulation. A characteristic of the contact with sliding is that contact forces can be either zero or oriented so as to press the object.

- Stiff contact. In this case, the contact point of the manipulator tip and object in the manipulation remain all the time fixed to the object envelope. Within this case one can distinguish four subcases:

- Stiff contact. This contact is characterized by the forces that can be directed both to and from the object, i.e. $F_c \leq 0$, $F_c \geq 0$, and that the force vector associated to the contact point can have the number of components different from zero, that is equal to the number of the object DOFs, $F_{ci} \in R^{\mathrm{DOF} \times 1}$.

- Stiff half-contact. A characteristic of this contact is that, in some directions, the forces can be directed both towards and from the object, whereas in some directions they must be pressure forces.

- Contact at holding. The characteristics of this contact are the same as in the contact with sliding, except for the fact that the manipulator-object contact point cannot move on the object envelope.

- Contact at half-holding. In contrast to the preceding contact, which can involve all the components of the force vector at the contact, in this case some components of the force vector cannot be prescribed at all, and they are equal to zero during the time of manipulation.

In the cooperative system control, it is possible to consider several sets from which (or to which) mapping is performed in the sense of input-output (Figure 41).

For the overall cooperative system as a control object, it can be adopted that the set of all driving torques τ is the set of inputs $\mathcal{D}_\tau = \{\tau\} \in R^{6m}$. The set of manipulator states $\mathcal{D}_q = \{q\} \in R^{6m}$ can be adopted as a set of the manipulator outputs. The set of contact forces F_c (elastic system drives) can be adopted as a set of elastic system inputs $\mathcal{D}_{Fc} = \{F_c\} \in R^{6m}$, whereas the set of states Y of the elastic system $\mathcal{D}_Y = \{Y\} \in R^{6m+6}$ can be adopted as the set of elastic system outputs. The set $\mathcal{D}_Y$ can be thought of as the union of the sets $\mathcal{D}_{Yc} = \{Y_c\} \in R^{6m}$ of states at contact points Y_c and the set $\mathcal{D}_{Y0} = \{Y_0\} \in R^6$ of states of the manipulated object Y_0. Between $6m$ components Y_{ci}^j, $i = 1, \ldots, m$, $j = 1, \ldots, 6$ of the vector Y_c, defining in the $6m$-dimensional space a point of the set of the elastic system states $\mathcal{D}_{Yc} = \{Y_c\} \in R^{6m} \subseteq \mathcal{D}_Y$ and $6m$ components q_i^j, $i = 1, \ldots, m$, $j = 1, \ldots, 6$ of the vector q, defining in the $6m$-dimensional space one point of the set of the states of six-DOF non-redundant manipulators $\mathcal{D}_q$, there is a mutually unique correspondence according to the law (172), for every state q.

Assuming that, for the known states, their derivatives are also known, and that the mapping from one set of states covers the mapping from the set of derivatives too, the cooperative system behavior can be represented by the mapping between these sets.

The set of manipulator states $\mathcal{D}_q$ is obtained by the mapping $(\tau, F_c) \rightarrow q$ of the pair (τ, F_c) to the state q. The set $\mathcal{D}_\tau$ of driving torques τ can be obtained by the mapping $(q, F_c) \rightarrow \tau$ of the pair (q, F_c) to the driving torques τ, and the

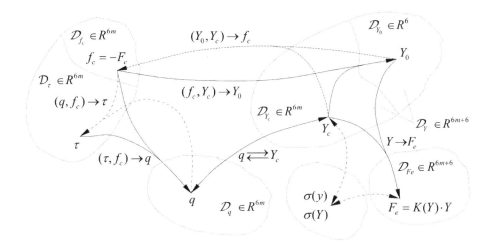

Figure 41. Mapping of the cooperative manipulation domain

set of contact forces $\mathcal{D}_{Fc}$ can be obtained by the mapping $(\tau, q) \rightarrow F_c$ of the pair (τ, q) to the contact force F_c. The mapping law is determined by the solution of the last two equations of (167). The set of elastic system states $\mathcal{D}_Y = \mathcal{D}_{Y_c} \cup \mathcal{D}_{Y_0}$ can be obtained by the mapping from the set of contact forces $\mathcal{D}_{Fc}$ to the set of elastic system states $\mathcal{D}_Y$ according to the law defined by the solution of the first three equations of the system of differential equations (100) or (113). As in the case of manipulators, other combinations of mapping are also possible, and some of them are presented in Figure 41. The set $\mathcal{D}_{Fe}$ of elastic forces F_e is obtained by the mapping $Y \rightarrow F_e$ of the elastic system state Y to these forces according to the law $F_e = K(Y)Y$. It should be noticed that between the set of states (y or Y) of the elastic system $\mathcal{D}_{\#} = \{\#\} \in R^{6m+6}$, $\# = y, Y$ or the set of elastic forces F_e, $\mathcal{D}_{Fe} = \{F_e\} \in R^{6m+6}$ and stresses ($\sigma = \sigma(y), \sigma = \sigma(Y), \sigma = \sigma(F_e)$) in the elastic system, there is a unique relation. Definition of that relation, however, is not the subject of this section.

Since the dimension of the space of the cooperative system inputs (driving torques) is $6m$, then $6m$ independent congruent requirements can be preset for the elements of any of the previous sets. This means that, at most, $6m$ independent congruent requirements can be met by the cooperative system for any of the previous sets expressed via the properties of the vector of controlled outputs.

Generally, in view of the set of quantities by which requirements are given, the control in cooperative manipulation can be classified as follows:

- control with the requirements imposed by the driving torques τ,

- control with the requirements imposed by the contact forces F_c,

- control with the requirements imposed by the manipulator states q, which is equivalent to the case when the requirements are imposed on the states at the contact points Y_c,

- control with the requirements imposed by the elasticity forces at contact points F_{ec}.

However, the main task is the control of the motion of the manipulated object. For such a task, it is necessary to directly preset six requirements for the state $Y_0 \in R^{6 \times 1}$ or indirectly via the requirement for the coordinates of some other node of the elastic system $Y_{ci} \in R^{6 \times 1}$. The remaining $6m - 6$ requirements can be preset for some other elements of the previous sets. Of all the possible cases we will single out only two:

- Control is imposed directly or indirectly on the manipulated object and the maintenance of elasticity forces is required, whereby the feedback loops can be closed

  - for one manipulator only on the basis of information about the state of the quantities,

  - for one manipulator on the basis of information about the state of quantities of one of the cooperation participants.

- Control is imposed directly or indirectly on the manipulated object and control of contact forces is required, whereby, as in the previous case, the feedback loops can be closed

  - for one manipulator only on the basis of information about the state of the quantities,

  - for one manipulator on the basis of information about the state of quantities of all the cooperation participants.

According to the conditions in the nominal motion, and in view of the requirements for gripping, cooperative manipulation can be classified as

- Nominal motion whose first phase is characterized by the initial gripping and the motion of the gripped elastic system is continued with the required contact forces that ensure the coordinated motion of the gripped elastic system.

- Nominal motion in which such change of contact forces is simultaneously required which, apart from the coordinated motion, will also realize the change in gripping conditions.

Each of the mentioned nominal conditions may be solvable with respect to the states of the MC or the states of any contact point of the manipulated object. In the previous section we considered the nominal motion in which no change in the gripping conditions was required.

In view of the function of the leader in the course of cooperative manipulation, the control can be performed

- without change of the leadership during the motion, or

- with simultaneous change in the leadership during the motion.

By combining the above conditions and tasks, we can obtain a very large number of tasks that need solving, which can be the subject of future research.

6.3   Choice of Control Tasks in Cooperative Manipulation

In Section 4.6 we modeled the dynamic behavior of a cooperative system consisted of m non-redundant, rigid, six-DOF manipulators with non-compliant joints, handling a rigid object, whereby the manipulator-object connections are elastic and the contact is rigid. The elastic interconnections possess the inertia and dissipation properties. The same model will describe the manipulation of an elastic object by rigid manipulators if the model can be split into elastically connected parts that can mimic previously described elastically connected rigid manipulators and rigid object. For the cooperative system thus defined, in Chapter 5 we adopted such nominal motion for which, in the first stage of motion, gripping is performed to the required force, and then the motion of the gripped elastic system is continued along the prescribed trajectory of the manipulated object MC, or of a contact point of the leader in cooperation, without requiring additional gripping. Coordinated motion under nominal conditions was ensured and all quantities were exactly calculated. For the cooperative system thus determined, such control laws should be selected that will guarantee tracking of the selected nominal quantities in the desired way. For the non-controlled quantities, it is necessary to give an estimate of their behavior.

It has already been said that for the overall cooperative system, the vector of driving torques $\tau \in R^{6m \times 1}$ can be adopted as the input with the domain $\mathcal{D}_\tau = \{\tau\} \in R^{6m}$. The set of controlled outputs must belong to the space of the same dimension as the domain of inputs ($6m$). As the vector of controlled outputs Y^u, for which certain requirements have been predefined, it is possible to choose some of the following vector types:

- *Vector of attitude of parts of the cooperative system*

- Part of the state vector $Y = \mathrm{col}(Y_v, Y_s, Y_0) = \mathrm{col}(Y_v, Y_{s0}) = \mathrm{col}(Y_c, Y_0) \in R^{(6m+6) \times 1}$ of the elastic system

$$Y^u = \mathrm{col}(Y_{s1}, \ldots, Y_{s(m-1)}, Y_0) = \mathrm{col}(Y_s, Y_0) = Y_{s0} \in R^{6m \times 1}. \quad (282)$$

- Part of the state vector Y of the elastic system equal to the position vector of contact points

$$Y^u = Y_c \in R^{6m \times 1}. \quad (283)$$

- In view of the one-to-one mapping of the internal coordinates q and position vector of contact points Y_c expressed by the relation (172) in the form $Y_c = \Theta(q)$, the choices equivalent to the previous ones are

$$Y^u = \mathrm{col}(q_s, Y_0) \in R^{6m \times 1}, \quad (284)$$

$$Y^u = q \in R^{6m \times 1}. \quad (285)$$

- *Vector of elasticity forces.* Between the vector of elasticity forces $F_e = \mathrm{col}(F_{ev}, F_{es}, F_{e0}) = \mathrm{col}(F_{ev}, F_{es0}) = \mathrm{col}(F_{ec}, F_{e0}) \in R^{(6m+6) \times 1}$ and the state vector of elastic system Y, there exists the relation (120) given by $F_e(Y) = K(Y) \cdot Y \in R^{(6m+6) \times 1}$, so that, instead of the part of the state vector Y of the elastic system, the controlled output can be part of the vector of elastic forces, that is

  - Part of the vector of elasticity forces acting at the contact points of the followers and manipulated object MC given by

$$Y^u = \mathrm{col}(F_{es}, F_0) = F_{es0} \in R^{6m \times 1}. \quad (286)$$

  - Part of the vector of elasticity forces equal to the vector of elasticity forces acting at the contact points

$$Y^u = F_{ec} \in R^{6m \times 1}. \quad (287)$$

- *Vector of contact forces.* In principle, the correctness of this choice can be corroborated in the following way. By solving the differential equations (115), describing the elastic system dynamics, the solution will be obtained in the form $Y = Y(F_c) \in R^{(6m+6) \times 1}$, and having (172) in mind, the relation $(q^T, Y_0^T)^T = (q^T, Y_0^T)^T(F_c) \in R^{(6m+6) \times 1}$. will be obtained. By solving the system of differential equations (167) that describe the manipulator dynamics, we get the solution $q = q(\tau, F_c) \in R^{6m \times 1}$ or, from

(172), $Y_c = Y_c(\tau, F_c) \in R^{6m \times 1}$. Elimination of the vector q, i.e. of the vector Y_c, will yield the dependence $F_c = F_c(\tau, Y_0) \in R^{6m \times 1}$, which can be written as a function of the selected input vector $\tau \in R^{6m \times 1}$ as $(F_c^T, Y_0^T)^T = (F_c^T, Y_0^T)^T(\tau) \in R^{(6m+6) \times 1}$. This means that the response to the drive $\tau \in R^{6m \times 1}$ is the contact forces $F_c \in R^{6m \times 1}$ and position of the manipulated object MC $Y_0 \in R^{6 \times 1}$. Thus, we have a total of $6m + 6$ quantities. The controlled output can be selected as

– The overall vector of contact forces

$$Y^u = F_c \in R^{6m \times 1}, \qquad (288)$$

whereby it should be borne in mind that in such choice of controlled output the position of the manipulated object MC in space can be arbitrary and, consequently, the position of the whole cooperative system too.

– Vector of attitude of the manipulated object MC $Y_0 \in R^{6 \times 1}$ and part of the vector of contact forces $F_{cs} = \mathrm{col}(F_{cs1}, \ldots, F_{cs(m-1)}) \in R^{(6(m-1) \times 1}$ acting at the contact points of the followers

$$Y^u = \mathrm{col}(Y_0, F_{cs}) \in R^{6m \times 1}. \qquad (289)$$

• *Part of the position vector of contact points $\bar{Y}_c$, i.e. the corresponding internal coordinates, $\bar{q}$, and part of the vector of contact forces $\bar{F}_c$.*

$$Y^u = \mathrm{col}(\bar{Y}_c, \bar{F}_c) \in R^{6m \times 1}, \qquad (290)$$

$$Y^u = \mathrm{col}(\bar{q}, \bar{F}_c) \in R^{6m \times 1}. \qquad (291)$$

In selecting such controlled outputs, care should be taken as to the congruence of the requirements to be fulfilled by the system, and that the dimension of the space $\mathcal{D}_y^u$ is $\dim\{\mathcal{D}_y^u\} = 6m$, i.e. to select the quantities that are mutually independent. Such a case is possible if the cooperative system can be decomposed in such a way that the controlled outputs are independent. A characteristic choice of the vector of controlled output is

$$Y^u = \mathrm{col}(Y_{cv}, F_{cs}) \in R^{6m \times 1}, \qquad (292)$$

$$Y^u = \mathrm{col}(q_v, F_{cs}) \in R^{6m \times 1}, \qquad (293)$$

which is structurally analogous to the vector (289), because, instead of the position of the manipulated object MC Y_0, the cooperative system in space is described in terms of the easily measurable position of one contact point (of the leader) Y_{cv} or q_v.

Above we gave some characteristic cases of choosing the controlled outputs. The choice of the controlled output implies the selection of external feedback loops, i.e. the selection of the appropriate sensors for furnishing information about the controlled outputs. From the point of view of engineering needs, the most suitable choice is the internal coordinates q as the output quantities of the actuators, which already possess sensors to measure them. Manipulators can be used to manipulate various objects. It is convenient that all the quantities needed for control are measured by the sensors with which the manipulators are equipped, so that, in addition to the internal coordinates, it is possible to use as feedback the contact forces of the manipulator tip and object measured by the sensors placed at the manipulators tips. For the needs of analysis, at least of a theoretical one, it is necessary to demonstrate that effective manipulation of the object is possible for the known current states, so that it is advisable to seek the control law with feedbacks in which the manipulated object states participate explicitly (as measured quantities).

From the point of view of the analysis, the choices of control laws in cooperative manipulation on the basis of position vectors of the parts of the cooperative system and vector of elasticity forces are equivalent. The choices of the controlled output $Y^u = Y_c$, $Y^u = q$ and $Y^u = F_{ec}$ are equivalent, so that it suffices to choose control laws for one of these cases, e.g. for $Y^u = q$. The choices of controlled outputs $Y^u = \mathrm{col}(Y_s, Y_0) = Y_{s0}$ and $Y^u = \mathrm{col}(F_{es}, F_{e0}) = F_{es0}$ are also equivalent, so that the choice of control laws can be carried out for $Y^u = Y_{s0}$, only, i.e. along with (172), for $Y^u = \mathrm{col}(q_s, Y_0)$.

Generally, all the above choices can be classified in two groups. One group consists of the control laws by which requirements are explicitly preset for the manipulated object MC and contact points of the followers. Controlled inputs are defined by (282) or (284), (286) and (289). To the other group belong the control laws by which the requirements are preset for the contact points, but without explicit requirements for the manipulated object MC, and the controlled outputs are determined by (283) or (285), (287) and (288).

In view of the above, the characteristic control tasks in cooperative manipulation are

- Tracking of the nominal trajectory of one point of the elastic system and tracking of the nominal trajectories of contact points of the followers, i.e. the nominal internal coordinates of the followers. Typical variants of such tracking are:

  - Tracking of the nominal trajectory $Y_0^0(t) \in R^6$ of the manipulated object MC and tracking of the nominal trajectories of contact points of the followers $Y_s^0 \in R^{(6m-6)}$, i.e. the nominal internal coordinates $q_s^0 \in R^{6m-6}$ of the followers.

The controlled output of the cooperative system is the $6m$-dimensional vector $Y^u = \text{col}(Y_s, Y_0)$, i.e. $Y^u = \text{col}(q_s, Y_0)$,

– Tracking of the nominal trajectory of the manipulated object MC without explicit tracking of its trajectory $Y_0^0(t) \in R^6$, but tracking of the trajectory $Y_v^0(t) \in R^6$ of the leader's contact point and tracking of the nominal trajectories of contact points of the followers $Y_s^0 \in R^{(6m-6)}$, i.e. of the nominal internal coordinates $q_s^0 \in R^{6m-6}$ of the followers. This means that direct tracking is performed of the nominal trajectory of all contact points given by the vector $Y_c^0 \in R^{6m}$ or by the vector of internal coordinates $q^0 \in R^{6m}$.

The controlled output of the cooperative system is the $6m$-dimensional vector $Y^u = \text{col}(Y_v, Y_s) = Y_c \in R^{6m}$, i.e. $Y^u = \text{col}(q_v, q_s) = q \in R^{6m}$.

The output quantities of the elastic system that are not directly tracked (non-controlled outputs) are the coordinates of the forces $F_c \in R^{6m}$ and position of one contact point ($\in R^6$).

- Tracking of the nominal trajectory of one node of the elastic system and tracking of the nominal contact forces at the contact points of the followers.

  Tracking of the nominal trajectory of one node of the elastic system ($Y_0^0(t) \in R^6$ of the manipulated object MC or $Y_v^0(t) \in R^6$ of the leader's contact point) and tracking of the nominal contact forces $F_{cs}^0 \in R^{6m-6}$ at the contact points of the followers.

  – Tracking of the nominal trajectory $Y_0^0(t) \in R^6$ of the manipulated object MC and tracking of the nominal contact forces $F_{cs}^0 \in R^{6m-6}$ at the contact points of the followers.

    The controlled output is the $6m$-dimensional vector $Y^u = \text{col}(F_{cs}, Y_0)$.

  – Tracking of the nominal trajectory of the manipulated object MC without explicit tracking of $Y_0^0(t)$, but with tracking the nominal trajectory of one (leader's) contact point $Y_v^0(t) \in R^6$ or $q_v^0 \in R^6$ and the nominal contact forces $F_{cs}^0 \in R^{6m-6}$ at the other contact points.

    The controlled output of the cooperative system is the $6m$-dimensional vector $Y^u = \text{col}(F_{cs}, Y_v)$, i.e. $Y^u = \text{col}(F_{cs}, q_v)$.

The output quantities of the elastic system that are not directly tracked (non-controlled outputs) are the positions of m nodes (when tracking Y_0^0, these are the positions of the contact points $Y_c \in R^{6m}$, whereas in tracking Y_v^0 these are positions of the followers' contact points and the manipulated object MC, i.e. the vector $Y_{s0} = \text{col}(Y_s, Y_0) \in R^{6m}$ or the vector $\text{col}(q_s, Y_0) \in R^{6m}$) and the contact force $F_{cv} \in R^6$ at the leader's contact point.

In this chapter, we will describe the synthesis of control laws for direct tracking of the nominal trajectory of the manipulated object.

6.4 Control Laws

The control laws are synthesized only for the directly tracked nominal trajectories of the manipulated object MC.

Before selecting the control laws, let us repeat in short the story about the mathematical model of cooperative manipulation with the emphasis on the properties that will be used later on.

6.4.1 Mathematical model

As we deal with the general motion, we shall consider the model given in the absolute coordinates. For the model in the coordinates of deviations of the immobile unloaded state of the elastic system, it is only necessary to introduce y instead of Y. The cooperative manipulation model for which the control laws will be selected was presented in Section 4.6 by Equations (113) or (115), (167) and (172). The combined form of the mathematical model is given by Equations (181) or (211).

Equation (115) represents the dynamic model of the elastic system that, under the action of the external forces F_c, performs the general motion. The model is of the form

$$W_{ca}(Y_c)\ddot{Y}_c + w_{ca}(Y, \dot{Y}) = F_c,$$

$$W_{0a}(Y_0)\ddot{Y}_0 + w_{0a}(Y, \dot{Y}) = 0.$$

The model of the dynamics of manipulators is given by (167) in the form

$$H(q)\ddot{q} + h(q, \dot{q}) = \tau + J^T f_c,$$

whereas the kinematic relations between the manipulator's internal and external coordinates are given by (172) in the form

$$Y_c = \Theta(q) \in R^{6m \times 1},$$

$$\dot{Y}_c = J(q)\dot{q} \in R^{6m \times 1},$$

$$\ddot{Y}_c = \dot{J}(q)\dot{q} + J(q)\ddot{q} \in R^{6m \times 1}.$$

By introducing the kinematic relations into the first equation, we obtain the description of the elastic system dynamics in terms of the internal coordinates q in the form

$$W_{ca}(\Theta(q))(\dot{J}(q)\dot{q} + J(q)\ddot{q}) + w_{ca}(\Theta(q), J(q)\dot{q}, Y_0, \dot{Y}_0) = F_c,$$

$$W_{0a}(Y_0)\ddot{Y}_0 + w_{0a}(\Theta(q), J(q)\dot{q}, Y_0, \dot{Y}_0) = 0. \quad (294)$$

By combining all the above equations, and taking that $F_c = -f_c$, we obtain the description of the cooperative system dynamics (181). Equations (181), together with the rearranged first of the above equations given in short form, represent the starting equations that describe the cooperative system's behavior, needed to introduce the control laws into the cooperative manipulation. Their form is

$$N(q)\ddot{q} + n(q, \dot{q}, Y_0, \dot{Y}_0) = \tau,$$

$$W(Y_0)\ddot{Y}_0 + w(q, \dot{q}, Y_0, \dot{Y}_0) = 0,$$

$$P(q)\ddot{q} + p(q, \dot{q}, Y_0, \dot{Y}_0) = F_c. \quad (295)$$

The first two equations of (295) are the repeated equations of the cooperative system's behavior (181), whereas the third equation determines the dependence of the contact forces on the internal coordinates.

Using the convention for the leader and followers, defined in Section 4.12, Equation (181) (i.e. (295)) was written in the form (211). The result is the mathematical model of the cooperative system dynamics in the form

$$N_v(q_v)\ddot{q}_v + n_v(q, \dot{q}, Y_0, \dot{Y}_0) = \tau_v,$$

$$N_s(q_s)\ddot{q}_s + n_s(q, \dot{q}, Y_0, \dot{Y}_0) = \tau_s,$$

$$W(Y_0)\ddot{Y}_0 + w(q, \dot{q}, Y_0, \dot{Y}_0) = 0,$$

$$P_v(q_v)\ddot{q}_v + p_v(q, \dot{q}, Y_0, \dot{Y}_0) = F_{cv},$$

$$P_s(q_s)\ddot{q}_s + p_s(q, \dot{q}, Y_0, \dot{Y}_0) = F_{cs}, \quad (296)$$

which represents the basic form of the model for introducing control into the cooperative system.

6.4.2 Illustration of the application of the input calculation method

The method of input calculation is a procedure of synthesizing the system input by solving a system of differential equations that describe the system's mathematical model and the control law error given in advance.

The procedure can be summarized as follows. For the system considered, the mathematical model is composed in the form (100), (100), (113), (115), (181), (183), (295) or (296). The quantities to be directly tracked are selected. The deviations of the directly controlled quantities from their nominal values are introduced and their higher derivatives are determined. The law of the behavior of deviations of the directly controlled quantities from their nominal values of the closed-loop system is selected in advance and given by the differential equation. This equation is solved with respect to the highest derivatives of deviations as a function of the lower derivatives of deviations as independent variables. The calculated highest derivatives are introduced into the differentiated equations and values of the highest derivatives of the directly tracked quantities are calculated. The values of the latter should be possessed by the controlled object in order that the deviation of the actual trajectory from its nominal value would satisfy the required differential equations of deviations. The calculated derivatives of the directly tracked quantities are introduced into the mathematical model and the inputs to be introduced are calculated.

The application of the input calculation method will be illustrated in the example of simple mechanical systems in which the number of inputs is equal to the number of equations of motion. In the equations of motion of mechanical systems, the highest derivative is the second one (acceleration), so that the simplest way is to choose that the deviations satisfy second-order differential equations. As an example, we consider a mechanical object with no stabilization loops ($\tau_{ob}(t) = \tau(t)$, Figure 42) that can be described by the following second-order differential equation:

$$M(y)\ddot{y} + m(y, \dot{y}) = J(y)\tau, \quad y \in R^1. \tag{297}$$

Let the nominal $y^0 \in R^1$, $\dot{y}^0 \in R^1$, $\ddot{y}^0 \in R^1$, $\tau^0 \in R^1$ to be described by the object be known. It is required that the object (297) follows the known nominal in an asymptotically stable manner. This will be realized if the deviations from the nominal trajectory converge to zero. By analogy to a linear regulation loop, it can be required that the deviations from the nominal trajectories in the closed-loop controlled system satisfy the differential equations with exactly determined properties in respect of stability of the indicators of the quality of behavior of their solution. One possible choice of differential equation is

$$\Delta\ddot{y} + 2\zeta\omega\Delta\dot{y} + \omega^2\Delta y = 0. \tag{298}$$

By adjusting the damping coefficient ζ and frequency ω, the stability properties and quality of nominal trajectory tracking, i.e. the properties of the closed-loop system for which the desired input is a zero deviation (Figure 42), are adjusted.

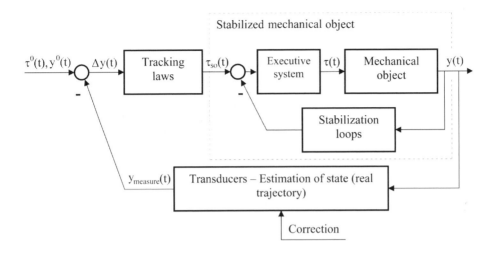

Figure 42. Global structure of the closed loop system

From (298) we determine the second derivative of deviations

$$\Delta \ddot{y} = -2\zeta \omega \Delta \dot{y} - \omega^2 \Delta y, \tag{299}$$

and since $\Delta \ddot{y} = \ddot{y}^0 - \ddot{y}$, $\Delta \dot{y} = \dot{y}^0 - \dot{y}$ and $\Delta y = y^0 - y$, the second derivative to be possessed by the real object is

$$\ddot{y} = \ddot{y}^0 - \Delta \ddot{y} = \ddot{y}^0 + 2\zeta \omega (\dot{y}^0 - \dot{y}) + \omega^2 (y^0 - y). \tag{300}$$

By introducing the necessary second derivative $\ddot{y}$ into the motion equation (297), we calculate the input to the object to realize that derivative

$$\tau = J^{-1}(y)\{M(y)[\ddot{y}^0 + 2\zeta \omega (\dot{y}^0 - \dot{y}) + \omega^2 (y^0 - y)] + m(y, \dot{y})\}. \tag{301}$$

The calculated input (301) represents the guiding law to be introduced into the real control object model in order to realize the asymptotically stable tracking of the nominal trajectory. Obviously, after introducing the calculated control law into the object model (297), the prescribed requirement (298) for the behavior of the deviation will be identically satisfied.

The application of the method of input calculation onto the objects having the number of inputs that is smaller than the number of motion equations, is more complex. With a cooperative system, the number of inputs (physical drives – driving torques) is smaller than the number of equations of motion.

6.4.3   Control laws for tracking the nominal trajectory of the manipulated object MC and nominal trajectories of contact points of the followers

In this case of tracking, the controlled input is the vector $Y^u = \mathrm{col}(q_s, Y_0)$. It is required that the controlled cooperative system is tracking the selected nominal trajectory $Y^0(t) = \mathrm{col}(q_s^0(t), Y_0^0(t))$ with a predefined quality, determined by the procedures given in Chapter 5. The output quantities of the cooperative system that are not directly tracked (non-controlled outputs) are the contact forces $F_c \in R^{6m}$ and the position of the leader's contact point, $Y_v \in R^6$. The character of deviation of non-controlled quantities in the system from their nominal values should be examined separately.

The procedure to synthesize the driving moments ensuring the error of con-trolled outputs has the properties determined in advance consists of the following.

Let

$$\overset{(k)}{\eta_s}(t) = \overset{(k)}{q_s^0}(t) - \overset{(k)}{q_s}(t), \quad k = 0, 1, 2, \dots,$$

$$\overset{(k)}{\Delta Y_0} = \overset{(k)}{Y_0^0}(t) - \overset{(k)}{Y_0}(t), \quad k = 0, 1, 2, \dots, \tag{302}$$

be the vectors of deviations and vectors of derivatives of deviations of the actual controlled trajectory from the nominal trajectory. If $\eta_s(t)$ and ΔY_0 are the solutions of the homogeneous differential equations

$$\chi_s(\overset{(l)}{\eta_s}, \overset{(l-1)}{\eta_s}, \dots, \overset{(0)}{\eta_s}) = 0, \quad \overset{(0)}{\eta_s} = \eta_s,$$

$$\chi_0(\overset{(k)}{\Delta Y_0}, \overset{(k-1)}{\Delta Y_0}, \dots, \overset{(0)}{\Delta Y_0}) = 0, \quad \overset{(0)}{\Delta Y_0} = \Delta Y_0, \tag{303}$$

obtained as the response to the initial states of deviations $\eta_s(t_0) = q_s^0(t_0) - q_s(t_0)$ and $\Delta Y_0(t_0) = Y_0^0(t_0) - Y_0(t_0)$, then a relationship can be established between the character of change of deviations $\eta_s(t)$ and realized deviations ΔY_0 from the nominal trajectories and the characteristics of the previous differential equations. It is required that the deviations from the nominal trajectories in the controlled closed-loop system satisfy differential equations with exactly determined proper-ties in respect of the stability and indicators of the quality of the behavior of their solution. By solving the previous differential equations with respect to the highest derivative, we obtain the functional relationships

$$\overset{(l)}{\eta_s} = \overset{(l)}{q_s^0}(t) - \overset{(l)}{q_s}(t) = Q_s(\overset{(l-1)}{\eta_s}, \overset{(l-2)}{\eta_s}, \dots, \eta_s)),$$

$$\overset{(k)}{\Delta Y_0} = \overset{(k)}{Y_0^0}(t) - \overset{(k)}{Y_0}(t) = Q_0(\overset{(k-1)}{\Delta Y_0}, \overset{(k-2)}{\Delta Y_0}, \dots, \Delta Y_0), \tag{304}$$

between the highest derivatives of deviations on their lower derivatives as indepen-dent variables. The calculation gives

$$\overset{(l)}{q_s} = \overset{(l)}{q_s^0}(t) - Q_s(\overset{(l-1)}{\eta_s}, \overset{(l-2)}{\eta_s}, \ldots, \eta_s),$$

$$\overset{(k)}{Y_0}(t) = \overset{(k)}{Y_0^0}(t) - Q_0(\Delta \overset{(k-1)}{Y_0}, \Delta \overset{(k-2)}{Y_0}, \ldots, \Delta Y_0), \tag{305}$$

the values of highest derivatives $\overset{(l)}{q_s}(t)$ and $\overset{(k)}{Y_0}(t)$ of the controlled quantities to be possessed by the controlled object in order that the deviation of the real trajectory from its nominal value would satisfy the sought differential equations. Based on the requirement for the realization of these derivatives, after introducing the calculated derivatives into (296), the driving torques τ are calculated. The proposed procedure represents the expansion into cooperative manipulation of the procedure based on the requirement that the deviations from the nominals satisfy linear differential equations, which are usually found in the open literature. This expansion has been given in [35] for a manipulator in contact with dynamic environment.

In this case of tracking, the calculated value $\overset{(k)}{Y_0}(t)$ should be introduced into the third equation of (296). If we choose, for example $k = 2$, we will obtain the dependence

$$W(Y_0)(\ddot{Y}_0^0 - Q_0(\Delta \dot{Y}_0, \Delta Y_0)) + w(q, \dot{q}, Y_0, \dot{Y}_0) = 0 \tag{306}$$

or, written differently,

$$\varphi_0(\ddot{Y}_0^0, \dot{Y}_0^0, Y_0^0, \dot{q}, q, \dot{Y}_0, Y_0) = 0 \tag{307}$$

which, for the rest of the controlled cooperative system, represents a non-holonomic relation. This relation defines six conditions and the same number of conditions is given to the vector of possible accelerations $\ddot{q}$, which has $6m$ compo-nents. These conditions may be associated to any component $\ddot{q}_i$ and, in this case of tracking, it has been chosen that these are the first six components, i.e. the vec-tor of the leader's acceleration. In order to obtain all possible accelerations of the leader, the above expression for φ_0 should be differentiated. The result will be the dependence on $\overset{...}{Y_0^0}$ that should be simultaneously determined in the course of con-trol on the basis of the known (prescribed) $\ddot{Y}_0^0$. Because of that, and for an easier proof of the stability of the closed-loop system, it is more convenient to differen-tiate the third equation of (296) prior to replacing the highest derivatives, and set the requirements via the third derivative of deviations ($k = 3$) of the real trajectory from the nominal trajectory of the manipulated object MC. By differentiating the third equation of (296), we obtain

$$\frac{d}{dt}W(Y_0)\ddot{Y}_0 + \frac{d}{dt}w(q,\dot{q},Y_0,\dot{Y}_0)$$

$$= \dot{W}(Y_0)\ddot{Y}_0 + W(Y_0)\,\dddot{Y}_0 + \frac{\partial w}{\partial q}\dot{q} + \frac{\partial w}{\partial \dot{q}}\ddot{q} + \frac{\partial w}{\partial Y_0}\dot{Y}_0 + \frac{\partial w}{\partial \dot{Y}_0}\ddot{Y}_0 = 0. \tag{308}$$

Using the sign convention for the leader and followers, we obtain

$$\frac{\partial w}{\partial \dot{q}}\ddot{q} = \frac{\partial w_v}{\partial \dot{q}_v}\ddot{q}_v + \frac{\partial w_s}{\partial \dot{q}_s}\ddot{q}_s, \quad \frac{\partial w_v}{\partial \dot{q}_v} \in R^{6\times6}, \quad \frac{\partial w_s}{\partial \dot{q}_s} \in R^{6\times(6m-6)}. \tag{309}$$

If the matrix $\partial w_v/\partial \dot{q}_v$ is non-singular, solving the last equation with respect to $\ddot{q}_v$ gives the possible accelerations of the leader as

$$\ddot{q}_v = -\alpha(\dddot{Y}_0, \ddot{Y}_0, \dot{Y}_0, Y_0, \dot{q}, q) - \beta(\dot{Y}_0, Y_0, \dot{q}, q)\cdot\ddot{q}_s \in R^{6\times1}, \quad \left|\frac{\partial w_v}{\partial \dot{q}_v}\right| \neq 0, \tag{310}$$

where

$$\alpha(\dddot{Y}_0, \ddot{Y}_0, \dot{Y}_0, Y_0, \dot{q}, q) = \left(\frac{\partial w_v}{\partial \dot{q}_v}\right)^{-1}$$

$$\times \left(\dot{W}(Y_0)\ddot{Y}_0 + W(Y_0)\,\dddot{Y}_0 + \frac{\partial w}{\partial q}\dot{q} + \frac{\partial w}{\partial Y_0}\dot{Y}_0 + \frac{\partial w}{\partial \dot{Y}_0}\ddot{Y}_0\right) \in R^{6\times1},$$

$$\beta(\dot{Y}_0, Y_0, \dot{q}, q) = \left(\frac{\partial w_v}{\partial \dot{q}_v}\right)^{-1}\frac{\partial w}{\partial \dot{q}_s} \in R^{6\times(6m-6)}. \tag{311}$$

The matrix $\partial w_v/\partial \dot{q}_v$ is singular for the conditions of the elastic system passing through the unloaded state $(1 - \|\rho_{ijo}\|/\|\rho_{ija}\| = 0)$. It is assumed that the object, in transferring, is gripped $(\|\rho_{ijo}\| \neq \|\rho_{ija}\|)$, otherwise it would not be held. If even then the matrix $\partial w_v/\partial \dot{q}_v$ is singular this means that a wrong choice of the leader was made, so that it has to be changed because the object dynamics has no direct influence on the acceleration of the manipulator selected as leader.

Let the differential equation

$$\Delta \dddot{Y}_0 = Q_0(\Delta \ddot{Y}_0, \Delta \dot{Y}_0, \Delta Y_0), \quad \Rightarrow \dddot{Y}_0 = \dddot{Y}_0^0 - Q_0(\Delta \ddot{Y}_0, \Delta \dot{Y}_0, \Delta Y_0) \tag{312}$$

have only the trivial asymptotically stable equilibrium state $\Delta Y_0 = 0$ (i.e... $Y_0^0 = Y_0$). Let this differential equation be chosen in the way that the solution obtained as the response to the initial deviation, $\Delta Y_0(t_0) = Y_0^0(t_0) - Y_0(t_0)$, is asymptotically stable with the desired indicators of the quality of dynamic behavior. Replacement of the calculated third derivative $\dddot{Y}_0\ (t)$ from (312) to (310) gives the possible accelerations of the leader as a function of the system state, preset requirement, and the followers' acceleration $\ddot{q}_s$ as

$$\ddot{q}_v = -\alpha(\dddot{Y}_0^0 - Q_0(\Delta\ddot{Y}_0, \Delta\dot{Y}_0, \Delta Y_0), \ddot{Y}_0, \dot{Y}_0, Y_0, \dot{q}, q) - \beta(\dot{Y}_0, Y_0, \dot{q}, q)\cdot\ddot{q}_s. \quad (313)$$

Introducing this acceleration into the first and fourth equations of (296) gives the driving torques and contact force of the leader, also as a function of the followers' acceleration $\ddot{q}_s$

$$
\begin{aligned}
\tau_v &= N_v(q_v)[-\alpha(\dddot{Y}_0^0 - Q_0(\Delta\ddot{Y}_0, \Delta\dot{Y}_0, \Delta Y_0), \ddot{Y}_0, \dot{Y}_0, Y_0, \dot{q}, q) \\
&\quad - \beta(\dot{Y}_0, Y_0, \dot{q}, q)\cdot\ddot{q}_s] + n_v(q, \dot{q}, Y_0, \dot{Y}_0) \\
&= \tau_v(\dddot{Y}_0^0 - Q_0(\Delta\ddot{Y}_0, \Delta\dot{Y}_0, \Delta Y_0), \ddot{Y}_0, \dot{Y}_0, Y_0, q, \dot{q}, \ddot{q}_s), \\
F_{cv} &= P_v(q_v)[-\alpha(\dddot{Y}_0^0 - Q_0(\Delta\ddot{Y}_0, \Delta\dot{Y}_0, \Delta Y_0), \ddot{Y}_0, \dot{Y}_0, Y_0, \dot{q}, q) \\
&\quad - \beta(\dot{Y}_0, Y_0, \dot{q}, q)\cdot\ddot{q}_s] + p_v(q, \dot{q}, Y_0, \dot{Y}_0) \\
&= F_{cv}(\dddot{Y}_0^0 - Q_0(\Delta\ddot{Y}_0, \Delta\dot{Y}_0, \Delta Y_0), \ddot{Y}_0, \dot{Y}_0, Y_0, q, \dot{q}, \ddot{q}_s). \quad (314)
\end{aligned}
$$

From the above, and in view of (296), it comes out that all driving torques and contact forces depend on the followers' acceleration $\ddot{q}_s$.

Let the differential equation

$$\dddot{\eta}_s(t) = Q_s(\dot{\eta}_s, \eta_s) \quad \Rightarrow \quad \dddot{q}_s(t) = \dddot{q}_s^0(t) - Q_s(\dot{\eta}_s, \eta_s) \quad (315)$$

have only the trivial asymptotically stable equilibrium state $\eta_s(t) = 0$ (i.e. $q_s^0(t_0) = q_s(t_0)$). Let this differential equation be chosen in a way that the solution obtained as the response to the initial deviation, $\eta_s(t_0) = q_s^0(t_0) - q_s(t_0)$, is asymptotically stable with the desired indicators of quality of the dynamic behavior.

By introducing the calculated acceleration of the followers from (315) into (314) and into the second equation of (296), we can calculate the driving torques

$$
\begin{aligned}
\tau_v &= N_v(q_v)[-\alpha(\dddot{Y}_0^0 - Q_0(\Delta\ddot{Y}_0, \Delta\dot{Y}_0, \Delta Y_0), \ddot{Y}_0, \dot{Y}_0, Y_0, \dot{q}, q) \\
&\quad - \beta(\dot{Y}_0, Y_0, \dot{q}, q)\cdot(\dddot{q}_s^0(t) - Q_s(\dot{\eta}_s, \eta_s))] + n_v(q, \dot{q}, Y_0, \dot{Y}_0) \\
&= \tau_v(\dddot{Y}_0^0 - Q_0(\Delta\ddot{Y}_0, \Delta\dot{Y}_0, \Delta Y_0), \ddot{Y}_0, \dot{Y}_0, Y_0, q, \dot{q}, \ddot{q}_s^0 - Q_s(\dot{\eta}_s, \eta_s)) \\
&= \tau_v(\dddot{Y}_0^0, \ddot{Y}_0, \dot{Y}_0, Y_0, q, \dot{q}, q_s^0, \dot{q}_s^0, \ddot{q}_s^0),
\end{aligned}
$$

$$\tau_s = N_s(q_s)(\ddot{q}_s^0(t) - Q_s(\dot{\eta}_s, \eta_s)) + n_s(q, \dot{q}, Y_0, \dot{Y}_0)$$

$$= \tau_s(\dot{Y}_0, Y_0, q, \dot{q}, \ddot{q}_s^0 - Q_s(\dot{\eta}_s, \eta_s)) = \tau_s(\dot{Y}_0, Y_0, q, \dot{q}, q_s^0, \dot{q}_s^0, \ddot{q}_s^0), (316)$$

that should be introduced at the joints of the manipulators in order to realize tracking of the controlled output $Y^{u0} = \mathrm{col}(Y_0^0, q_s^0)$ with the quality given indirectly by (312) and (315). The introduction of the driving torques (316) allows the realization of the contact forces

$$F_{cv} = P_v(q_v)[-\alpha(\dddot{Y}_0^0 - Q_0(\Delta \ddot{Y}_0, \Delta \dot{Y}_0, \Delta Y_0), \ddot{Y}_0, \dot{Y}_0, Y_0, \dot{q}, q)$$

$$- \beta(\dot{Y}_0, Y_0, \dot{q}, q) \cdot (\ddot{q}_s^0(t) - Q_s(\dot{\eta}_s, \eta_s))] + p_v(q, \dot{q}, Y_0, \dot{Y}_0)$$

$$= F_{cv}(\dddot{Y}_0^0 - Q_0(\Delta \ddot{Y}_0, \Delta \dot{Y}_0, \Delta Y_0), \ddot{Y}_0, \dot{Y}_0, Y_0, q, \dot{q}, \ddot{q}_s^0 - Q_s(\dot{\eta}_s, \eta_s))$$

$$= F_{cv}(\dddot{Y}_0^0, \ddot{Y}_0, \dot{Y}_0, Y_0, q, \dot{q}, q_s^0, \dot{q}_s^0, \ddot{q}_s^0),$$

$$F_{cs} = P_s(q_s)(\ddot{q}_s^0(t) - Q_s(\dot{\eta}_s, \eta_s)) + p_s(q, \dot{q}, Y_0, \dot{Y}_0)$$

$$= F_{cs}(\dot{Y}_0, Y_0, q, \dot{q}, \ddot{q}_s^0 - Q_s(\dot{\eta}_s, \eta_s))$$

$$= F_{cs}(\dot{Y}_0, Y_0, q, \dot{q}, q_s^0, \dot{q}_s^0, \ddot{q}_s^0). \tag{317}$$

To form the driving torques, it is necessary to have information about all instantaneous kinematic quantities $\ddot{Y}_0, \dot{Y}_0, Y_0$ of the manipulated object MC, information about all internal coordinates q and their derivatives $\dot{q}$, and information about the nominal outputs $Y_0^0, \dddot{Y}_0^0, \dot{Y}_0^0, Y_0^0, q_s^0, \dot{q}_s^0$ and $\ddot{q}_s^0$.

Let us introduce the driving torques (316) as the input to the model of cooperative manipulation (296) and let us prove that the preset requirements will be realized.

$$N_v(q_v)\ddot{q}_v + n_v(q, \dot{q}, Y_0, \dot{Y}_0)$$

$$= N_v(q_v)[-\alpha(\dddot{Y}_0^0 - Q_0(\Delta \ddot{Y}_0, \Delta \dot{Y}_0, \Delta Y_0), \ddot{Y}_0, \dot{Y}_0, Y_0, \dot{q}, q)$$

$$- \beta(\dot{Y}_0, Y_0, \dot{q}, q) \cdot (\ddot{q}_s^0(t) - Q_s(\dot{\eta}_s, \eta_s))] + n_v(q, \dot{q}, Y_0, \dot{Y}_0),$$

$$N_s(q_s)\ddot{q}_s + n_s(q, \dot{q}, Y_0, \dot{Y}_0) = N_s(q_s)(\ddot{q}_s^0(t) - Q_s(\dot{\eta}_s, \eta_s)) + n_s(q, \dot{q}, Y_0, \dot{Y}_0),$$

$$W(Y_0)\ddot{Y}_0 + w(q, \dot{q}, Y_0, \dot{Y}_0) = 0, \tag{318}$$

i.e.

$$N_v(q_v)[\ddot{q}_v + \alpha(\dddot{Y}_0^0 - Q_0(\Delta \ddot{Y}_0, \Delta \dot{Y}_0, \Delta Y_0), \ddot{Y}_0, \dot{Y}_0, Y_0, \dot{q}, q)$$

$$+ \beta(\dot{Y}_0, Y_0, \dot{q}, q) \cdot (\ddot{q}_s^0(t) - Q_s(\dot{\eta}_s, \eta_s))] = 0,$$

$$N_s(q_s)[\ddot{q}_s^0(t) - \ddot{q}_s - Q_s(\dot{\eta}_s, \eta_s)] = 0,$$

$$W(Y_0)\ddot{Y}_0 + w(q, \dot{q}, Y_0, \dot{Y}_0) = 0. \tag{319}$$

Because of the non-singularity of the matrix $N_s(q_s)$, it follows from the second equation that the driving torque τ_s introduced from (316) realizes

$$\ddot{q}_s(t) = \ddot{q}_s^0(t) - Q_s(\dot{\eta}_s, \eta_s) \quad \Rightarrow \ddot{\eta}_s(t) = Q_s(\dot{\eta}_s, \eta_s), \tag{320}$$

which is identical to the preset requirement (315). As the matrix $N_v(q_v)$ is also non-singular, then

$$\ddot{q}_v + \alpha(\ddot{Y}_0^0 - Q_0(\Delta \ddot{Y}_0, \Delta \dot{Y}_0, \Delta Y_0), \ddot{Y}_0, \dot{Y}_0, Y_0, \dot{q}, q)$$
$$+ \beta(\dot{Y}_0, Y_0, \dot{q}, q) \cdot (\ddot{q}_s^0(t) - Q_s(\dot{\eta}_s, \eta_s)) = 0. \tag{321}$$

The leader's acceleration $\ddot{q}_v$ must satisfy (310), so that the introduction of that acceleration into the last expression, along with the relation for the realized followers' acceleration, after rearrangement gives

$$\alpha(\ddot{Y}_0^0 - Q_0(\Delta \ddot{Y}_0, \Delta \dot{Y}_0, \Delta Y_0), \ddot{Y}_0, \dot{Y}_0, Y_0, \dot{q}, q) - \alpha(\ddot{Y}_0, \ddot{Y}_0, \dot{Y}_0, Y_0, \dot{q}, q) = 0. \tag{322}$$

Replacement of the values for $\alpha(\dddot{Y}_0, \ddot{Y}_0, \dot{Y}_0, Y_0, \dot{q}, q)$ according to (311) and rearrangement gives

$$\left(\frac{\partial w_v}{\partial \dot{q}_v}\right)^{-1} W(Y_0) \cdot [\ddot{Y}_0^0 - \ddot{Y}_0 - Q_0(\Delta \ddot{Y}_0, \Delta \dot{Y}_0, \Delta Y_0)] = 0. \tag{323}$$

As the matrices $\partial w_v/\partial \dot{q}_v$ and $W(Y_0)$ are non-singular, it finally comes out that the driving torques introduced realize

$$\ddot{Y}_0 = \ddot{Y}_0^0 - Q_0(\Delta \ddot{Y}_0, \Delta \dot{Y}_0, \Delta Y_0) \quad \Rightarrow \Delta \ddot{Y}_0 = Q_0(\Delta \ddot{Y}_0, \Delta \dot{Y}_0, \Delta Y_0), \tag{324}$$

which is just the starting requirement (312).

It has been shown that the introduction of control laws, represented by the relations for the synthesized driving torques (316), ensure the controlled cooperative system follows the nominal controlled outputs $Y^{u0} = \mathrm{col}(Y_0^0, q_s^0)$ in a stable manner and with the prescribed quality requirements, given indirectly by (312) and (315). As the dependence of deviation of the third and second derivatives $\Delta \dddot{Y}_0$ and $\ddot{\eta}_s = \ddot{q}_s^0 - \ddot{q}_s$ of the controlled outputs, adopted through the control laws, is realized, the deviation of the lower derivatives and the realized lower derivative of the controlled output Y_0 will be

$$\Delta \ddot{Y}_0 = \int_{t_0}^{t} Q_0(\Delta \ddot{Y}_0, \Delta \dot{Y}_0, \Delta Y_0) \, dt$$

$$\Rightarrow \ddot{Y}_0 = \ddot{Y}_0^0 - \int_{t_0}^{t} Q_0(\Delta \ddot{Y}_0, \Delta \dot{Y}_0, \Delta Y_0) \, dt,$$

$$\Delta \dot{Y}_0 = \int_{t_0}^{t} \int_{t_0}^{t} Q_0(\Delta \ddot{Y}_0, \Delta \dot{Y}_0, \Delta Y_0) \, dt \, dt$$

$$\Rightarrow \dot{Y}_0 = \dot{Y}_0^0 - \int_{t_0}^{t} \int_{t_0}^{t} Q_0(\Delta \ddot{Y}_0, \Delta \dot{Y}_0, \Delta Y_0) \, dt \, dt,$$

$$\Delta Y_0 = \int_{t_0}^{t} \int_{t_0}^{t} \int_{t_0}^{t} Q_0(\Delta \ddot{Y}_0, \Delta \dot{Y}_0, \Delta Y_0) \, dt \, dt \, dt$$

$$\Rightarrow Y_0 = Y_0^0 - \int_{t_0}^{t} \int_{t_0}^{t} \int_{t_0}^{t} Q_0(\Delta \ddot{Y}_0, \Delta \dot{Y}_0, \Delta Y_0) \, dt \, dt \, dt, \qquad (325)$$

whereas the deviation of the lower derivatives and the realized lower derivative of the controlled output q_s will be

$$\dot{\eta}_s = \int_{t_0}^{t} Q_s(\dot{\eta}_s, \eta_s) \, dt \qquad \Rightarrow \dot{q}_s = \dot{q}_s^0 - \int_{t_0}^{t} Q_s(\dot{\eta}_s, \eta_s) \, dt,$$

$$\eta_s = \int_{t_0}^{t} \int_{t_0}^{t} Q_s(\dot{\eta}_s, \eta_s) \, dt \, dt \Rightarrow q_s = q_s^0 - \int_{t_0}^{t} \int_{t_0}^{t} Q_s(\dot{\eta}_s, \eta_s) \, dt \, dt. \qquad (326)$$

The controlled outputs are tracked in an asymptotically stable manner, which implies that, after the initial deviation, they converge in time to their nominal values

$$\lim_{t \to \infty} \ddot{Y}_0 = \lim_{t \to \infty} \left[\ddot{Y}_0^0 - \int_{t_0}^{t} Q_0(\Delta \ddot{Y}_0, \Delta \dot{Y}_0, \Delta Y_0) \, dt \right] = \ddot{Y}_0^0,$$

$$\lim_{t \to \infty} \dot{Y}_0 = \lim_{t \to \infty} \left[\dot{Y}_0^0 - \int_{t_0}^{t} \int_{t_0}^{t} Q_0(\Delta \ddot{Y}_0, \Delta \dot{Y}_0, \Delta Y_0) \, dt \, dt \right] = \dot{Y}_0^0,$$

$$\lim_{t \to \infty} Y_0 = \lim_{t \to \infty} \left[Y_0^0 - \int_{t_0}^{t} \int_{t_0}^{t} \int_{t_0}^{t} Q_0(\Delta \ddot{Y}_0, \Delta \dot{Y}_0, \Delta Y_0) \, dt \, dt \, dt \right] = Y_0^0,$$

$$\lim_{t \to \infty} \dot{q}_s = \lim_{t \to \infty} \left[\dot{q}_s^0 - \int_{t_0}^{t} Q_s(\dot{\eta}_s, \eta_s) \, dt \right] = \dot{q}_s^0,$$

$$\lim_{t \to \infty} q_s = \lim_{t \to \infty} \left[q_s^0 - \int_{t_0}^{t} \int_{t_0}^{t} Q_s(\dot{\eta}_s, \eta_s) \, dt \, dt \right] = q_s^0. \tag{327}$$

Because of the existence of the functional dependence $Y_c = \Theta(q)$ all the expressions related to the internal coordinates q, after the coordinates transformation, also hold for the coordinates of contact points Y_c, which will be used to assess the behavior of the non-controlled quantities q_v, F_{cv} and F_{cs} and calculated driving torques τ.

The conclusion about the behavior of the non-controlled quantities will be derived on the basis of the analysis of the physical laws in the elastic system, at the moment when the control-realized asymptotic tracking of the controlled quantities Y_0 and q_s takes place. The goal is to estimate or determine the deviations of the non-controlled quantities from their nominal values on the basis of considering the physical laws in the elastic system. The solution should answer the following question: Starting from the known deviations of the controlled quantities from their nominals, is it possible to exactly determine the deviations of non-controlled quantities from their nominal values, and what will be their character?

The analysis of the behavior of non-controlled quantities will be performed using the physical properties of the mobile elastic system.

6.4.4 Behavior of the non-controlled quantities in tracking the manipulated object MC and nominal trajectories of contact points of the followers

To derive conclusions about the behavior of non-controlled quantities, it is possible to apply the same reasoning as in defining the coordinated nominal motion, which will be illustrated first with a simple example.

To illustrate the way of reasoning we will first analyze the motion in the vertical plane of a simple elastic structure (elastic system), composed of two rigid bodies

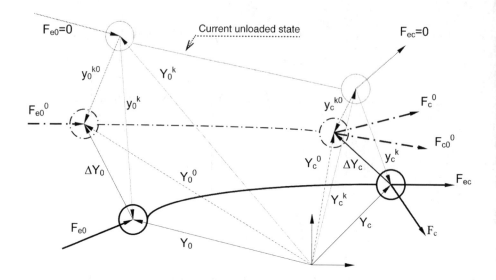

Figure 43. Motion in the plane of the loaded elastic system

with the MCs at the nodes, interconnected by the non-inertial elastic insertions (Figure 43).

Let some external load $F_c = F_c(t)$ act at one of the nodes. Let us consider what happens at an arbitrary moment t. Let the nominal trajectories of nodes $Y_0^0 = Y_0^0(t)$, $Y_c^0 = Y_c^0(t)$ and nominal contact forces $F_c^0 = F_c^0(t)$ be determined in advance. For legibility's sake, the explanation will be given without mentioning time dependence, i.e. by assuming that all the quantities refer to the instant t. In the motion, it is not known where the elastic structure of the unloaded state is. Let its position be determined by the coordinates Y_0^k and Y_c^k. If the nodes move along the nominal trajectories, the unloaded state displacements will be y_0^{k0} and y_c^{k0}. If the trajectories of the nodes deviate from the nominal trajectories by ΔY_0 and ΔY_c, the total displacements of the nodes from the mobile unloaded state will be y_0^k and y_c^k. Hence, the kinematic relations

$$y_0^k = Y_0 - Y_0^k, \qquad\qquad y_c^k = Y_c - Y_c^k,$$

$$y_0^{k0} = Y_0^0 - Y_0^k, \qquad\qquad y_c^{k0} = Y_c^0 - Y_c^k,$$

$$\Delta Y_0 = Y_0^0 - Y_0 = y_0^{k0} - y_0^k, \quad \Delta Y_c = Y_c^0 - Y_c = y_c^{k0} - y_c^k \qquad (328)$$

will hold. A consequence of these displacements is the appearance of the elastic forces acting at the elastic structure nodes. In the real motion, these forces are $F_e = \mathrm{col}(F_{e0}, F_{ec})$, and in the nominal motion, they are $F_e^0 = \mathrm{col}(F_{e0}^0, F_{ec}^0)$. The elasticity properties are preserved irrespective of the character and origin of the forces acting at the elastic structure nodes. Hence, if certain masses exist at the nodes and there act some external forces, and if the non-inertial elastic connection has damping properties, then, according to (186) to (195), the elastic forces F_e and F_e^0 balance the resultant of the dynamic $F_{d*}^{\#}$, gravitational G_* and contact $F_{c*}^{\#}$, $* = 0, c$, $\# = -, 0$ forces. To make the picture more intelligible, damping properties will be omitted from further discussions, so that the dynamic forces at one point of the elastic structure will depend only on the kinematic quantities describing the state of this node. For the domain of a linear relationship between the stress and dilatation for the unperturbed (nominal) and perturbed motions the following relations hold:

$$
\begin{aligned}
F_e^0 &= \begin{pmatrix} F_{ec}^0 \\ F_{e0}^0 \end{pmatrix} \\
&= \begin{pmatrix} F_{dc}^0 + G_c + F_c^0 \\ F_{d0}^0 + G_0 \end{pmatrix} = \begin{pmatrix} A^k & b^k \\ c^k & d^k \end{pmatrix} \cdot \begin{pmatrix} y_c^{k0} \\ y_0^{k0} \end{pmatrix} \in R^{6 \times 1}, \\
F_e &= \begin{pmatrix} F_{ec} \\ F_{e0} \end{pmatrix} \\
&= \begin{pmatrix} F_{dc} + G_c + F_c \\ F_{d0} + G_0 \end{pmatrix} = \begin{pmatrix} A^k & b^k \\ c^k & d^k \end{pmatrix} \cdot \begin{pmatrix} y_c^k \\ y_0^k \end{pmatrix} \in R^{6 \times 1}, \quad (329)
\end{aligned}
$$

where $A^k \in R^{3 \times 3}$, $b^k \in R^{3 \times 3}$, $c^k \in R^{3 \times 3}$, $d^k \in R^{3 \times 3}$ are some constant submatrices of the constant stiffness matrix K^k. If the position of the mobile unloaded state 0 is known and if the known stiffness matrix K has been determined for the immobile unloaded state 0, then, according to (245), the stiffness matrix K^k for the moment t can be obtained from the expression $K^k = A_r^T (A_i^k - A_{i0}) K A_r (A_i^k - A_{i0})$. The matrix $A_r (A_i^k - A_{i0})$ is defined by the expression (150) for the change of orientation $a(t) = A_i^k(t) - A_{i0}$, which is determined by the difference of the orientation $A_i^k(t)$ of the mobile unloaded state 0 at the moment t and the orientation A_{i0} of the immobile unloaded state 0. Although the position of the mobile unloaded state 0 is not known and, consequently, neither is the orientation $A_i^k(t)$ of some of its nodes, it is essential in further inference that, for each moment t, there exists a constant matrix K^k.

Subtracting the second equation from the first one gives

$$
\begin{aligned}
\Delta F_e &= F_e^0 - F_e = \begin{pmatrix} \Delta F_{ec}^0 \\ \Delta F_{e0}^0 \end{pmatrix} \\
&= \begin{pmatrix} \Delta F_{dc} + \Delta F_c \\ \Delta F_{d0} \end{pmatrix} = \begin{pmatrix} A^k & b^k \\ c^k & d^k \end{pmatrix} \cdot \begin{pmatrix} y_c^{k0} - y_c^k \\ y_0^{k0} - y_0^k \end{pmatrix} \in R^{6 \times 1}, \quad (330)
\end{aligned}
$$

where $F_{d*} = F^0_{d*} - \Delta F_{d*}$, $* = c, 0$ and $F_c = F^0_c - \Delta F_c$ are the deviations of dynamic and contact forces from their nominal values. Having in mind (328), the last equation can be formulated with respect to the deviations from the nominal trajectory by the expression

$$\begin{pmatrix} \Delta F_{dc} + \Delta F_c \\ \Delta F_{d0} \end{pmatrix} = \begin{pmatrix} A^k & b^k \\ c^k & d^k \end{pmatrix} \cdot \begin{pmatrix} \Delta Y_c \\ \Delta Y_0 \end{pmatrix} \in R^{6\times 1} \tag{331}$$

or, in scalar form,

$$\Delta F_{dc} + \Delta F_c = A^k \cdot \Delta Y_c + b^k \cdot \Delta Y_0,$$

$$\Delta F_{d0} = c^k \cdot \Delta Y_c + d^k \cdot \Delta Y_0. \tag{332}$$

Equations (332) describe the additional load of the elastic structure, but they can also be treated as the equilibrium equations of a fictitious space grid loaded at the nodes by the forces $\Delta F_{dc} + \Delta F_c$ and ΔF_{d0} that produce the node displacements ΔY_c i ΔY_0. Let us adopt a point with the coordinates Y_0 as the support of the fictitious space grid. In that case, the displacement ΔY_0 determines the support displacement, whereas ΔF_{d0} determines the support reaction.

If the trajectories Y^0_0 and Y_0 are known (i.e. the character of their displacement $\Delta Y_0 = Y^0_0 - Y_0$), then it is possible to explicitly determine the character of the forces F^0_{d0}, F_{d0}, ΔF_{d0}. From the second equation of (332), we can determine the displacement ΔY_c of the other node that must be realized in order that the support displacement ΔY_0 would produce the support reaction ΔF_{d0}. For the known Y_c and ΔY_c, the values of Y^0_c are determined and, on the basis of them, F^0_{dc}, F_{dc} and ΔF_{dc}, so that from the first equation of (332), it is easy to determine ΔF_c. Hence, on the basis of the known ΔY_0, it is possible to exactly determine ΔY_c, and on the basis of this, also ΔF_c.

On the basis of the considered displacements from the unknown mobile un-loaded state, it is possible to form, for each moment, the equations of the fictitious space grid at the nodes of which act the deviations of dynamic and contact forces from their nominal values, causing displacements equal to the displacement of the elastic system nodes from the nominal trajectories. To derive conclusions about what is possible to calculate, one can rely upon the methodology used in statics.

For the elastic system, on the basis of Equations (247), from which damping forces have been omitted, for the nominal conditions at the instant t and using the sign conventions (206), (209), (203) and (164), we get the relations

$$F^0_{dv} + G_v + F^0_{cv} = u^k_v Y^0_v + u^k_s Y^0_s + u^k_0 Y^0_0,$$

$$F^0_{ds} + G_s + F^0_{cs} = A^k_v Y^0_v + A^k_s Y^0_s + A^k_0 Y^0_0,$$

$$F^0_{d0} + G_0 = c^k_v Y^0_v + c^k_s Y^0_s + c^k_0 Y^0_0, \tag{333}$$

where the superscript '0' denotes the nominal quantities and where $F^0_{d*} = -W_*(Y^0_*)\ddot{Y}^0_* - F_{b*}(Y^0_*, \dot{Y}^0_*)$, $* = v, s, 0$.

For the perturbed motion of the elastic system, it holds that

$$F_{dv} + G_v + F_{cv} = u^k_v Y_v + u^k_s Y_s + u^k_0 Y_0,$$

$$F_{ds} + G_s + F_{cs} = A^k_v Y_v + A^k_s Y_s + A^k_0 Y_0,$$

$$F_{d0} + G_0 = c^k_v Y_v + c^k_s Y_s + c^k_0 Y_0, \tag{334}$$

where $F_{d*} = -W_*(Y_*)\ddot{Y}_* - F_{b*}(Y_*, \dot{Y}_*)$, $* = v, s, 0$.

Subtracting (334) from (333) gives the equations

$$\Delta F_{dv} + \Delta F_{cv} = u^k_v \Delta Y_v + u^k_s \Delta Y_s + u^k_0 \Delta Y_0,$$

$$\Delta F_{ds} + \Delta F_{cs} = A^k_v \Delta Y_v + A^k_s \Delta Y_s + A^k_0 \Delta Y_0,$$

$$\Delta F_{d0} = c^k_v \Delta Y_v + c^k_s \Delta Y_s + c^k_0 \Delta Y_0, \tag{335}$$

that describe equilibrium of the fictitious space grid loaded at the nodes by the forces $\Delta F_{dv} + \Delta F_{cv}$, $\Delta F_{ds} + \Delta F_{cs}$ and ΔF_{d0} that produce the node displacements ΔY_v, ΔY_s and ΔY_0.

Bearing in mind that the physical features of the elastic system impose the dependence (335), conclusion should be derived as to the behavior of the displacements from the nominal quantities that are not directly controlled, i.e. the contact forces $F_c \in R^{6m}$ and position of the leader's contact points, $Y_v \in R^6$.

In the preceding section, the control laws were chosen for the $6m$-dimensional vector of controlled outputs $Y^u(t) = \text{col}(q_s(t), Y_0(t))$. For that choice of control laws, the vector of finite nominal positions of the nodes $Y^0_{s0} = \text{col}(Y^0_s(t), Y^0_0(t))$ and vectors of its derivatives are known. In the course of motion, the vector of the realized positions of the nodes $Y_{s0} = \text{col}(Y_s(t), Y_0(t))$ and vectors of its derivatives are also known. On the basis of them, $\Delta F_{ds0} = \text{col}(\Delta F_{ds}, \Delta F_{d0})$, so that the vector ΔF_{ds0} can be considered as a known quantity.

The control laws synthesized in the preceding section ensure the vector of increments $\Delta Y_{s0} = \mathrm{col}(\Delta Y_s, \Delta Y_0)$ and its derivatives have an exponentially descending character. Because of the character of the change of ΔY_{s0} and its derivatives, the vector ΔF_{ds0} also has an exponentially descending character, converging to a zero value.

For the non-singular matrix c_v^k, the value ΔY_v is calculated from the last equation of (335). Because of a linear dependence on the quantities ΔF_{d0}, ΔY_s and ΔY_0, the deviation ΔY_v will also have an exponentially descending character, converging to zero. Because of that, the non-controlled output quantity Y_v will asymptotically converge to the nominal trajectory of the leader's contact point Y_v^0. By an analogous procedure, on the basis of the second equation of (335), it can be concluded that the increments of the contact forces of the followers ΔF_{cs} have also an exponentially descending character, converging to the zero values, i.e. the contact forces of the followers converge asymptotically to their nominal values F_{cs}^0.

Thus, we have proved the asymptotic tracking of all the non-controlled quantities of the elastic system in the case of tracking the nominal trajectories of the manipulated object MC and nominal trajectories of the followers' contact points. Exactly the same conclusion could be derived if the trajectories of the other m nodes were selected as nominals, e.g. the nominal trajectories of the contact points $Y_c^0 = \mathrm{col}(Y_v^0(t), Y_s^0(t))$, only, when selecting the vector of controlled outputs $Y^u = q$.

Estimation of the driving torques constraints. Constraints on the driving torques can be estimated by norming some expression by an expression of the cooperative system dynamics in which they are explicitly contained (316), (295), (296), or by norming the expression for the description of the manipulator dynamics (167). The simplest way is to norm (167)

$$\|\tau\| \le \|H(q)\|\|\ddot{q}\| + \|h(q,\dot{q})\| + \|J^T(q)\|\|f_c\|.$$

Since, in the course of time, after the initial deviation, q_s^0 is realized and since q_v and $f_c = -F_c = -\mathrm{col}(F_v, F_s)$ are constrained, then for the constrained arguments, all the expressions on the right-hand side are constrained so that the driving torques are also constrained.

Let us conclude that the introduction of the control laws defined by the expressions (316) for the driving torques that are to be realized at the manipulators' joints ensures tracking of the nominal controlled outputs $Y^u = \mathrm{col}(Y_0^0, q_s^0)$ in the required way, indirectly given by (312) and (315), whereby the non-controlled quantities (kinematic quantities of the leader, contact forces and driving torques) will not be unconstrained after the transient process caused by the initial deviation of the controlled outputs from their nominal values, but they will asymptotically converge to their nominal values.

6.4.5 Control laws to track the nominal trajectory of the manipulated object MC and nominal contact forces of the followers

In this case of tracking, the controlled output is the vector $Y^u = \mathrm{col}(F_{cs}, Y_0) \in R^{6m}$, composed of the nominal contact forces of the followers $F_{cs}^0 \in R^{6m-6}$ and the nominal trajectory $Y_0^0(t) \in R^6$ of the manipulated object MC. The output quantities of the cooperative system that are not directly tracked (non-controlled outputs) are the positions of m contact points $Y_c \in R^{6m}$ and the contact force $F_{cv} \in R^6$ at the leader's contact point.

The task of the control law synthesis is to determine the driving torques that are to be introduced at the manipulators' joints in order that the cooperative system would follow the output $Y^u = \mathrm{col}(F_{cs}, Y_0)$ with the indicators of the quality of dynamic behavior given in advance. The behavior of the deviations of the quantities that are not directly tracked from their nominal values should be estimated separately.

Let the requirement for the character of tracking the manipulated object MC be given by the relation (312). Then, on the basis of (310), we obtain the dependence (314) for the driving torques τ_v of the leader and the contact force F_{cv} at the leader's contact point. From (314) and (296), it can be concluded that all the driving torques and all contact forces depend on the followers' accelerations $\ddot{q}_s$. In the case of tracking the controlled outputs $Y^u = \mathrm{col}(q_s, Y_0)$, the required character for tracking nominal trajectories of the followers' internal coordinates is given by (315), from which the necessary accelerations of the followers in the real motion are determined. On the basis of the necessary accelerations of the followers, the driving torques to be introduced to the manipulators' joints are calculated from (310).

The followers' accelerations $\ddot{q}_s$ can be also determined from the last equality in (296) depending on the contact forces F_{cs} at the contact points of the followers. Hence, the requirement for the quality of tracking can be given by the contact forces F_{cs}.

Let

$$\overset{(k)}{\mu_s}(t) = \overset{(k)}{F_{cs}^0}(t) - \overset{(k)}{F_{cs}}(t), \quad i = 0, 1, \ldots, \quad \overset{(0)}{\mu_s}(t) = \mu_s(t), \tag{336}$$

be the vectors of deviations and vectors of derivatives of the deviation of the realized controlled contact forces from their nominal values. Let $\mu_s(t) = F_{cs}^0(t) -$

$F_{cs}(t)$ be the solution of the homogeneous differential equation

$$\chi_s(\overset{(k)}{\mu_s}, \overset{(k-1)}{\mu_s}, \ldots, \overset{(0)}{\mu_s}) = 0, \qquad \overset{(0)}{\mu_s} = \mu_s, \tag{337}$$

obtained as the response to the initial deviation $\mu_s(t_0) = F_{cs}^0(t_0) - F_{cs}(t_0)$. Let $\mu_s(t) = 0$ (which is equivalent to $F_{cs}^0(t) = F_{cs}(t)$) be the equilibrium state of the above differential equation. Let the highest derivative be one ($k = 1$) and let the differential equation be chosen in the way that each solution, obtained as the response to the initial deviation $\mu_s(t_0) = F_{cs}^0(t_0) - F_{cs}(t_0)$, be asymptotically stable with the desired indicators of the quality of dynamic behavior, which is mathematically described by the expression

$$\dot{\mu}_s(t) = S(\mu_s(t)) \quad \Rightarrow \quad \dot{F}_{cs}^0(t) = \dot{F}_{cs}(t) - S(\mu_s(t)), \tag{338}$$

whose integration gives the values of contact forces $F_{cs}(t)$ that should be realized at the moment t in order that the above law (338) be fulfilled.

$$F_{cs}(t) = F_{cs}^0(t) - \int_{t_0}^{t} S(\mu_s)\, dt. \tag{339}$$

After introducing this value of contact force into the last equation of (296), the followers' accelerations are obtained as

$$\ddot{q}_s = P_s^{-1}(q_s) \left[F_{cs}^0(t) - \int_{t_0}^{t} S(\mu_s) dt - p_s(q, \dot{q}, Y_0, \dot{Y}_0) \right], \tag{340}$$

since the inertia matrix $P_s(q_s)$ is non-singular. The driving torques τ_v at the leader's joints are found by introducing the followers' accelerations $\ddot{q}_s$ from (340) into (314), whereas the driving torques at the followers joints are obtained by introducing the accelerations $\ddot{q}_s$ from (340) into the second equality of (296). Thus, we obtain

$$
\begin{aligned}
\tau_v &= N_v(q_v)[-\alpha(\ddot{Y}_0^0 - Q_0(\Delta\ddot{Y}_0, \Delta\dot{Y}_0, \Delta Y_0), \ddot{Y}_0, \dot{Y}_0, Y_0, \dot{q}, q)] \\[2mm]
&\quad - \beta(\dot{Y}_0, Y_0, \dot{q}, q) P_s^{-1}(q_s) \left[F_{cs}^0(t) - \int_{t_0}^{t} S(\mu_s)\, dt - p_s(q, \dot{q}, Y_0, \dot{Y}_0) \right] \\[2mm]
&\quad + n_v(q, \dot{q}, Y_0, \dot{Y}_0) \\[2mm]
&= \tau_v\left(\ddot{Y}_0^0 - Q_0(\Delta\ddot{Y}_0, \Delta\dot{Y}_0, \Delta Y_0), \ddot{Y}_0, \dot{Y}_0, Y_0, q, \dot{q}, F_{cs}^0(t) - \int_{t_0}^{t} S(\mu_s)\, dt \right)
\end{aligned}
$$

$$= \tau_v(\ddot{Y}^0_0, \ddot{Y}_0, \dot{Y}_0, Y_0, q, \dot{q}, F_{cs}, F^0_{cs}),$$

$$\tau_s = N_s(q_s)P_s^{-1}(q_s)\left[F^0_{cs}(t) - \int_{t_0}^t S(\mu_s)\,dt - p_s(q, \dot{q}, Y_0, \dot{Y}_0)\right]$$

$$+ n_s(q, \dot{q}, Y_0, \dot{Y}_0)$$

$$= \tau_s\left(\dot{Y}_0, Y_0, q, \dot{q}, F^0_{cs}(t) - \int_{t_0}^t S(\mu_s)\,dt\right)$$

$$= \tau_s(\dot{Y}_0, Y_0, q, \dot{q}, F_{cs}, F^0_{cs}). \qquad (341)$$

The calculated driving torques should be introduced at the manipulators' joints in order to realize the tracking of the controlled output $Y^u = \mathrm{col}(Y^0_0, F^0_{cs})$ with the quality of dynamic behavior given indirectly in advance by (312) and (338). To determine the driving torques, it is necessary to have information about all instantaneous kinematic quantities $\ddot{Y}_0$, $\dot{Y}_0$ and Y_0 of the manipulated object MC, information about the instantaneous values of the internal coordinates q and their derivatives $\dot{q}$, information about the nominal output Y^0_0 and its derivatives $\dot{Y}^0_0$, $\ddot{Y}^0_0$, $\dddot{Y}^0_0$, and information about the real F_{cs} and nominal F^0_{cs} contact forces at the contact points of the followers.

The introduction of driving torques into (341) ensures the realization of the contact force.

$$F_{cv} = P_v(q_v)[-\alpha(\ddot{Y}^0_0 - Q_0(\Delta\ddot{Y}_0, \Delta\dot{Y}_0, \Delta Y_0), \ddot{Y}_0, \dot{Y}_0, Y_0, \dot{q}, q)]$$

$$- \beta(\dot{Y}_0, Y_0, \dot{q}, q)P_s^{-1}(q_s)\left[F^0_{cs}(t) - \int_{t_0}^t S(\mu_s)\,dt - p_s(q, \dot{q}, Y_0, \dot{Y}_0)\right]$$

$$+ p_v(q, \dot{q}, Y_0, \dot{Y}_0)$$

$$= F_{cv}\left(\ddot{Y}^0_0 - Q_0(\Delta\ddot{Y}_0, \Delta\dot{Y}_0, \Delta Y_0), \ddot{Y}_0, \dot{Y}_0, Y_0, q, \dot{q}, F^0_{cs}(t) - \int_{t_0}^t S(\mu_s)\,dt\right)$$

$$= F_{cv}(\ddot{Y}^0_0, \ddot{Y}_0, \dot{Y}_0, Y_0, q, \dot{q}, F_{cs}, F^0_{cs}),$$

$$F_{cs} = F^0_{cs}(t) - \int_{t_0}^t S(\mu_s)\,dt. \qquad (342)$$

Let us introduce the calculated driving torques (341) into the model of cooperative manipulation (296) and let us prove that the prescribed requirements will be fulfilled:

$$N_v(q_v)\ddot{q}_v + n_v(q, \dot{q}, Y_0, \dot{Y}_0)$$

$$= N_v(q_v)[-\alpha(\ddot{Y}_0^0 - Q_0(\Delta\ddot{Y}_0, \Delta\dot{Y}_0, \Delta Y_0), \ddot{Y}_0, \dot{Y}_0, Y_0, \dot{q}, q)]$$

$$- \beta(\dot{Y}_0, Y_0, \dot{q}, q) P_s^{-1}(q_s) \left[F_{cs}^0(t) - \int\limits_{t_0}^{t} S(\mu_s)\,dt - p_s(q, \dot{q}, Y_0, \dot{Y}_0) \right]$$

$$+ n_v(q, \dot{q}, Y_0, \dot{Y}_0),$$

$$N_s(q_s)\ddot{q}_s + n_s(q, \dot{q}, Y_0, \dot{Y}_0)$$

$$= N_s(q_s) P_s^{-1}(q_s) \left[F_{cs}^0(t) - \int\limits_{t_0}^{t} S(\mu_s)\,dt - p_s(q, \dot{q}, Y_0, \dot{Y}_0) \right]$$

$$+ n_s(q, \dot{q}, Y_0, \dot{Y}_0),$$

$$W(Y_0)\ddot{Y}_0 + w(q, \dot{q}, Y_0, \dot{Y}_0) = 0, \tag{343}$$

i.e. after rearranging

$$N_v(q_v)\left\{ \ddot{q}_v + \alpha(\ddot{Y}_0^0 - Q_0(\Delta\ddot{Y}_0, \Delta\dot{Y}_0, \Delta Y_0), \ddot{Y}_0, \dot{Y}_0, Y_0, \dot{q}, q) \right.$$

$$\left. + \beta(\dot{Y}_0, Y_0, \dot{q}, q) P_s^{-1}(q_s) \left[F_{cs}^0(t) - \int\limits_{t_0}^{t} S(\mu_s)\,dt - p_s(q, \dot{q}, Y_0, \dot{Y}_0) \right] \right\} = 0,$$

$$N_s(q_s)\left\{ \ddot{q}_s - P_s^{-1}(q_s) \left[F_{cs}^0(t) - \int\limits_{t_0}^{t} S(\mu_s)\,dt - p_s(q, \dot{q}, Y_0, \dot{Y}_0) \right] \right\} = 0,$$

$$W(Y_0)\ddot{Y}_0 + w(q, \dot{q}, Y_0, \dot{Y}_0) = 0. \tag{344}$$

The inertia matrices $N_v(q_v)$ and $N_s(q_s)$ are non-singular, and the last equation after differentiation is transformed into (310). Hence,

$$\ddot{q}_v + \alpha(\ddot{Y}_0^0 - Q_0(\Delta\ddot{Y}_0, \Delta\dot{Y}_0, \Delta Y_0), \ddot{Y}_0, \dot{Y}_0, Y_0, \dot{q}, q)$$

$$+ \beta(\dot{Y}_0, Y_0, \dot{q}, q) P_s^{-1}(q_s) \left[F_{cs}^0(t) - \int\limits_{t_0}^{t} S(\mu_s)\,dt - p_s(q, \dot{q}, Y_0, \dot{Y}_0) \right] = 0,$$

$$\ddot{q}_s - P_s^{-1}(q_s) \left[F_{cs}^0(t) - \int_{t_0}^{t} S(\mu_s)\, dt - p_s(q, \dot{q}, Y_0, \dot{Y}_0) \right] = 0. \tag{345}$$

By calculating the accelerations from the last equation of (296) and introducing it into the second equality of (345), it follows that

$$P_s^{-1}(q_s) \left[F_{cs}(t) - p_s(q, \dot{q}, Y_0, \dot{Y}_0) + F_{cs}^0(t) \right.$$

$$\left. - \int_{t_0}^{t} S(\mu_s)\, dt + p_s(q, \dot{q}, Y_0, \dot{Y}_0) \right] = 0. \tag{346}$$

Since the inertia matrix $P_s(q_s)$ is non-singular, the following relation is realized:

$$F_{cs}(t) - F_{cs}^0(t) - \int_{t_0}^{t} S(\mu_s)\, dt = 0, \tag{347}$$

which is identical to the relation (339) resulting from the integration for the preset requirements (338) for tracking the followers' contact forces.

By introducing $\ddot{q}_v$ from (310) to the first equality of (345), we get

$$-\alpha(\ddot{Y}_0, \ddot{Y}_0, \dot{Y}_0, Y_0, \dot{q}, q) + \alpha(\ddot{Y}_0^0 - Q_0(\Delta \ddot{Y}_0, \Delta \dot{Y}_0, \Delta Y_0), \ddot{Y}_0, \dot{Y}_0, Y_0, \dot{q}, q)$$

$$- \beta(\dot{Y}_0, Y_0, \dot{q}, q) \left\{ \ddot{q}_s - P_s^{-1}(q_s) \left[F_{cs}^0(t) - \int_{t_0}^{t} S(\mu_s)\, dt - p_s(q, \dot{q}, Y_0, \dot{Y}_0) \right] \right\}$$

$$= 0. \tag{348}$$

Having in mind the second equality of (345), the last equality becomes identical to the equality (322) from which, according to (311), follow the equalities (323) and (324), which demonstrate the realization of the initially prescribed requirements (312). Thus, it has been shown that the introduction of the control laws presented by the relations for the calculated driving torques (341) allows the controlled cooperative system (296) to follow the nominal controlled outputs $Y^{u0} = \text{col}(F_{cs}^0, Y_0^0)$ in a stable manner and with the quality requirements indirectly prescribed by (312) and (338).

Since the required laws of deviation of the derivatives $\Delta \dddot{Y}_0$ and $\dot{\mu}_s = \dot{F}_{cs}^0 - \dot{F}_{cs}$ of the controlled outputs Y_0 and F_{cs} adopted by the control laws (341) are realized, then, according to (325), the deviation of the lower derivatives of the controlled output Y_0 will be realized too, whereas the followers' contact force will be realized according to (339). The controlled outputs are tracked in an asymptotically stable manner so that, in the course of time, the relation (327) will be fulfilled for the controlled output Y_0 and

$$\lim_{t\to\infty} F_{cs} = \lim_{t\to\infty} \left(F_{cs}^0 - \int_{t_0}^{t} S(\mu_s)\, dt \right) = F_{cs}^0 \qquad (349)$$

for the controlled output F_{cs}.

6.4.6 Behavior of the non-controlled quantities in tracking the trajectory of the manipulated object MC and nominal contact forces of the followers

The discussion concerning the behavior of the mobile elastic structure given in Section 6.4.4 will be used to examine the properties of the non-controlled quantities $F_{cv}, q, \dot{q}, \ddot{q}$ (i.e. $Y_c, \dot{Y}_c, \ddot{Y}_c$) and calculated driving torques τ in considering the elastic system with controlled trajectories of contact points and controlled contact forces.

The starting equations for the analysis are (335), written as

$$\Delta F_{dv} + \Delta F_{cv} = u_v^k \Delta Y_v + u_s^k \Delta Y_s + u_0^k \Delta Y_0,$$

$$\Delta F_{ds} + \Delta F_{cs} = A_v^k \Delta Y_v + A_s^k \Delta Y_s + A_0^k \Delta Y_0,$$

$$\Delta F_{d0} = c_v^k \Delta Y_v + c_s^k \Delta Y_s + c_0^k \Delta Y_0, \qquad (350)$$

to describe the equilibrium of a fictitious space grid loaded at the nodes by the forces $\Delta F_{dv} + \Delta F_{cv}$, $\Delta F_{ds} + \Delta F_{cs}$ and ΔF_{d0} that produce the node displacements ΔY_v, ΔY_s and ΔY_0.

In the previous section, the choice of control laws was made for the $6m$-dimensional vector of the controlled outputs $Y^u = \mathrm{col}(F_{cs}, Y_0)$. For such a choice of control laws, the known quantities are the nominal trajectory of the manipulated object MC, Y_0^0, nominal contact forces of the followers F_{cs}^0 and derivative of the nominal quantities. During the motion, the known quantities are the vector of the realized position of the object MC, Y_0, vector of the realized contact force of the follower F_{cs}, and the derivatives of the realized outputs. Hence, the corresponding vectors of deviations ΔY_0 and ΔF_{cs}, and their derivatives are also known. The value ΔF_{d0} is determined on the basis of the known nominal and realized trajectory, so that the increment of dynamic force ΔF_{d0} in (350) can be considered as being known. All other quantities in (350) are unknown. The equation of equilibrium (350) of the fictitious space grid is defined by $6m+6$ conditions, of which $6m$

are independent. To find the instantaneous configuration of acting forces and grid displacement in the course of control (motion), there are at our disposal six components of the vector ΔY_0, $6m-1$ components of the vector ΔF_{cs} and six components of the vector ΔF_{d0}. The unknowns are ΔF_{dv}, ΔF_{ds}, ΔF_{cv}, ΔY_v and ΔY_s, i.e. in total $6+(6m-6)+6+6+(6m-6) = 12m+6$ unknown quantities. Obviously, there exist an infinite number of combinations of the unknown quantities that, together with the known quantities, determine the configuration of the fictitious space grid. Equilibrium equations can be satisfied for arbitrary values of the unknown quantities from the set of real numbers, and thus for the unconstrained values. This is straightforward for the case of zero values of the increments $\Delta Y_0 = 0$, $\Delta F_{cs} = 0$ and ΔF_{d0} that appear in the ideally realized nominal conditions. In that case, the equilibrium conditions (350) reduce to

$$\Delta F_{dv} + \Delta F_{cv} = u_v^k \Delta Y_v + u_s^k \Delta Y_s,$$

$$\Delta F_{ds} = A_v^k \Delta Y_v + A_s^k \Delta Y_s,$$

$$0 = c_v^k \Delta Y_v + c_s^k \Delta Y_s. \tag{351}$$

If the matrix c_v^k is non-singular, from the third equality, we can determine the deviation ΔY_v as a function of the deviation ΔY_s. Replacing the determined deviation into the first two equation yields

$$\Delta F_{dv} + \Delta F_{cv} = [-u_v^k (c_v^k)^{-1} c_s^k + u_s^k] \Delta Y_s,$$

$$\Delta F_{ds} = [-A_v^k (c_v^k)^{-1} c_s^k + A_s^k] \Delta Y_s,$$

$$\Delta Y_v = -(c_v^k)^{-1} c_s^k \Delta Y_s. \tag{352}$$

For an arbitrarily chosen value of the realized deviation ΔY_s from the nominal trajectory of the followers' contact points, it is possible to determine the corresponding deviation ΔY_v of the realized trajectory from the nominal trajectory of the leader's contact points and the necessary deviation of the nominal force ΔF_{cv} at the leader's contact point that will balance the increments of the elastic and dynamic forces.

In other words, even when the control satisfies the preset requirements in respect of the input quantities, the deviations of the non-controlled nominal values can be unconstrained. Hence, the realized non-controlled quantities can be, but not necessarily, unconstrained.

By introducing into the equation of elastic behavior (247) the realized control behaviors (327) and (349) of the controlled outputs, we obtain

$$F_{dv}(Y_v, \dot{Y}_v, \ddot{Y}_v) + D_{uvs}(Y_c, Y_0^0)\dot{Y}_c + D_{u0}(Y_c, Y_0^0)\dot{Y}_0^0$$

$$+ u_{vs}(Y_c, Y_0^0)Y_c + u_0(Y_c, Y_0^0)Y_0^0 = G_v + F_{cv}(Y),$$

$$F_{ds}(Y_s, \dot{Y}_s, \ddot{Y}_s) + D_{Avs}(Y_c, Y_0^0)\dot{Y}_c + D_{A0}(Y_c, Y_0^0)\dot{Y}_0^0$$

$$+ A_{vs}(Y_c, Y_0^0)Y_c + A_0(Y_c, Y_0^0)Y_0^0 = G_s + F_{cs}^0(Y),$$

$$F_{d0}(Y_0^0, \dot{Y}_0^0, \ddot{Y}_0^0) + D_c(Y_c, Y_0^0)\dot{Y}_v + D_d(Y_c, Y_0^0)\dot{Y}_0^0$$

$$+ c(Y_c, Y_0^0)Y_c + d(Y_c, Y_0^0)Y_0^0 = G_0, \qquad (353)$$

where $F_{d*} = W_*(Y_*)\ddot{Y}_* + F_{b*}(Y_*, \dot{Y}_*)$, $* = v, s, 0$. These non-linear differential equations describe the elastic system motion along the trajectory Y_0^0 with the controlled excitation F_{cs}^0 during the motion. Properties of the solutions of Equations (353) as a function of the system parameters and character of the drives, are subject to the theory of oscillations and dynamics of constructions [6, 7, 23] in the frame of the analysis of forced oscillations with an arbitrary finite number of DOFs.

From the point of view of cooperative manipulation and practical application, it can be concluded that the control laws (341) follow in an asymptotically stable manner, the nominal trajectory of the manipulated object MC and nominal trajectories of the followers' contact forces. The elastic system will behave as a mobile elastic structure excited in a controlled manner. The response of such a structure depends on the characteristics of the elastic structure and character of the nominal (required) contact forces. The excited elastic structure can assume any state, including the resonant one.

6.5   Examples of Selected Control Laws

The synthesis of the control laws for the phase of gripping and general motion of the cooperative system will be illustrated on the example of the 'linear' cooperative system (Figure 26), considered in Chapter 3 (Figures 8 and 9). The synthesis of control laws will be illustrated for guiding along the nominal trajectories the 'linear' cooperative system (Figure 26), by which is approximated the cooperative manipulation of the object by two manipulators along a vertical straight line. The model of the non-controlled system is given in Chapter 5 by the relations (260), (261), (262) and (263). It is assumed that the masses of the connections of the object and manipulators are smaller than the mass of the manipulated object, so that they can be neglected.

Control laws are introduced on the basis of the dynamic model of cooperative manipulation for the mobile loaded state given in the form of (296). For this example, this form is obtained by uniting the relations (261), (262) and (263)

$$m_1\ddot{Y}_1 + d_{pm}\dot{Y}_1 - d_p\dot{Y}_2 + d_{ps}\dot{Y}_3 + c_{pm}Y_1 - c_pY_2 + c_{ps}Y_3 + m_1g + c_ps_1 = \tau_1,$$

$$m_2 \ddot{Y}_3 - d_{ks} \dot{Y}_1 - d_k \dot{Y}_2 + d_{km} \dot{Y}_3 + c_{ks} Y_1 - c_k Y_2 + c_{km} Y_3 + m_2 g - c_k s_3 = \tau_2,$$

$$m \ddot{Y}_2 - d_p \dot{Y}_1 + (d_p + d_k) \dot{Y}_2 - d_k \dot{Y}_3$$
$$- c_p Y_1 + (c_p + c_k) Y_2 - c_k Y_3 + mg - c_p s_1 + c_k s_3 = 0,$$

$$d_p \dot{Y}_1 - d_p \dot{Y}_2 + c_p Y_1 - c_p Y_2 + c_p s_1 = F_{c1},$$

$$-d_k \dot{Y}_2 + d_k \dot{Y}_3 - c_k Y_2 + c_k Y_3 - c_k s_3 = F_{c2}. \tag{354}$$

Values of the damping coefficients $d_{pm} = d_p + d_1$, d_{ps}, $d_{km} = d_k + d_2$, d_{ks} and stiffness coefficients $c_{pm} = c_p + c_1$, c_{ps}, $c_{km} = c_k + c_2$, c_{ks} are adjusted within the local stabilization of the cooperative system. For example, if $d_1 \neq 0$, $d_{ps} = 0$, $d_2 \neq 0$, $d_{ks} = 0$, $c_1 \neq 0$, $c_{ps} = 0$, $c_2 \neq 0$, $c_{ks} = 0$, the local stabilization is performed individually for each manipulator based on the information from the given manipulator only. For this example, the control laws will be selected for the non-stabilized cooperative system with the coefficients having the following values: $d_{pm} = d_p + d_1 = d_p$, $d_{ps} = 0$, $d_{km} = d_k + d_2 = d_k$, $d_{ks} = 0$, $c_{pm} = c_p + c_1 = c_p$, $c_{ps} = 0$, $c_{km} = c_k + c_2 = c_k$ and $c_{ks} = 0$.

Having in mind the kinematic relations (263) and adopting the first manipulator as a leader, a comparison of (354) with (296) yields the conclusions that

$$q_v = q_1 = Y_1, \quad q_s = q_2 = Y_3, \quad Y_0 = Y_2, \quad \tau_v = \tau_1, \quad \tau_s = \tau_2,$$

$$N_v(q_v) = m_1, \quad N_s(q_s) = m_2, \quad W(Y_0) = m, \quad P_v(q_v) = 0, \quad P_s(q_s) = 0,$$

$$\begin{aligned}
n_v(q, \dot{q}, Y_0, \dot{Y}_0) &= d_{pm} \dot{Y}_1 - d_p \dot{Y}_2 + d_{ps} \dot{Y}_3 \\
&\quad + c_{pm} Y_1 - c_p Y_2 + c_{ps} Y_3 + m_1 g + c_{ps} s_1, \\
n_s(q, \dot{q}, Y_0, \dot{Y}_0) &= -d_{ks} \dot{Y}_1 - d_{km} \dot{Y}_2 + d_k \dot{Y}_3 \\
&\quad + c_{ks} Y_1 - c_k Y_2 + c_{km} Y_3 + m_2 g - c_k s_3, \\
w(q, \dot{q}, Y_0, \dot{Y}_0) &= -d_p \dot{Y}_1 + (d_p + d_k) \dot{Y}_2 - d_k \dot{Y}_3 \\
&\quad - c_p Y_1 + (c_p + c_k) Y_2 - c_k Y_3 + mg - c_p s_1 + c_k s_3, \\
p_v(q, \dot{q}, Y_0, \dot{Y}_0) &= d_p \dot{Y}_1 - d_p \dot{Y}_2 + c_p Y_1 - c_p Y_2 + c_p s_1, \\
p_s(q, \dot{q}, Y_0, \dot{Y}_0) &= -d_k \dot{Y}_2 + d_k \dot{Y}_3 - c_k Y_2 + c_k Y_3 - c_k s_3. \tag{355}
\end{aligned}$$

The selected control laws will track the nominal trajectories of the manipulated object MC and nominal trajectories of the followers' contact points, considered in Section 6.4.3, or the nominal trajectories of the manipulated object MC and the followers' nominal contact forces, considered in Section 6.4.5.

In choosing the control laws for tracking nominal trajectories of the manipulated object MC and nominal trajectories of the followers' contact forces, a concrete form of differential equations should be selected for the law of deviation of
the realized trajectory from the nominal trajectory of the manipulated object MC
(312) and the law of deviations of the realized trajectories from the nominal trajectories of the followers' contact points (315). These equations are taken in the
linear forms

$$\Delta \ddot{Y}_2 + b_2 \Delta \ddot{Y}_2 + b_1 \Delta \dot{Y}_2 + b_0 \Delta Y_2 = k_{y2} u_{y2}|_{u_{y2}=0} = 0, \qquad \Delta Y_2 = Y_2^0 - Y_2,$$

$$\ddot{\eta}_s + a_1 \dot{\eta}_s + b_0 \eta_s = k_\eta u_\eta|_{u_\eta=0} = 0, \qquad \eta_s = Y_3^0 - Y_3. \tag{356}$$

By comparing these equations with (312) and (315) it can be concluded that
$Q_0(\Delta \ddot{Y}_2, \Delta \dot{Y}_2, \Delta Y_2) = -b_2 \Delta \ddot{Y}_2 - b_1 \Delta \dot{Y}_2 - b_0 \Delta Y_2 + k_{y2} u_{y2}|_{u_{y2}=0}$ and $Q_s(\dot{\eta}_s, \eta_s)$
$= -a_1 \dot{\eta}_s + b_0 \eta_s + k_\eta u_\eta|_{u_\eta=0}$. The selected numerical values of the coefficients
are $b_0 = 7106.118$ [s^{-3}], $b_1 = 8883.0936$ [s^{-2}], $b_2 = 46.38938$ [s^{-1}], $a_0 = 355.3059$ [s^{-2}] and $a_1 = 26.38938$ [s^{-1}]. The leader's acceleration is determined
from the relation (310), after determining from (311) the auxiliary expressions

$$\alpha = -\frac{1}{d_p}[m \ddot{Y}_2 + (d_p + d_k)\ddot{Y}_2 - c_p \dot{Y}_1 + (c_p + c_k)\dot{Y}_2 - c_k \dot{Y}_3],$$

$$\beta = \frac{d_k}{d_p}, \tag{357}$$

into which has already been introduced $\partial w_v/\partial \dot{q}_v = \partial w/\partial \dot{Y}_1 = -d_p$ and
$\partial w_s/\partial \dot{q}_s = \partial w/\partial \dot{Y}_3 = -d_k$. The control laws are determined from (316) by
the expressions

$$\tau_v = N_v[-\alpha - \beta(\ddot{q}_s^0 - Q_s)] + n_v \Rightarrow \tau_1 = m_1[-\alpha^0 - \beta(\ddot{Y}_3^0 - Q_s^0)] + n_v,$$

$$\tau_s = N_s[\ddot{q}_s^0 - Q_s] + n_s \qquad \Rightarrow \tau_1 = m_2[\ddot{Y}_3^0 - Q_s^0] + n_s, \tag{358}$$

where N_v, N_s, n_v and n_s are shortened notations for the functions (355) (without
designating independent variables), whereas the new notations with the superscript
0 determine the developed expressions with the same basic notation, i.e.

$$\alpha^0 = -\frac{1}{d_p}[m(\ddot{Y}_2 - Q_0^0) + (d_p + d_k)\ddot{Y}_2 - c_p \dot{Y}_1 + (c_p + c_k)\dot{Y}_2 - c_k \dot{Y}_3],$$

$$Q_0^0 = -b_2(\ddot{Y}_2^0 - \ddot{Y}_2) - b_1(\dot{Y}_2^0 - \dot{Y}_2) - b_0(Y_2^0 - Y_2),$$

$$Q_s^0 = -a_1(\dot{Y}_3^0 - \dot{Y}_3) - b_0(Y_3^0 - Y_3), \tag{359}$$

where Y_2^0, $\dot{Y}_2^0$, $\ddot{Y}_2^0$ and $\dddot{Y}_2^0$ represent the nominal trajectory of the manipulated object MC and its derivatives, and Y_3^0, $\dot{Y}_3^0$, $\ddot{Y}_3^0$ are the nominal trajectory of the followers' contact points and its derivatives, which are in this example identical to the followers' internal coordinates and derivatives.

For the control law to track the nominal trajectory of the manipulated object MC and nominal contact forces of the followers, a concrete form of differential equations should be selected for the law of deviations of the realized trajectory of the object MC from its nominal value (312) and for the law of deviations of the realized contact forces of the followers from their nominal values (338). For tracking the nominal trajectory of the object MC and nominal contact forces of the followers the law of deviations of the realized trajectory of the object MC from its nominal is defined by (356). The law of deviations of the realized followers' contact forces from their nominals selected in the linear form, i.e.

$$\dot{\mu}_s = -\sigma\mu_s, \qquad \mu_s = F_{c2}^0 - F_{c2}. \tag{360}$$

By comparing this with (338), it is obvious that $S(\mu_s) = -\sigma\mu_s$. The selected numerical value is $\sigma = 20$ [s^{-1}].

Since the elastic interconnections have no masses, the acceleration of the followers' contact points, instead from (340), is obtained by differentiating the last equation of (354)

$$\ddot{Y}_3 = \frac{1}{d_k}\dot{F}_{c2} + \ddot{Y}_2 + \frac{c_k}{d_k}\dot{Y}_2 - \frac{c_k}{d_k}\dot{Y}_3. \tag{361}$$

By introducing into this equation the derivative of the followers' contact force $\dot{F}_{c2} = \dot{F}_{c2}^0 - \sigma(F_{c2}^0 - F_{c2})$ needed to satisfy the law (360), we obtain

$$\ddot{Y}_3 = \frac{1}{d_k}[\dot{F}_{c2}^0 - \sigma(F_{c2}^0 - F_{c2})] + \ddot{Y}_2 + \frac{c_k}{d_k}\dot{Y}_2 - \frac{c_k}{d_k}\dot{Y}_3. \tag{362}$$

The control laws are determined, according to (341), by the expressions

$$\tau_v = N_v[-\alpha - \beta\ddot{q}_s] + n_v$$

$$\Rightarrow \tau_1 = m_1\left[-\alpha^0 - \beta\frac{1}{d_k}[\dot{F}_{c2}^0 - \sigma(F_{c2}^0 - F_{c2}) + d_k\ddot{Y}_2 + c_k\dot{Y}_2 - c_k\dot{Y}_3]\right] + n_v,$$

$$\tau_s = N_s\ddot{q}_s + n_s$$

$$\Rightarrow \tau_1 = m_2\frac{1}{d_k}[\dot{F}_{c2}^0 - \sigma(F_{c2}^0 - F_{c2}) + d_k\ddot{Y}_2 + c_k\dot{Y}_2 - c_k\dot{Y}_3] + n_s, \tag{363}$$

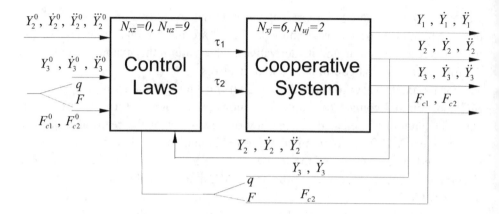

Figure 44. Block diagram of the closed-loop cooperative system

where F_{c2}^0 and $\dot{F}_{c2}^0$ are the followers' nominal contact forces and derivatives.

By comparing the control laws (358) for tracking the nominal trajectory of the manipulated object MC and nominal trajectory of the followers' contact points with the control laws (363) for tracking the nominal trajectory of the manipulated object MC and the followers' nominal contact forces, it can be concluded that the difference is only in the determination of the necessary acceleration of the followers. This gives the possibility of representing both control laws by one block diagram (Figure 44) and realizing them by one program. Hence, to the input of the controlled system the necessary input data for tracking both the nominal trajectory of contact point ($\ddot{Y}_3^0$, $\dot{Y}_3^0$ and Y_3^0) and nominal contact forces ($\dot{F}_{c3}^0$ and F_{c3}^0) of the followers are introduced simultaneously. The switches F and q in the main branch and feedback branch are switched on simultaneously and they show what the quantities are to be used for the selected control law from the main branch and feedback branch.

The selected control laws have been tested on the example of tracking nominal trajectories in the case of the existence of the initial deviation of the real trajectories from their nominals. The nominal trajectories along which the system is guided, were determined for the same example as in Chapter 5 and are given in Figure 28 for the gripping phase and in Figure 31 for the general motion. The results of simulation of the controlled cooperative system are presented in Figures 45–48. The results of the simulated work of the cooperative system with the control laws for tracking the nominal trajectory of the object MC and nominal trajectory of the followers' contact points are given under the title 'NPZU_Q'. The corresponding results are presented graphically in Figures 45a and 45b for the gripping phase and in Figures 47a and 47b for the general motion. Results of the simulated work of the

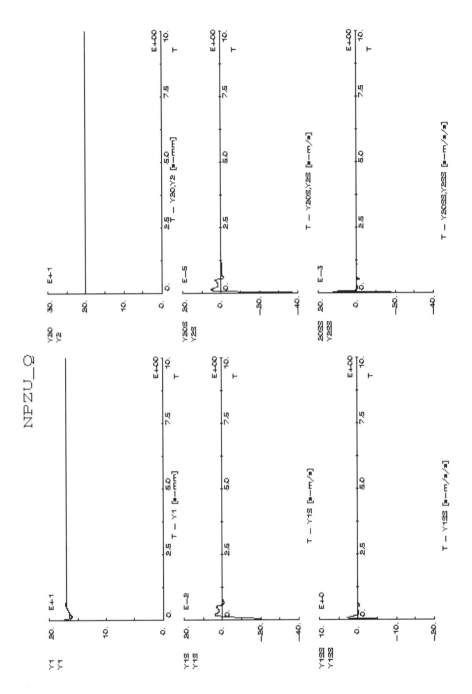

Figure 45a. Gripping – tracking Y_2^0 and Y_3^0

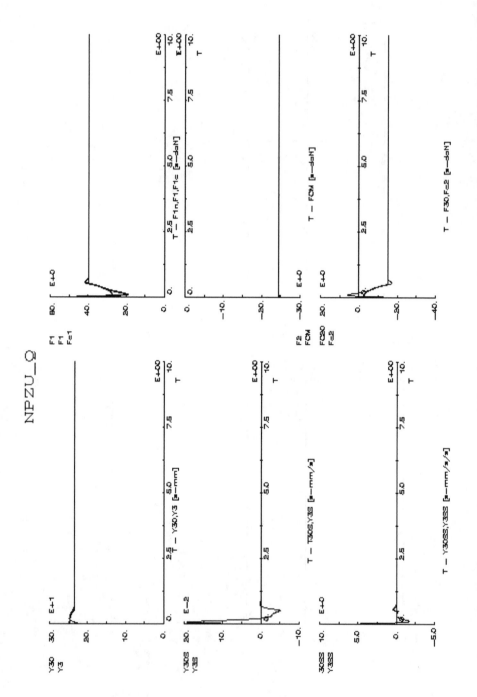

Figure 45b. Gripping – tracking Y_2^0 and Y_3^0

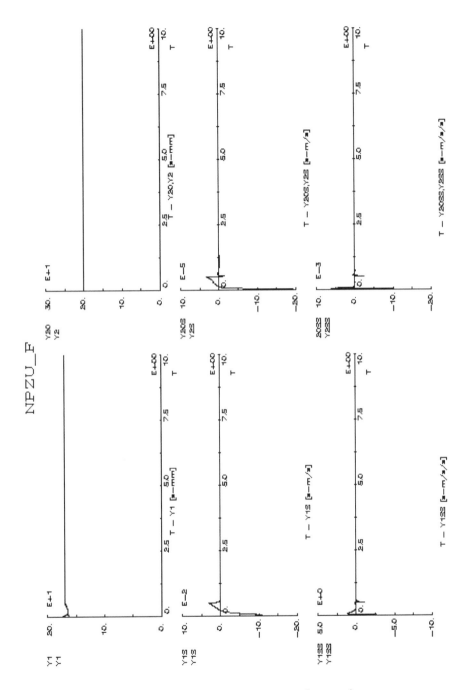

Figure 46a. Gripping – tracking Y_2^0 and F_{c2}^0

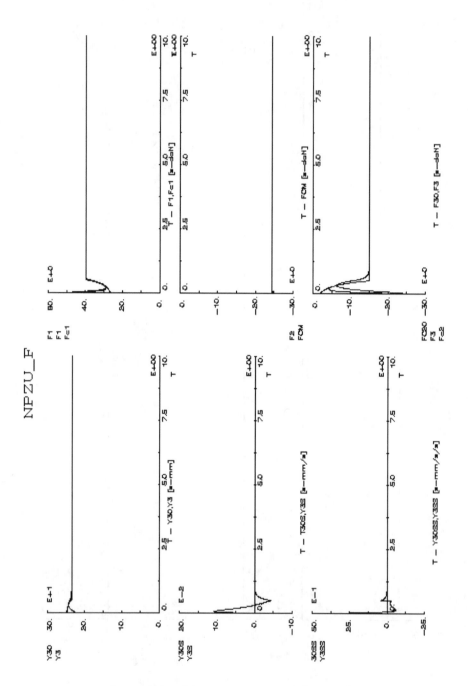

Figure 46b. Gripping – tracking Y_2^0 and F_{c2}^0

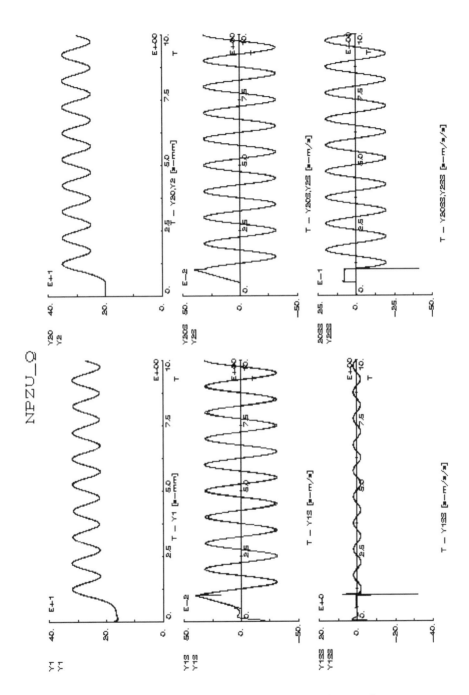

Figure 47a. General motion – tracking Y_2^0 and Y_3^0

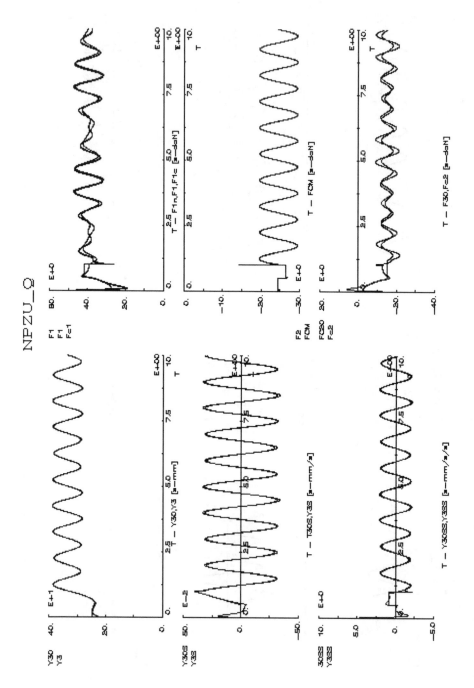

Figure 47b. General motion – tracking Y_2^0 and Y_3^0

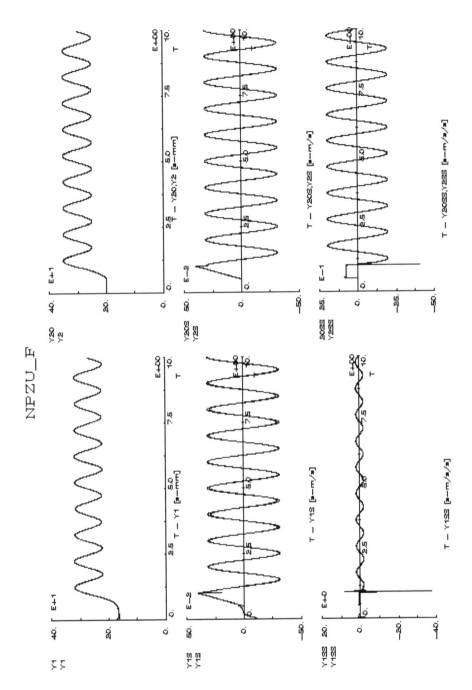

Figure 48a. General motion – tracking Y_2^0 and F_{c2}^0

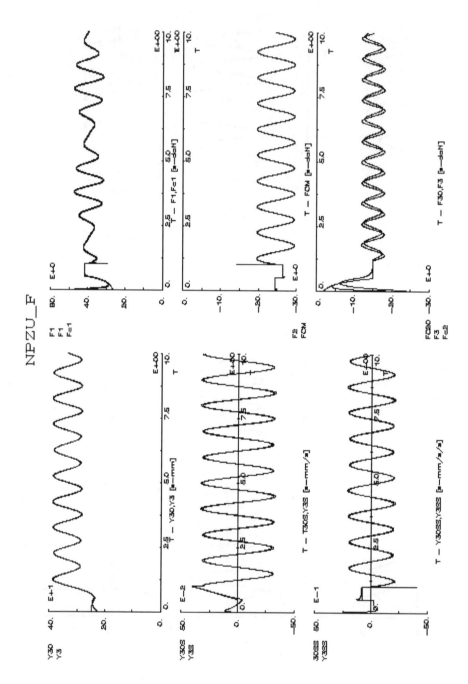

Figure 48b. General motion – tracking Y_2^0 and F_{c2}^0

cooperative system controlled by applying the control laws for tracking the nominal trajectory of the manipulated object MC and the followers' nominal contact force are given under the title 'NPZU_F' and presented graphically in Figures 46a and 46b for the gripping phase and in Figures 48a and 48b for the general motion. In the diagrams, the required and realized trajectories are superimposed. It can be concluded that both control laws ensured high-quality tracking of the nominal trajectories.

7 CONCLUSION: LOOKING BACK ON THE PRESENTED RESULTS

7.1 An Overview of the Introductory Considerations

In the introductory chapters of this monograph (Chapters 1, 2 and 3), we have shown that kinematic uncertainty is not essentially a problem of cooperative manipulation, whereas the problem of determining the real distribution of the contact forces that arise as a consequence of the action of the weight and dynamic forces of the manipulated object, known in the literature as force uncertainty, is practically the basic problem of cooperative manipulation.

By analyzing the cooperative system's properties in the state of rest, it was concluded that the cooperative system at rest can be considered as a statically undetermined spatial grid. Distribution of forces in such a grid cannot be solved if its elastic properties are neglected.

The impossibility of finding the real load distribution from the object onto contact forces is a consequence of neglecting the elastic properties of the cooperative system in the part where this distribution is realized [8]. None of the proposed criteria for solving force uncertainty in rigid [12–17, 19, 20] or elastic [1, 3–5, 18] cooperative systems could yield the solution of the real distribution of the loads involved.

It should be noticed that kinematic uncertainty and contact force uncertainty do not exist as a problem if the manipulated object position and force are determined on the basis of the positions of contacts and contact forces. Hence, the uncertainty problems are not essentially problems of cooperative manipulation any more, but it is the cooperative system control. Control schemes 'leader/follower control' and 'master/slave control' [14, 50], 'motion/force control' [51, 52], 'coordinated control' [15, 53, 54], are synthesized for the mathematical model of cooperative system dynamics with the problem of force uncertainty unresolved, so that they are of no practical importance. Besides, the control scheme 'hybrid position/force control' [12, 54, 55], based on [56, 57], involved inadequate realization of control decoupling with respect to position and force [35, 37–39].

The procedure for solving the problem of load distribution was presented for a cooperative system consisting of two fingers (manipulators) handling an object that is translated along a vertical straight line. From the cooperative system, we single out the part between the manipulators tips where the load distribution should be carried out. That part is approximated by three elastically interconnected concentrated masses placed at the contact points and at the object MC. The procedure of modeling dynamics of the elastic system for the case of its general motion using Lagrange equations, was consistently applied. It was shown that, in an attempt to separate the macro and micro motions of the elastic system, i.e. to separate the transfer and relative motion represented by elastic displacements [1, 3–5], the number of independent quantities exceeded the number of state quantities needed and was sufficient for a description of the elastic system's dynamics. This problem does not appear when the elastic system's dynamics is described by using the absolute coordinates of the nodes and information about the mutual positions of the nodes prior to the beginning of deformation, while the elastic system was still immobile.

Let us point out once more that by abandoning the assumption of the rigid cooperative system in the contact surroundings system enables us to uniquely solve the problem of load distribution, i.e. the problem of force uncertainty in cooperative manipulation.

The overall model of dynamics of the above cooperative system was obtained by uniting the kinematic relations between the internal and absolute coordinates of the contact, model of the elastic system dynamics, and the model of the manipulators' dynamics.

7.2 On Mathematical Modeling

The mathematical model was derived for an arbitrary system of cooperative manipulation of an object, involving an arbitrary number of manipulators (m). The model was derived for the case of complex motion of the object, i.e. when the object performs translation and rotation.

To make understanding the properties of the cooperative system easier, certain assumptions were introduced to significantly simplify the problem of modeling. The cooperative system was split into its 'rigid' (manipulators) and 'elastic' (the object and elastic contacts) parts. Namely, it is adopted that the manipulators are rigid, whereas the connections of the object and manipulators are elastic. The manipulated object is either rigid or can be divided into a rigid part and an elastic part, the latter being characteristic of the contact surroundings.

In modeling, the elastic part of the object is considered as part of the elastic interconnections of the object and manipulators. The object and elastic interconnections make the elastic system, which is approximated by a spatial grid having m external nodes and one internal node, connected by non-inertial elastic connections. At each node is placed a fictitious rigid body to represent the inertial properties of the elastic system in the surroundings of that node, i.e. inertial properties of the elastic interconnection of the rigid part of the object and manipulators. The bodies are placed in such a way that their MCs coincide with the grid nodes, so that the external loads (vectors of forces and moments) act at the MCs of these bodies. The manipulated object is placed at the internal node. Each node has six DOFs so that the total number of DOFs of the elastic system motion is equal to $6m + 6$. Elastic displacements of the nodes are equivalent to the displacements of the bodies placed at the grid nodes. Gravitational, dissipative, and contact forces are adopted as external loads of the elastic system.

In the modeling of elastic system dynamics, two case of motion can be distinguished.

In the first case, the unloaded state of the elastic system is immobile throughout the duration of the cooperative work, and the results of the theory of elasticity related to finding the acting load as a function of elastic displacements are directly applied, whereby it should be noted that all elastic displacements of nodes are treated as independent variables.

In the second case, the unloaded state of elastic system is mobile throughout the duration of the cooperative work. The general motion of the elastic system is described in terms of absolute coordinates of the loaded elastic system nodes. To form the motion equations, it is necessary to express deformation work as a function of absolute coordinates. As an expansion of the method of finite elements it was proposed to express elastic displacements as a function of the absolute coordinates of the nodes of the loaded elastic system and modules of the difference of absolute coordinates of the nodes of the unloaded elastic system that existed in the beginning of the gripping phase (before deformation of the elastic system), when the elastic system was in the state of rest. The determination of the dependence of elastic displacements on absolute coordinates is thus avoided, as the dependence of the forces at the nodes on the position vectors of the loaded elastic system nodes is achieved by expressing deformation work as a sum of the products of internal forces and mutual displacements of the nodes.

As a consequence of the choice that all node displacements are independent variables, the stiffness matrix in the models of elastic system is singular irrespective of whether the unloaded state is mobile or immobile. Because of the singularity of the stiffness matrix, the elastic system model must also contain the modes of motion of the elastic system as a rigid body. In the general case, it is necessary to know at least six independent position quantities (positions and orientations) in order to define the elastic system's position in space. These quantities can be arbitrarily selected, but if they are selected so as to describe the position of only one contact point (as was done in this monograph), then the manipulator that forms this contact is the leader, the other manipulators being followers. The quantities selected to define the elastic system's position in space can be changed simultaneously, without any discontinuity of forces or positions. In this way, a solution is also given for a simultaneous exchange of the leader's role in cooperative manipulation.

Using the same procedure as for deriving the model of the cooperative system in the introductory chapter, two complete models of cooperative system dynamics were derived, whereby the dynamic model of non-redundant manipulators and kinematic relations between internal coordinates and absolute coordinates of contacts were taken from the available literature [32]. The first model describes the dynamics of cooperative manipulation in which the elastic system unloaded state is immobile all the time of duration of the cooperative work. other model of cooperative manipulation dynamics covers the general case of motion of the manipulated object. This model is described via the absolute coordinates of the contact points and manipulated object MC, whereby it is supposed that the mutual position of the manipulators' tips at the moment of forming contact (at the beginning of gripping) between the manipulators and object is known.

The derived mathematical models give a sufficiently exact description of the cooperative system under static and dynamic conditions. It is concluded that the number of state quantities of the cooperative system must be greater than the number of driving torques. The results obtained in model testing on selected examples showed that our approach to the modeling of cooperative manipulation is consistent and the conclusions derived are correct.

7.3   Cooperative System Nominals

The preset requirement is that the nominal motion in cooperative manipulation must be coordinated and realizable.

Coordinated motion of a cooperative system is the motion in which the object gripping is carried out first, and then the manipulated object motion is continued. In the course of coordinated motion of the cooperative system, the manipulators perform such motion that ensures that the gripping conditions are not essentially disturbed, i.e. the geometric configuration of the elastic system nodes established at the end of the gripping phase is essentially preserved.

The nominals were synthesized for two cases. In the first case, the starting point was the assumption that contact forces were independent variables. This means that during the motion the distribution and absolute values of the contact forces realized at the end of the gripping phase change with respect to the object only as a consequence of the change of object orientation on its path and additional dynamic forces, and/or due to the additional changes of contact forces required at the contact points of the followers. The other approach starts from the assumption that there is no change in the geometric configuration of contact points that was established at the end of the gripping phase.

Using the proposed procedures, the elastic system nominals are synthesized first. It is assumed that the elastic displacements are not large and that the position of the elastic system nodes in the 'static' displacement (neglecting dynamic forces) and in the general motion along the trajectory given for the manipulated object, cannot essentially change. The procedure used is a quasi-static one.

The gripped object is being statically transferred to a series of selected points on the trajectory. Thereby, it is assumed that there are no changes in the direction and load intensity with respect to the loaded elastic system realized at the end of the gripping phase, that is, no real loads are considered. In the transferred positions, the load intensity realized at the end of gripping is reduced to zero. As a result, the current unloaded the states of the elastic system at selected points on the obtained trajectory. The current loaded state of the elastic system is obtained by the static action of the node load from the previously defined current unloaded state.

The node load is assumed to be the result of the real gravitational forces, rotated contact forces at the end of gripping, and dynamic forces. Dynamic forces are determined by using velocities and accelerations of nodes obtained from the condition that the elastic system at the end of the gripping phase moves as a rigid body. If, in addition to the motion of the manipulated object along the nominal trajectory, simultaneous changes of gripping conditions are required, then the resulting node loads are calculated using the required contact forces instead of rotated contact forces existing at the end of the gripping phase. The approximate values of the contact forces are determined from the condition that the nodes in the elastic system motion are on the calculated trajectories and that the motion equations from which dissipative forces of elastic interconnections are omitted, are satisfied. These contact forces are adopted as the coordinated motion nominals. It is assumed that, at the end of the motion between selected points on the trajectory, the change of contact forces can be described by a monotonous function.

The realizable trajectories are determined by numerically solving the system of differential equations describing the dynamics of the elastic interconnections of the followers, whereby the system is excited by the synthesized nominal contact forces for the coordinated motion. The nominals of elastic interconnection of the leader are determined as functions of the nominals for the manipulated object and elastic interconnections of the followers.

The nominals obtained for the elastic system external nodes (contacts with the manipulators) are the nominals of the manipulators' tips and, based on them, the nominals for the manipulators in cooperation are synthesized.

The proposed procedure of nominals synthesis ensures a realizable coordinated motion of the cooperative system. The obtained nominals are input quantities to the controlled cooperative system.

7.4   Cooperative System Control Laws

The dynamic model of cooperative manipulation with the force uncertainty resolved, serves as the basis for the synthesis of cooperative system control laws.

First, it is necessary to analyze the number and nature of the requirements from the cooperative system and then consider the laws which should be introduced to realize a coordinated motion of the cooperative system.

The analysis of definitions and criteria for assessing the controllability and observability of linear systems showed that these definitions and criteria determine the conditions of direct and inverse mapping between the domains of inputs, states, and outputs of the system (the control object). The results of this analysis are then applied to non-linear systems. The following conclusions are essential for cooperative manipulation.

First, the number of independently controlled output quantities (directly tracked quantities, quantities closing feedback loops) cannot be larger than the number of independent input variables.

Second, the natural space of the system output is the union of the space of inputs and space of states. The controlled output can be any quantity from the natural space of system outputs or the 'image' of the selected controlled output from the natural space in another space. This means that the controlled output may be either state (or the real output as the 'image' of the state in the output space) or input to the system and/or even a combination of both. The natural space of cooperative system outputs considered in this monograph is the union of the space of driving torques, space of manipulated object MC position, and space of positions of contact points, or, instead of them, the space of internal coordinates. Total dimension of this space is $6m + 6m + 6$. Of interest are also the spaces of the contact forces and space of elastic forces, which are obtained by the mapping from the natural space of outputs.

Third, the consistent approach to modeling cooperative system dynamics automatically resolves the problem of its controllability.

As the cooperative system is excited by $6m$ independent inputs (driving torques), it is possible to select $6m$ independent controlled outputs from its natural space of outputs.

An analysis is carried out of the possible choices of controlled outputs of the cooperative system. To have a controlled motion of the cooperative system in space, the vector of controlled outputs should contain at least such a number of independent position coordinates that is equal to the dimension of the space in which the cooperative system moves. In this monograph, this role is assigned to the position vector of one node of the elastic system. This node is either the manipulated object MC or the leader's contact point. It was demonstrated that for the remaining $6m - 6$ controlled outputs there are only two qualitatively different choices. One is the choice of the position vectors of the contact positions of the followers, and the other is the choice of the vector of the contact forces at the followers' contacts. Control laws were selected for both of the controlled outputs. It was shown that the cooperative system that is controlled using the selected laws follows the controlled outputs in an asymptotically stable manner.

The analysis of the behavior of non-controlled quantities proved that the choice of position vectors of the contact points of the followers as controlled outputs ensures the stability of all the system's quantities. For the choice of the vector of contact forces at the followers' contact points, it was proved that, even when the control ensures the realization of all the required nominal output quantities, the deviations of non-controlled quantities from their nominal values can be, but not necessarily, unconstrained. The elastic system will behave as a mobile elastic structure excited by the controlled external loads.

7.5 General Conclusions about the Study of Cooperative Manipulation

On the basis of the properties of the controlled cooperative system, it can be concluded that

- it is not suitable to control contact forces in cooperative manipulation, but the control should be exclusively concerned with the manipulators' positions, i.e. positions of the manipulator-object contact points, and

- manipulators can control contact forces irrespective of the instantaneous positions of contact points of the manipulators and object, and this can be used to examine the dynamics of mobile elastic structures using controlled loads.

It should be noted that the derived model of cooperative manipulation is modular. Thus, the methodology and results of this monograph can also be used to control elastic structures in which executive organs are not manipulators but isolated one-DOF actuators. The models of elastic system and cooperative manipulation, synthesized nominals, along with the selected control laws and conclusions about the properties of the controlled cooperative system, can easily be expanded and adapted to control arbitrary mobile elastic structures.

7.6   Possible Directions of Further Research

The results obtained in this monograph could, in a certain way, determine the direction of future research concerning cooperative manipulation and contact tasks in general.

In cooperative manipulation, the problem of force uncertainty cannot be resolved without introducing elastic elements between the object and manipulators. Hence, it would be necessary to equip the grippers by elastic elements. For such elements, it would be possible to determine in advance their individual characteristic stiffness matrices. The gripper's elastic tips themselves (without the manipulated object) can be adopted as an elastic system. Using the direct stiffness method, mathematical modeling of an elastic system can be simplified and automated. The outcome would be the usable information about elastic displacements and elastic forces at the grippers's elastic tips. During the motion in cooperative manipulation, the geometric configuration of manipulator tips, i.e. their mutual geometrical arrangement, should be such that, in any instant, are realized desired elastic displacement and/or desired elastic forces, which is equivalent to the existence of the external load produced by the object. At that, it is not necessary to join the object itself to the elastic system, so that it is not necessary to know its characteristics.

The methodology presented in this monograph solves, in an exact way, the problem of force uncertainty in cooperative manipulation and determines the appropriate laws to control cooperative manipulation. This methodology is a result of the awareness of the necessity of considering cooperative system elasticity, at least in the part where force uncertainty arises, which allows an exact (to a desired limit) calculation of loads in all parts of the cooperative system for all its motions. These loads are the necessary input data to calculate the solidity of the cooperative system and input parameters for defining servoactuators. An important conclusion is that it is necessary to control the position of the manipulators if the goal is to realize a stable motion and load of the cooperative system. The application of the proposed methodology requires a comprehensive knowledge of the user in several theoretical areas, and can induce some problems in practical realizations. The application of this methodology also requires an exact model of the cooperative system, simultaneous consideration of 'fast' and 'slow' dynamics (of all characteristic oscillation modes) of the cooperative system, and the use of force servoactuators to control the driving torques at the manipulators joints during the motion. The bandwidth of such servoactuators should be larger than the characteristic frequency of the cooperative (i.e. elastic) system.

The conclusion that one should control manipulator positions implies that it is possible to form a new model of cooperative manipulation that will also take into account the dynamics of the actuators in which feedback involves velocity and/or position, and not force or moment. The external load of the actuators whose feedback is velocity and/or position, is its natural local feedback. The dynamics of manipulators and cooperative systems as a whole can be considered as a local feedback of the actuators, which, being low-bandwidth filters, filtrate out the real 'fast' dynamics of the cooperative system, leaving the possibility to control its 'slow' dynamics. This means that for the purpose of the synthesis of cooperative system control, it is useful to reduce the complex and exact model of the elastic system, so that it contains only its characteristic oscillation modes within the bandwidth of the servoactutors. Such an approach to solving the task of cooperative system control is suitable for engineering practice and opens up some new possibilities to modeling and selecting appropriate control laws.

APPENDIX A: ELASTIC SYSTEM MODEL FOR THE IMMOBILE UNLOADED STATE

Here we give the entire procedure for deriving the mathematical model of dynamics of an elastic system as a part of a cooperative system composed of m six-DOF rigid manipulators, handling an object whose motion in three-dimensional space proceeds without any constraint (Figure 12).

It is assumed that the connections of the object and manipulators are elastic and the object is either rigid or elastic. For both cases, we assume that each connection or part of the manipulated object in the neighborhood of the contact point, can be represented by a rigid body at the MC where the forces of contact, gravitation, damping, and elasticity act. The object with the connections forms an elastic system of $m + 1$ elastically interconnected rigid bodies (Figure 12). It is assumed that elastic properties are such that a linear relationship can be established between each relative displacement of any part of the elastic system. Each body is allowed to have six DOFs of motion. Gravitational and contact forces are considered as a system of the external forces acting at the MC.

The coordinates of the MC displacements with respect to the unloaded state y_i defined by (69), are adopted as generalized coordinates.

Potential energy Π of the elastic system is equal to the deformation work

$$2\Pi = y^T K y = (\delta^T \mathcal{A}^T) K \begin{pmatrix} \delta \\ \mathcal{A} \end{pmatrix},$$

$$K = K^T \in R^{(6m+6)\times(6m+6)}, \quad \text{rank } K \leq 6m, \tag{364}$$

so that the derivative with respect to the coordinate is equal to elastic force and is given by

$$F_e = \frac{\partial \Pi}{\partial y} = K y \in R^{6m+6}, \tag{365}$$

261

$$\frac{\partial \Pi}{\partial y} = \begin{bmatrix} \dfrac{\partial \Pi}{\partial y_o} \\ \cdots \\ \dfrac{\partial \Pi}{\partial y_i} \\ \cdots \\ \dfrac{\partial \Pi}{\partial y_m} \end{bmatrix} = \begin{bmatrix} K_0 y \\ \cdots \\ K_i y \\ \cdots \\ K_m y \end{bmatrix} \Rightarrow \frac{\partial \Pi}{\partial y_i} = K_i y \in R^6, \quad K_i \in R^{6 \times (6m+6)},$$

where K_i are the submatrices composed of $6i + 1$ to $6i + 6$ rows of the matrix K.

Relations between the angular velocity $\omega_i = \mathrm{col}(p_i, q_i, r_i)$ measured along the body i main inertia axes and the derivatives of orientation given by $\dot{\mathcal{A}}_i = \mathrm{col}(\dot{\psi}, \dot{\theta}, \dot{\varphi})$ are

$$\begin{bmatrix} p_i \\ q_i \\ r_i \end{bmatrix} = \begin{bmatrix} \dot{\psi} \sin \theta_i \sin \varphi_i + \dot{\theta}_i \cos \varphi_i \\ \dot{\psi} \sin \theta_i \cos \varphi_i - \dot{\theta}_i \sin \varphi_i \\ \dot{\psi} \cos \theta_i + \dot{\varphi}_i \end{bmatrix} = \begin{bmatrix} \sin \theta_i \sin \varphi_i & \cos \varphi_i & 0 \\ \sin \theta_i \cos \varphi_i & -\sin \varphi_i & 0 \\ \cos \theta_i & 0 & 1 \end{bmatrix} \begin{bmatrix} \dot{\psi} \\ \dot{\theta}_i \\ \dot{\varphi}_i \end{bmatrix},$$

$$\omega_i = \mathrm{col}(p_i, \ q_i, \ r_i) = L_\omega(\mathcal{A}_i) \cdot \dot{\mathcal{A}}_i.$$

Motion velocity in expanded form is

$$v_{ia} = \begin{bmatrix} \dot{x}_i \\ \dot{y}_i \\ \dot{z}_i \\ p_i \\ q_i \\ r_i \end{bmatrix} = \begin{bmatrix} \dot{r}_{ia} \\ \omega_{ia} \end{bmatrix} = \begin{bmatrix} I_{3 \times 3} & 0_{3 \times 3} \\ 0_{3 \times 3} & L_\omega(\mathcal{A}_i) \end{bmatrix} \cdot \begin{bmatrix} \dot{\delta}_i \\ \dot{\mathcal{A}}_i \end{bmatrix} = L_v(\mathcal{A}_i) \dot{y}_i = L_v(y_i) \dot{y}_i,$$

i.e.

$$v_i = \begin{bmatrix} 1 & 0 & 0 & 0 & 0 & 0 \\ 0 & 1 & 0 & 0 & 0 & 0 \\ 0 & 0 & 1 & 0 & 0 & 0 \\ 0 & 0 & 0 & \sin \theta_i \sin \varphi_i & \cos \varphi_i & 0 \\ 0 & 0 & 0 & \sin \theta_i \cos \varphi_i & -\sin \varphi_i & 0 \\ 0 & 0 & 0 & \cos \theta_i & 0 & 1 \end{bmatrix} \cdot \begin{bmatrix} \dot{x}_i \\ \dot{y}_i \\ \dot{z}_i \\ \dot{\psi} \\ \dot{\theta}_i \\ \dot{\varphi}_i \end{bmatrix} = L_v(y_i) \dot{y}_i.$$

The total kinetic energy is defined by

$$T = T_0 + T_1 + \cdots + T_i + \cdots + T_m,$$

while the kinetic energy of the ith part is

$$2T_i = m_i \dot{\delta}_i^T \dot{\delta}_i + \omega_i^T I_i \omega_i = m_i \dot{\delta}_i^{x^2} + m_i \dot{\delta}_i^{y^2} + m_i \dot{\delta}_i^{z^2} + A \dot{p}_i^2 + B \dot{q}_i^2 + C \dot{r}_i^2,$$

where m_i is the mass and A_i, B_i, C_i are the ith object's main moments of inertia. By substituting the angular velocity, we obtain

$$2T_i = m_i \dot{\delta}_i^T \dot{\delta}_i + \dot{\mathcal{A}}_i^T L_\omega^T(\mathcal{A}_i) I_i L_\omega(\mathcal{A}_i) \dot{\mathcal{A}}_i$$

or together with

$$2T_i = \dot{y}_i^T L_v^T(y_i) M_i L_v(y_i) \dot{y}_i = \dot{y}_i W_i(y_i) \dot{y}_i,$$

$$W_i(y_i) = L_v^T(y_i) M_i L_v(y_i) \in R^{6 \times 6}, \quad W_i(y_i) = W_i^T(y_i), \quad \det W_i(y_i) \neq 0,$$

where $I_i = \mathrm{diag}(A_i, B_i, C_i)$ and $M_i = \mathrm{diag}(m_i, m_i, m_i, A_i, B_i, C_i)$. In an expanded form $W_i(y_i)$ is given by

$W_i(y_i) =$

$$\begin{bmatrix} m_i & 0 & 0 & 0 & 0 & 0 \\ 0 & m_i & 0 & 0 & 0 & 0 \\ 0 & 0 & m_i & 0 & 0 & 0 \\ 0 & 0 & 0 & \sin^2\theta_i(A_i\sin^2\varphi_i + B_i\cos^2\varphi_i) + C_i\cos^2\theta_i & \frac{1}{2}(A_i - B_i)\sin\theta_i\sin2\varphi_i & C_i\cos\theta_i \\ 0 & 0 & 0 & \frac{1}{2}(A_i - B_i)\sin\theta_i\sin2\varphi_i & A_i\cos^2\varphi_i + B_i\sin^2\varphi_i & 0 \\ 0 & 0 & 0 & C_i\cos\theta_i & 0 & C_i \end{bmatrix}$$

or, in scalar form,

$$2T_i = m_i\dot{\delta}_i^{x^2} + m_i\dot{\delta}_i^{y^2} + m_i\dot{\delta}_i^{z^2}$$

$$+\dot{\psi}_i^2(\sin^2\theta_i(A_i\sin^2\varphi_i + B_i\cos^2\varphi_i) + C_i\cos^2\theta_i)$$

$$+\dot{\psi}_i\dot{\theta}_i(A_i - B_i)\sin\theta_i\sin2\varphi_i + +2\dot{\psi}_i\dot{\varphi}_iC_i\cos\theta_i$$

$$+\dot{\theta}_i^2(A_i\cos^2\varphi_i + B_i\sin^2\varphi_i) + C_i\dot{\varphi}_i^2.$$

Derivatives of kinetic energy with respect to velocity are defined by

$$\frac{\partial T_i}{\partial \dot{y}_i} = L_v^T(y_i) M_i L_v(y_i) \dot{y}_i = W_i(y_i) \dot{y}_i,$$

$$\frac{d}{dt}\frac{\partial T_i}{\partial \dot{y}_i} = \dot{W}_i(y_i) \dot{y}_i + W_i(y_i) \ddot{y}_i.$$

The member $\dot{W}_i(y_i) \dot{y}_i$ is defined in the following way:

$$\dot{W}_i(y_i) \dot{y}_i =$$

$$
\begin{bmatrix}
0 \\
0 \\
0 \\
\{\dot{\psi}_i\dot{\theta}_i \sin 2\theta_i (A_i \sin^2 \varphi_i + B_i \cos^2 \varphi_i - C_i) + \dot{\psi}_i\dot{\varphi}_i (A_i - B_i) \sin^2 \theta_i \sin 2\varphi_i \\
+\dot{\theta}_i\dot{\varphi}_i \sin \theta_i ((A_i - B_i) \cos 2\varphi_i - C_i) + \frac{1}{2}\dot{\theta}_i^2 (A_i - B_i) \cos \theta_i \sin 2\varphi_i\} \\
\frac{1}{2}\dot{\psi}_i\dot{\theta}_i (A_i - B_i) \cos \theta_i \sin 2\varphi_i + \dot{\psi}_i\dot{\varphi}_i (A_i - B_i) \sin \theta_i \cos 2\varphi_i \\
-\dot{\theta}_i\dot{\varphi}_i (A_i - B_i) \sin 2\varphi_i - \dot{\psi}_i\dot{\theta}_i C_i \sin \theta_i
\end{bmatrix}.
$$

The partial derivative of kinetic energy with respect to the coordinate is the function of that coordinate and its derivative, and is given by

$$
\frac{\partial T_i}{\partial y_i}(y_i, \dot{y}_i) =
\begin{bmatrix}
0 \\
0 \\
0 \\
0 \\
\frac{1}{2}\dot{\psi}_i^2 \sin 2\theta_i (A_i \sin^2 \varphi_i + B_i \cos^2 \varphi_i - C_i) \\
+\frac{1}{2}\dot{\psi}_i\dot{\theta}_i (A_i - B_i) \cos \theta_i \sin 2\varphi_i - \dot{\psi}_i\dot{\varphi}_i C_i \sin \theta_i \\
\frac{1}{2}\dot{\psi}_i^2 (A_i - B_i) \sin^2 \theta_i \sin 2\varphi_i + \dot{\psi}_i\dot{\theta}_i (A_i - B_i) \sin \theta_i \cos 2\varphi_i \\
-\frac{1}{2}\dot{\theta}_i^2 (A_i - B_i) \sin 2\varphi_i
\end{bmatrix}.
$$

The preceding equations for kinetic energy can be given concisely as

$$
2T = \sum_{i=0}^{m} \left(\sum_{j=1}^{\infty} dm_j v_j \right) = \sum_{i=0}^{m} \left(\sum_{j=1}^{\infty} dm_j \frac{d}{dt}(y_i + b_j) \right),
$$

$$
2T = \sum_{i=0}^{m} m_i \delta_i^2 + \sum_{i=0}^{m} I_i \omega_i^2
$$

$$
= \sum_{i=0}^{m} v_i^T M_i v_i = v^T M v = \dot{y}^T L_v^T(y) M L_v(y) \dot{y}, \qquad (366)
$$

where

$$
M = \operatorname{diag}(M_0, M_1, \ldots, M_n) \in R^{(6m+6)\times(6m+6)},
$$

$$
M_i = \operatorname{diag}(m_i, m_i, m_i, A_i, B_i, C_i) \in R^{6\times6},
$$

$$
v = \operatorname{col}(v_0, v_1, \ldots, v_m) \in R^{(6m+6)\times1},
$$

$$
v_i = \operatorname{col}(\dot{\delta}_i, \omega_i(\mathcal{A}_i)) = L_{vi}(y_i)\dot{y}_i \in R^{6\times1},
$$

$$
\omega_i = L_{\omega i}(\mathcal{A}_i)\dot{\mathcal{A}}_i \in R^{3\times1},
$$

$$
L_{vi}(y_i) = \operatorname{diag}(I_{3\times3}, L_{\omega_i}(\mathcal{A}_i)) \in R^{6\times6},
$$

$$
L_v(y) = \operatorname{diag}(L_{v0}, L_{v1} \ldots L_{vm}) \in R^{(6m+6)\times(6m+6)}. \qquad (367)
$$

If the connections have dissipative properties, the dissipation energy can be expressed as

$$2\mathcal{D} = -\dot{y}^T D \dot{y}, \quad D = D^T \geq 0, \quad D \in R^{(6m+6)\times(6m+6)}, \tag{368}$$

where $D \in R^{6(m+1)\times6(m+1)}$ is the matrix with damping coefficients corresponding to velocities. The derivative of the dissipation energy with respect to velocity is given by

$$\frac{\partial \mathcal{D}}{\partial \dot{y}} = -D\dot{y}, \quad \frac{\partial \mathcal{D}}{\partial \dot{y}} = \begin{bmatrix} \dfrac{\partial \mathcal{D}}{\partial \dot{y}_o} \\ \cdots \\ \dfrac{\partial \mathcal{D}}{\partial \dot{y}_i} \\ \cdots \\ \dfrac{\partial \mathcal{D}}{\partial \dot{y}_m} \end{bmatrix} = \begin{bmatrix} D_0\dot{y} \\ \cdots \\ D_i\dot{y} \\ \cdots \\ D_m\dot{y} \end{bmatrix} \Rightarrow \frac{\partial \mathcal{D}}{\partial \dot{y}_i} = -D_i\dot{y}, \quad D_i \in R^{6\times(6m+6)},$$

where D_i are the submatrices composed of the rows atarting from $6i + 1$ to $6i + 6$ inclusive of the matrix D.

By substituting these expressions into the Langrange equations we obtain

$$\frac{d}{dt}\frac{\partial T}{\partial \dot{y}_i} - \frac{\partial T}{\partial y_i} - \frac{\partial \mathcal{D}}{\partial \dot{y}_i} + \frac{\partial \Pi}{\partial y_i} = Q_i, \quad i = 0, 1, \ldots, m, \tag{369}$$

where

$$Q_i = \text{col}(Q_i^1, \ldots, Q_i^6) = G_i(m_i g) + F_i,$$

$$Q_i^j = \sum_{k=0}^{m} \frac{\partial y_k}{\partial y_i^j} Q_k = \sum_{k=0}^{m} \frac{\partial y_k}{\partial y_i^j}(G_k(m_k g) + F_k), \quad i = 0, 1, \ldots, m, \quad j = 1, \ldots, 6,$$

$$Q_i^j = \sum_{k=0}^{m} \left(F_{ku}^x \frac{\partial \delta_k^x}{\partial y_i^j} + F_{ku}^y \frac{\partial \delta_k^y}{\partial y_i^j} + F_{ku}^z \frac{\partial \delta_k^z}{\partial y_i^j} + F_{ku}^\psi \frac{\partial \psi_k}{\partial y_i^j} + F_{ku}^\theta \frac{\partial \theta_k}{\partial y_i^j} + F_{ku}^\varphi \frac{\partial \varphi_k}{\partial y_i^j} \right)$$

$$= F_{iu}^j$$

$$\Rightarrow F_{iu} = G_i(m_i g) + F_i = \text{col}(F_{iu}^1, \ldots, F_{iu}^6) \in R^{6\times1}, \quad j = x, y, z, \psi_i, \theta_i, \varphi_i,$$

and y_i^j, $j = 1, \ldots, 6$ are the individual components of the vector $y_i = \text{col}(y_i^1, y_i^2, y_i^3, y_i^4, y_i^5, y_i^6) = \text{col}(\delta_i^x, \delta_i^y, \delta_i^z, \psi_i, \theta_i, \varphi_i)$.

For the elastic system performing general motion about the immobile unloaded state 0 under the action of the external forces F, a general form of the model is obtained as

$$W_i(y_i)\ddot{y}_i + \dot{W}_i(y_i)\dot{y}_i - \frac{\partial T_i}{\partial y_i}(y_i, \dot{y}_i) + D_i\dot{y} + K_i y = G_i(m_i g) + F_i, \quad i = 0, 1, \ldots, m$$

or, in short form,

$$W_i(y_i)\ddot{y}_i + w_i(y, \dot{y}) = F_i, \quad i = 0, 1, \ldots, m,$$

where

$$w_i(y, \dot{y}) = \dot{W}_i(y_i)\dot{y}_i - \frac{\partial T_i}{\partial y_i}(y_i, \dot{y}_i)$$

$$+ D_i\dot{y} + K_i y - G_i(m_i g) \in R^{6 \times 1}, \quad i = 0, 1, \ldots, m.$$

Taking into account that $G_i(m_i g) = (0, 0, -m_i g, 0, 0, 0)^T$, the member $w_i(y, \dot{y})$ has the expanded form

$$w_i(y, \dot{y}) =$$

$$\begin{bmatrix} \sum_{j=0}^{m} D_{(6i+1)j}\dot{y} + \sum_{j=0}^{m} K_{(6i+1)j}y \\ \cdots \\ \sum_{j=0}^{m} D_{(6i+2)j}\dot{y} + \sum_{j=0}^{m} K_{(6i+2)j}y \\ \cdots \\ \sum_{j=0}^{m} D_{(6i+3)j}\dot{y} + \sum_{j=0}^{m} K_{(6i+3)j}y + m_i g \\ \cdots \\ \dot{\psi}_i\dot{\theta}_i \sin 2\theta_i (A_i \sin^2\varphi_i + B_i \cos^2\varphi_i - C_i) + \dot{\psi}_i\dot{\varphi}_i(A_i - B_i)\sin^2\theta_i \sin 2\varphi_i + \\ \dot{\theta}_i\dot{\varphi}_i \sin\theta_i((A_i - B_i)\cos 2\varphi_i - C_i) + \frac{1}{2}\dot{\theta}_i^2(A_i - B_i)\cos\theta_i \sin 2\varphi_i \\ + \sum_{j=0}^{m} D_{(6i+4)j}\dot{y} + \sum_{j=0}^{m} K_{(6i+4)j}y \\ \cdots \\ -\frac{1}{2}\dot{\psi}_i^2 \sin 2\theta_i (A_i \sin^2\varphi_i + B_i \cos^2\varphi_i C_i) + \dot{\psi}_i\dot{\varphi}_i \sin\theta_i((A_i - B_i)\cos 2\varphi_i + C_i) \\ -\dot{\theta}_i\dot{\varphi}_i(A_i - B_i)\sin 2\varphi_i + \sum_{j=0}^{m} D_{(6i+5)j}\dot{y} + \sum_{j=0}^{m} K_{(6i+5)j}y \\ \cdots \\ -\frac{1}{2}\dot{\psi}_i^2(A_i - B_i)\sin^2\theta_i \sin 2\varphi_i - \dot{\psi}_i\dot{\theta}_i \sin\theta_i((A_i - B_i)\cos 2\varphi_i + C_i) \\ +\frac{1}{2}\dot{\theta}_i^2(A_i - B_i)\sin 2\varphi_i + \sum_{j=0}^{m} D_{6(i+1)j}\dot{y} + \sum_{j=0}^{m} K_{6(i+1)j}y \end{bmatrix}.$$

By uniting all $6m + 6$ equations, we obtain

$$W(y)\ddot{y} + w(y, \dot{y}) = F, \tag{370}$$

where

$$W(y) = \text{diag}(W_0(y_0)W_1(y_1) \ldots W_m(y_m)) \in R^{(6m+6) \times (6m+6)}, \quad W(y) = W^T(y),$$

$$\det W(y) \neq 0, \quad w(y, \dot{y}) = \text{col}(w_o(y, \dot{y}), \ldots, w_m(y, \dot{y})) \in R^{(6m+6) \times 1}.$$

From $6m+6$ equations (370), the number of independent equations is exactly equal to the stiffness matrix rank (rank K).

Equation (370) can be presented in such a way that the descriptions of connections motion and manipulation object are separated

$$W_c(y_c)\ddot{y}_c + w_c(y, \dot{y}) = F_c,$$

$$W_0(y_0)\ddot{y}_0 + w_0(y, \dot{y}) = 0, \tag{371}$$

where the subscript c designates the quantities related to the contact points, and the subscript 0 designates the quantities related to the manipulated object. Hereby

$$y_c = \text{col}(y_1, y_2, \ldots, y_m) \in R^{6m \times 1}, \, y_o \in R^{6 \times 1},$$

$$F_c = \text{col}(F_1, F_2, \ldots, F_m) \in R^{6m \times 1}, \, F_0 = 0 \in R^{6 \times 1},$$

$$W_c(y_c) = \text{diag}(W_1(y_1) \ldots W_m(y_m)) \in R^{6m \times 6m},$$

$$W_c(y_c) = W_c^T(y_c), \, \det W_c(y_c) \neq 0,$$

$$w_c(y, \dot{y}) = \text{col}(w_1(y, \dot{y}), \ldots, w_m(y, \dot{y})) \in R^{6m \times 1},$$

where y_c denotes the expanded contact position vector in the $6m$-dimensional space and F_c is the expanded vector of the contact forces, adjoint to that vector. Let us mention that at the manipulated object MC, no contact force acts directly, hence $F_0 = 0$. Equations (370) and (371) represent the final form of equations of the elastic system behavior which, under the action of external contact forces F_c, performs the general motion about the immobile unloaded state 0.

The result would be also obtained by using the d'Alembert principle by replacing the components of inertial, damping and gravitational forces on the left-hand side of Equation (365).

APPENDIX B: ELASTIC SYSTEM MODEL FOR THE MOBILE UNLOADED STATE

Let the geometrical figure move from the state 0. Stress of the elastic system takes place in the same way as when the state 0 is at rest. In other words, the elastic system stress is still regarded only with respect to the mobile state 0, while this motion of state 0 influences only the members which do not reflect the elastic properties of the elastic system. Simply, the potential energy and dissipation energy of the elastic system connections are determined by displacement of the system relative to the mobile unloaded state, while the other quantities are defined for the elastic system absolute coordinates.

Kinetic energy is defined by the absolute velocities as

$$
2T_a = \sum_{i=0}^{m}\left(\sum_{j=1}^{\infty} dm_j v_j\right) = \sum_{i=0}^{m}\left(\sum_{j=1}^{\infty} dm_j \frac{d}{dt}(r_{ia} + b_j)\right),
$$

$$
2T_a = \sum_{i=0}^{m} m_i \dot{r}_{ia}^2 + \sum_{i=0}^{m} I_i \omega_{ia}^2
$$

$$
= \sum_{i=0}^{m} v_{ia}^T M_i v_{ia} = v_a^T M v_a = \dot{Y}^T L_{va}^T(Y) M L_{va}(Y)\dot{Y},
$$

$$
2T_a = \dot{Y}^T W_a(Y)\dot{Y}, \tag{372}
$$

where

$$
Y_i = \mathrm{col}(r_{ia}, \mathcal{A}_{ia}) \in R^{6\times1}, \quad Y = \mathrm{col}(Y_0, Y_1, \ldots, Y_m) \in R^{(6m+6)\times1},
$$

$$
v_{ia} = \mathrm{col}(\dot{r}_{ia}, \omega_{ia}(\mathcal{A}_{ia})) = L_{via}(Y_i)\dot{Y}_i \in R^{6\times1},
$$

$$
\omega_{ia} = L_{\omega ia}(\mathcal{A}_{ia})\dot{\mathcal{A}}_{ia} \in R^{3\times1},
$$

269

$$v_a = \mathrm{col}(v_{0a}, v_{1a}, \ldots, v_{ma}) \in R^{(6m+6) \times 1},$$

$$L_{via}(Y_i) = \mathrm{diag}(I_{3 \times 3}, L_{\omega ia}(\mathcal{A}_{ia})) \in R^{6 \times 6},$$

$$L_{va}(Y) = \mathrm{diag}(L_{v0a}, L_{v1a} \ldots, L_{vma}) \in R^{(6m+6) \times (6m+6)},$$

$$W_a(Y) = \mathrm{diag}(W_{0a}, W_{1a}, \ldots, W_{ma})$$

$$= L_{va}^T(Y) M L_{va}(Y) \in R^{(6m+6) \times (6m+6)}. \tag{373}$$

The mathematical form of W_{ia} is identical to the mathematical form of W_i, whereby instead of the subscript i, the subscript ia is used as the designation of absolute coordinates. Hence, the derivatives of the kinetic energy

$$\frac{\partial T_{ia}}{\partial \dot{Y}_i} = W_{ia}(Y_i) \dot{Y}_i, \quad \frac{d}{dt} \frac{\partial T_{ia}}{\partial \dot{Y}_i} = \dot{W}_{ia}(Y_i) \dot{Y}_i + W_{ia}(Y_i) \ddot{Y}_i, \quad \frac{\partial T_{ia}}{\partial Y_i}(Y_i, \dot{Y}_i),$$

are also identical to the expressions obtained in Appendix A, in which the subscripts should be changed, i.e. instead of the displacement coordinates marked by i, the absolute coordinates with the subscript ia should be used.

It has already been mentioned that the elastic system potential energy is equal to the elastic system deformation work.

Total potential energy due to the linear and rotational displacement of the body i relative to the body j, $i, j = 0, 1, \ldots, m$ is defined by

$$\Pi_a = \Pi_{01a} + \Pi_{02a} + \Pi_{03a} + \cdots + \Pi_{0ma} +$$

$$+ \Pi_{12a} + \Pi_{13a} + \cdots + \Pi_{1ma} +$$

$$+ \Pi_{23a} + \cdots + \Pi_{2ma} +$$

$$\ldots \ldots$$

$$+ \Pi_{(m-1)m}.$$

The arbitrary member Π_{ija} of this sum for $y_{ij} = y_{ij}^D$ is defined by

$$2\Pi_{ija} = (y_{ij}^D)^T K_{ija} y_{ij}^D = (Y_i - Y_j)^T \Lambda_{ij}(Y_i, Y_j) K_{ija} \Lambda_{ij}(Y_i, Y_j)(Y_i - Y_j),$$

$$2\Pi_{ija} = (Y_i - Y_j)^T \pi_{ij}(Y_i - Y_j) = Y_i^T \pi_{ij} Y_i - 2Y_i^T \pi_{ij} Y_j + Y_j^T \pi_{ij} Y_j,$$

where $\pi_{ij} = \Lambda_{ij}(Y_i, Y_j) K_{ija} \Lambda_{ij}$,

$$\pi_{ij} = \pi_{ji}$$

$$= \begin{bmatrix} c_{ij}^x \left(1 - \frac{\|\rho_{ij0}\|}{\|r_{ia}-r_{ja}\|}\right)^2 & 0 & 0 & 0 & 0 & 0 \\ 0 & c_{ij}^y \left(1 - \frac{\|\rho_{ij0}\|}{\|r_{ia}-r_{ja}\|}\right)^2 & 0 & 0 & 0 & 0 \\ 0 & 0 & c_{ij}^z \left(1 - \frac{\|\rho_{ij0}\|}{\|r_{ia}-r_{ja}\|}\right)^2 & 0 & 0 & 0 \\ 0 & 0 & 0 & c_{ij}^\psi & 0 & 0 \\ 0 & 0 & 0 & 0 & c_{ij}^\theta & 0 \\ 0 & 0 & 0 & 0 & 0 & c_{ij}^\varphi \end{bmatrix}.$$

Therefore, after substitution, the total potential energy is

$$2\Pi_a = Y_0^T(\pi_{01} + \pi_{02} + \cdots + \pi_{0m})Y_0 - Y_0^T\pi_{01}Y_1 - Y_0^T\pi_{02}Y_2 - \cdots$$

$$- Y_0^T\pi_{0m}Y_m - Y_1^T\pi_{10}Y_0 + Y_1^T(\pi_{10} + \pi_{12} + \cdots + \pi_{1m})Y_1$$

$$- Y_1^T\pi_{12}Y_2 - \cdots - Y_1^T\pi_{1m}Y_m$$

$$\cdots\cdots\cdots\cdots$$

$$- Y_m^T\pi_{m0}Y_0 - Y_m^T\pi_{m1}Y_1 - \cdots + Y_m^T(\pi_{m0} + \pi_{m1} + \cdots + \pi_{m(m-1)})Y_m$$

or, in comprised form,

$$2\Pi_a = Y^T\pi_a(Y)Y, \tag{374}$$

where, due to $\pi_{ij} = \pi_{ji}$,

$$\pi_a(Y) = \pi_a^T(Y)$$

$$= \begin{bmatrix} \sum_{k=0,k\neq0}^n \pi_{0k} & -\pi_{01} & -\pi_{02} & \cdots & -\pi_{0m} \\ -\pi_{01} & \sum_{k=0,k\neq1}^n \pi_{1k} & -\pi_{12} & \cdots & -\pi_{1m} \\ \cdots & \cdots & \cdots & \cdots & \cdots \\ -\pi_{0m} & -\pi_{1m} & -\pi_{2m} & \cdots & \sum_{k=0,k\neq m}^n \pi_{km} \end{bmatrix}.$$

From this, the derivative of potential energy with respect to the coordinate is

$$\frac{\partial\Pi_a}{\partial Y} = \frac{1}{2}\frac{\partial Y^T\bar{\pi}_aY}{\partial Y} + \pi_a(Y)Y,$$

$$\frac{\partial\Pi_a}{\partial Y} = \begin{bmatrix} \frac{\partial\Pi_a}{\partial Y_0} \\ \cdots \\ \frac{\partial\Pi_a}{\partial Y_i} \\ \cdots \\ \frac{\partial\Pi_a}{\partial Y_m} \end{bmatrix} = \begin{bmatrix} \frac{1}{2}\frac{\partial Y^T\bar{\pi}_aY}{\partial Y_0} + \pi_{0a}(Y)Y \\ \cdots \\ \frac{1}{2}\frac{\partial Y^T\bar{\pi}_aY}{\partial Y_i} + \pi_{ia}(Y)Y \\ \cdots \\ \frac{1}{2}\frac{\partial Y^T\bar{\pi}_aY}{\partial Y_m} + \pi_{ma}(Y)Y \end{bmatrix},$$

$$\Rightarrow \frac{\partial\Pi_a}{\partial Y_i} = \frac{1}{2}\frac{\partial Y^T\bar{\pi}_aY}{\partial Y_i} + \pi_{ia}(Y)Y \in R^6,$$

where $\pi_{ia}(Y) \in R^{6 \times (6m+6)}$ are the submatrices composed of the rows starting from $6i + 1$ to $6i + 6$ inclusive of the matrix $\pi_a(Y)$, and $\partial(Y^T \bar{\pi}_a Y)/\partial Y_i$ is the vector of the quadratic form (scalar) derivative $Y^T \pi_a Y$ with respect to the vector Y_i, whereby the macron designates that partial derivation is carried out over the matrix π_a. In expanded form, this vector is

$$
\frac{\partial Y^T \bar{\pi}_a Y}{\partial Y_i} =
\begin{bmatrix}
\dfrac{\partial Y^T \bar{\pi}_a Y}{\partial Y_i^1} \\
\dfrac{\partial Y^T \bar{\pi}_a Y}{\partial Y_i^2} \\
\cdots \\
\dfrac{\partial Y^T \bar{\pi}_a Y}{\partial Y_i^6}
\end{bmatrix}
=
\begin{bmatrix}
Y^T \dfrac{\partial \pi_a}{\partial Y_i^1} Y \\
Y^T \dfrac{\partial \pi_a}{\partial Y_i^2} Y \\
\cdots \\
Y^T \dfrac{\partial \pi_a}{\partial Y_i^6} Y
\end{bmatrix}.
$$

Total dissipation energy consumed in the course of linear and rotational displacement of the body i relative to the body j, $i, j = 0, 1, \ldots, m$ is defined by

$$
\mathcal{D}_a = \mathcal{D}_{01a} + \mathcal{D}_{02a} + \mathcal{D}_{03a} + \cdots + \mathcal{D}_{0ma} +
$$

$$
+ \mathcal{D}_{12a} + \mathcal{D}_{13a} + \cdots + \mathcal{D}_{1ma} +
$$

$$
+ \mathcal{D}_{23a} + \cdots + \mathcal{D}_{2ma} +
$$

$$
\cdots \cdots
$$

$$
+ \mathcal{D}_{(m-1)m}.
$$

An arbitrary member $\mathcal{D}_{ija}$ of that sum is given by

$$
\begin{aligned}
-2\mathcal{D}_{ija} &= (\dot{\delta}_{ij}^D)^T D_{ij}^{\delta} \dot{\delta}_{ij}^D + (\dot{\mathcal{A}}_{ia} - \dot{\mathcal{A}}_{ja})^T D_{ij}^{\mathcal{A}} (\dot{\mathcal{A}}_{ia} - \dot{\mathcal{A}}_{ja}) \\
&= (\dot{r}_{ia} - \dot{r}_{ja})^T \mathcal{G}_{ija}(r_{ia}, r_{ja}) D_{ij}^{\delta} \mathcal{G}_{ija}(r_{ia}, r_{ja})(\dot{r}_{ia} - \dot{r}_{ja}) \\
&\quad + (\dot{\mathcal{A}}_{ia} - \dot{\mathcal{A}}_{ja})^T D_{ij}^{\mathcal{A}} (\dot{\mathcal{A}}_{ia} - \dot{\mathcal{A}}_{ja}) \\
&= ((\dot{r}_{ia} - \dot{r}_{ja})^T \mid (\dot{\mathcal{A}}_{ia} - \dot{\mathcal{A}}_{ja})^T) \\
&\quad \times
\begin{bmatrix}
\mathcal{G}_{ija}(r_{ia}, r_{ja}) D_{ij}^{\delta} \mathcal{G}_{ija}(r_{ia}, r_{ja}) & 0_{3 \times 3} \\
0_{3 \times 3} & D_{ij}^{\mathcal{A}}
\end{bmatrix}
\begin{bmatrix}
\dot{r}_{ia} - \dot{r}_{ja} \\
\dot{\mathcal{A}}_{ia} - \dot{\mathcal{A}}_{ja}
\end{bmatrix} \\
&= (\dot{Y}_i - \dot{Y}_j)^T D_{ij}(\dot{Y}_i - \dot{Y}_j) \\
&= \dot{Y}_i^T D_{ij} \dot{Y}_i - 2\dot{Y}_i^T D_{ij} \dot{Y}_j + \dot{Y}_j^T D_{ij} \dot{Y}_j,
\end{aligned}
$$

where

$$
D_{ij} = D_{ij}^T = D_{ji} = \operatorname{diag}(\mathcal{G}_{ija}(r_{ia}, r_{ja}) D_{ij}^{\delta} \mathcal{G}_{ija}(r_{ia}, r_{ja}), D_{ij}^{\mathcal{A}}) \in R^{6 \times 6},
$$

wherefrom, after substitution, the total dissipation energy is determined by

$$-2\mathcal{D}_a = \dot{Y}_0^T (D_{01} + D_{02} + \cdots + D_{0m}) \dot{Y}_0 - \dot{Y}_0^T D_{01} \dot{Y}_1 - \cdots - \dot{Y}_0^T D_{0m} \dot{Y}_m$$

$$-\dot{Y}_1^T D_{10} \dot{Y}_0 + \dot{Y}_1^T (D_{10} + D_{12} + \cdots + D_{1m}) \dot{Y}_1 - \cdots - \dot{Y}_1^T D_{1m} \dot{Y}_m$$

$$\cdots \cdots \cdots \cdots$$

$$-\dot{Y}_m^T D_{m0} \dot{Y}_0 - \dot{Y}_m^T D_{m1} \dot{Y}_1 - \cdots + \dot{Y}_m^T (D_{m0} + D_{m1} + \cdots + D_{m(m-1)}) \dot{Y}_m.$$

In united quadratic form with respect to the absolute coordinates derivatives, the dissipation energy is given by

$$2\mathcal{D}_a = -\dot{Y}^T D_a(Y) \dot{Y}, \tag{375}$$

where, because of $D_{ij} = D_{ji}$

$$D_a(Y) = D_a^T(Y)$$

$$= \begin{bmatrix} \sum_{k=0, k \neq 0}^n D_{0k} & -D_{01} & -D_{02} & \cdots & -D_{0m} \\ -D_{01} & \sum_{k=0, k \neq 1}^n D_{1k} & -D_{12} & \cdots & -D_{1m} \\ \cdots & \cdots & \cdots & \cdots & \cdots \\ -D_{0m} & -D_{1m} & -D_{2m} & \cdots & \sum_{k=0, k \neq m}^n D_{km} \end{bmatrix}.$$

From this, the derivative with respect to velocity is defined by

$$\frac{\partial \mathcal{D}_a}{\partial \dot{Y}} = -D_a(Y) \dot{Y},$$

$$\frac{\partial \mathcal{D}_a}{\partial \dot{Y}} = \begin{bmatrix} \frac{\partial \mathcal{D}_a}{\partial \dot{Y}_o} \\ \cdots \\ \frac{\partial \mathcal{D}_a}{\partial \dot{Y}_i} \\ \cdots \\ \frac{\partial \mathcal{D}_a}{\partial \dot{Y}_m} \end{bmatrix} = \begin{bmatrix} D_{0a}(Y) \dot{Y} \\ \cdots \\ D_{ia}(Y) \dot{Y} \\ \cdots \\ D_{ma}(Y) \dot{Y} \end{bmatrix} \Rightarrow \frac{\partial \mathcal{D}_a}{\partial \dot{Y}_i} = -D_{ia}(Y) \dot{Y} \in R^{6 \times 1},$$

where $D_{ia}(Y) \in R^{6 \times 6(m+1)}$ are the submatrices composed of the rows starting from $6i + 1$ to $6i + 6$ inclusive of the matrix $D_a(Y)$.

Substituting the obtained expressions into Langrange's equations

$$\frac{d}{dt} \frac{\partial T_a}{\partial \dot{Y}_i^j} - \frac{\partial T_a}{\partial Y_i^j} - \frac{\partial \mathcal{D}_a}{\partial \dot{Y}_i^j} + \frac{\partial \Pi_a}{\partial Y_i^j} = Q_{ia}^j, \quad i = 0, 1, \ldots, m, \quad j = 1, \ldots, 6, \tag{376}$$

where the generalized forces for the individual components Y_i^j of the vector Y_i are given by

$$Q_{ia}^j = \sum_{k=0}^{m} \frac{\partial Y_k}{\partial Y_i^j} (G_k(m_k g) + F_k)$$

$$= G_i^j(m_i g) + F_i^j, \quad i = 0, 1, \ldots, m, \quad j = 1, \ldots, 6,$$

for the elastic system performing general motion under the action of a system of external forces about the mobile unloaded state 0, which is also performing general motion, the general form is obtained as

$$W_{ia}(Y_i)\ddot{Y}_i + \dot{W}_{ia}(Y_i)\dot{Y}_i - \frac{\partial T_{ia}}{\partial Y_i}(Y_i, \dot{Y}_i) + D_{ia}\dot{Y}$$

$$+ \frac{1}{2}\frac{\partial Y^T \bar{\pi}_a Y}{\partial Y_i} + \pi_{ia}(Y)Y = G_i(m_i g) + F_i,$$

for $i = 0, 1, \ldots, m$ or, in short form,

$$W_{ia}(Y_i)\ddot{Y}_i + w_{ia}(Y, \dot{Y}) = F_i, \quad i = 0, 1, \ldots, m,$$

where

$$w_{ia}(Y, \dot{Y}) = \dot{W}_{ia}(Y_i)\dot{Y}_i - \frac{\partial T_{ia}}{\partial Y_i}(Y_i, \dot{Y}_i) + D_{ia}\dot{Y}$$

$$+ \frac{1}{2}\frac{\partial Y^T \bar{\pi}_a Y}{\partial Y_i} + \pi_{ia}(Y)Y - G_i(m_i g) \in R^{6 \times 1},$$

for $i = 0, 1, \ldots, m$. By putting together all $6m + 6$ equations, we obtain

$$W_a(Y)\ddot{Y} + w_a(Y, \dot{Y}) = F, \tag{377}$$

where

$$W_a(Y) = \text{diag}(W_{0a}(Y_0), W_{1a}(Y_1), \ldots, W_{ma}(Y_m)) \in R^{(6m+6) \times (6m+6)},$$

$$W_a(Y) = W_a^T(Y), \quad \det W_a(Y) \neq 0,$$

$$w_a(Y, \dot{Y}) = \text{col}(w_{0a}(Y, \dot{Y}), \ldots, w_{ma}(Y, \dot{Y})) \in R^{(6m+6) \times 1}.$$

From $6m + 6$ equations (377), only rank K equations are independent.

Equation (377) can be presented in such way that the descriptions of connections and manipulated object motion are divided:

$$W_{ca}(Y_c)\ddot{Y}_c + w_{ca}(Y, \dot{Y}) = F_{ca},$$

$$W_{0a}(Y_0)\ddot{Y}_0 + w_{0a}(Y, \dot{Y}) = 0, \tag{378}$$

where the subscript c designates the quantities related to the contact points, and the subscript 0 quantities related to the manipulated object. Here

$$Y_{ca} = \text{col}(Y_{1a}, Y_{2a}, \ldots, Y_{ma}) \in R^{6m \times 1}, Y_{0a} \in R^{6 \times 1},$$

$$F_{ca} = \text{col}(F_{1a}, F_{2a}, \ldots, F_{ma}) \in R^{6m \times 1}, F_{0a} = 0 \in R^{6 \times 1},$$

$$W_{ca}(Y_{ca}) = \text{diag}(W_{1a}(Y_1) \ldots W_{ma}(Y_m)) \in R^{6m \times 6m},$$

$$W_{ca}(Y_c) = W_{ca}^T(Y_c), \det W_{ca}(Y_c) \neq 0,$$

$$w_{ca}(Y, \dot{Y}) = \text{col}(w_{1a}(Y, \dot{Y}), \ldots, w_{ma}(Y, \dot{Y})) \in R^{6m \times 1},$$

where the expanded position vector of the contact position in the $6m$-dimensional space is denoted by Y_{ca} and the expanded vector of the contact forces acting at the contact point in that space is denoted by F_{ca}. It should be noted that no contact force acts directly at the manipulated object MC, hence $F_{0a} = 0$. Equations (377) and (378) represent the final form of the equations of elastic system behavior under the action of external forces F_{ca}, while the system performs general motion about the unloaded state 0, which also performs general motion.

The general motion of elastic system motion is described by (377), i.e. $6m + 6$ relations are defined, of which rank K are independent. This means that for the unique definition of elastic system position during the motion, it is necessary to prescribe $6m + 6 - \text{rank } K$ absolute generalized coordinates and their derivatives.

REFERENCES

[1] Z. Luo, Y. Uematsu, K. Ito, A. Kato, and M. Ito, "On cooperative manipulation of dynamic object," in *Proceedings of IROS*, pp. 1423–1431, 1994.

[2] Z. Luo and M. Ito, "Control design of robot for compliant manipulation on dynamic environments," *IEEE Trans. of Robotics and Automation*, Vol. 9, pp. 286–296, June 1993.

[3] Z. Luo, K. Ito, and M. Ito, "Multiple robot manipulators' compliant manipulation on dynamical environments," in *Proceedings of IROS*, pp. 1927–1934, 1993.

[4] D. Sun, X. Shi, and Y. Liu, "Modeling and cooperation of two-arm robotic system manipulating a deformable object," in *Proceedings of IEEE Conference on Robotics and Automation* (Minneapolis, Minnesota), pp. 2346–2351, April 1996.

[5] D. Sun and Y. Liu, "Modeling and impedance control of a two-manipulator system handling a flexible beam," *Journal of Dynamic Systems, Measurement, and Control, Transactions of the ASME*, Vol. 119, pp. 736–742, December 1997.

[6] S. Rao, *The Finite Element Method in Engineering*. Oxford: Pergamon Press, 1982.

[7] R. Gallagher, *Finite Element Analysis* (in Russian). Englewood Cliffs, New Jersey: Prentice-Hall, 1975.

[8] M. Živanović and M. Vukobratović, "General mathematical model of multi-arm cooperating robots with elastic interconnection at the contact," *Trans. of the ASME: Journal on Dynamic Systems, Measurement, and Control*, Vol. 119, pp. 707–117, December 1997.

[9] M. Živanović, *Contribution to the Research of Multi-Arm Robots Cooperative Work* (in Serbian). PhD Thesis, Faculty of Technical Sciences, Novi Sad, July 1997.

[10] M. Živanović and M. Vukobratović, "Synthesis of nominal motion of the multi-arm cooperating robots with elastic interconnections at the contacts," *Trans. of the ASME: Journal on Dynamic Systems, Measurement, and Control*, Vol. 126, pp. 336–346, June 2004.

[11] M. Živanović and M. Vukobratović, "Control of multi-arm cooperating robots with elastic interconnection at the contacts," *ROBOTICA*, Vol. 18, No. 2, pp. 183–193, 2000.

[12] S. Hayati, "Hybrid position/force control of multi-arm cooperating robots," in *Proceedings of IEEE Conference on Robotics and Automation* (San Francisco, USA), pp. 82–89, April 1986.

[13] Y. Nakamura, K. Nagai, and T. Yoshikawa, "Mechanics of coordinative manipulation by multiple robotic mechanisms," in *Proceedings of IEEE Conference on Robotics and Automation* (Raleigh, USA), pp. 991–998, March–April 1987.

[14] K. Kosuge, J. Ishikawa, K. Furuta, and M. Sakai, "Control of single-master multi-slave manipulator system using vim," in *Proceedings of IEEE Conference on Robotics and Automation* (Cincinnati, Ohio), pp. 1172–1177, May 1990.

[15] A. B. Cole, J. E. Hauser, and S. S. Sastry, "Kinematics and control of multifingered hands with rolling contact," *IEEE Trans. of Automatic Control*, Vol. 34, pp. 398–404, April 1989.

[16] Y. F. Zheng and J. Y. S. Luh, "Optimal load distribution for two industrial robots handling a single object," in *Proceedings of IEEE Conference on Robotics and Automation* (Philadelphia, USA), pp. 344–349, April 1988.

[17] A. Ramadorai, T. Tarn, and A. Bejczy, "Task definition, decoupling and redundancy resolution by nonlinear feedback in multi-robot object handling," in *Proceedings of IEEE Conference on Robotics and Automation* (Nice, France), pp. 467–474, May 1992.

[18] K. Jankowski, H. ElMaragfy, and W. ElMaragfy, "Dynamic coordination of multiple robot arms with flexible joints," *Int. J. Robotic Research*, Vol. 12, pp. 505–528, December 1993.

[19] M. Djurović and M. Vukobratović, "A contribution to dynamic modeling of cooperative manipulation," *Mechanism and Machine Theory*, Vol. 25, No. 4, pp. 407–416, 1990.

[20] I. Walker, S. Marcus, and R. Freeman, "Distribution of dynamic loads for multiple cooperating robot manipulators," *Journal of Robotic Systems*, Vol. 6, No. 1, pp. 35–47, 1989.

[21] I. Prokofyev, *Theory of Constructions* (in Russian). Moscow: Nauka, 1963.

[22] K. Bathe and E. Wilson, *Numerical Methods in Finite Element Analysis*. Englewood Cliffs, New Jersey: Prentice-Hall, Inc., 1976.

[23] K. Bathe, *Finite Element Procedures in Engineering Analysis*. Englewood Cliffs, New Jersey: Prentice-Hall, Inc., 1984.

[24] O. Zienkiewicz, *The Finite Element Method in Engineering Sciences, 3rd ed.,*. London: McGraw-Hill, 1993.

[25] P. Akella and M. Cutkosky, "Contact transition control with semiactive soft fingertips," *IEEE Trans. of Robotics and Automation*, Vol. 11, pp. 859–867, December 1995.

[26] O. Al-Jarrah, Y. Zheng, and K. Yi, "Trajectory planning for two manipulators to deform flexible materials using compliant motion," in *Proceedings IEEE Int. Conference on Robotics and Automation* (Nagoya, Japan), pp. 1517–1522, 1995.

[27] A. Krasovski, *Systems of Automatic Flight Control and Their Analitical Design* (in Russian). Moscow: Nauka, 1973.

[28] V. Velichenko, *Matrichno-geometricheskie metod v mekhanike s prilozheniyami k zadacham robototekhniki*. Moscow: Nauka, 1988.

[29] C. Harris and C. Crede, *Shock and Vibration Handbook*. New York: McGraw-Hill, 1976.

[30] V. Jahno, *Obratniye zadachi dlya differencial'nykh uravnenii uprugosti*. Novosibirsk: Nauka, Sibirskoe Otdelenie, 1990.

[31] G. Korn and T. Korn, *Matematical Handbook for Scientists and Engineers* (in Russian). Moscow: Nauka, 1984.

[32] M. Vukobratović, *Applied Dynamics of Manipulation Robots*. New York: Springer-Verlag, 1989.

[33] M. Vukobratović and D. Stokić, *Control of Manipulation Robots*. New York: Springer-Verlag, 1989.

[34] M. Vukobratović and et al., *Introduction to Robotics* (in Serbian). Belgrade: Mihailo Pupin Institute, 1986.

[35] M. Vukobratović and Y. Ekalo, "Unified approach to control laws synthesis for robotic manipulators in contact with dynamic environment," in *Tutorial S5: Force and Contact Control in Robotic Systems, IEEE Int. Conference on Robotics and Automation* (Atlanta, USA), pp. 213–229, 1993.

[36] M. Vukobratović, R. Stojić, and Y. Ekalo, "Contribution to the problem solution of position/force control of manipulation robots in contact with dynamic environments - a generalization," *IFAC Automatica*, Vol. 34, No. 10, 1998.

[37] Y. Ekalo and M. Vukobratović, "Adaptive stabilization of motion and forces in contact tasks for robotic manipulators with non-stationary dynamics," *Int. Journal on Robotics and Automation*, Vol. 9, No. 3, pp. 91–98, 1994.

[38] M. Vukobratović and Y. Ekalo, "New approach to control of robotic manipulators interacting with dynamic environment," *Robotica*, Vol. 14, pp. 31–39, 1996.

[39] Y. Ekalo and M. Vukobratović, "Quality of stabilization of robots interacting with dynamic environment," *Journal of Intelligent and Robotic Systems*, Vol. 14, No. 2, pp. 155–178, 1995.

[40] F. Gantmaher, *Teoriya Matric*. Moscow: Gosudarstvennoe izdatel'stvo tekhniko-teoreticheskoi literatur, 1954.

[41] R. Bellman, *Introduction to Matrix Analysis*. New York Toronto London: McGraw-Hill Inc., 1960.

[42] M. Erdmann and T. Lozano-Perez, "On multiple moving objects," in *Proceedings of IEEE Conference on Robotics and Automation* (San Francisco, USA), pp. 1419–1424, April 1986.

[43] Y. Koga and J.-C. Latombe, "Experiments in dual-arm manipulation planning," in *Proceedings of IEEE Conference on Robotics and Automation* (Nice, France), pp. 2238–2245, May 1992.

[44] D. Henrich and X. Cheng, "Fast distance computation for on-line collision detection with multi-arm robots," in *Proceedings of IEEE Conference on Robotics and Automation* (Nice, France), pp. 2514–2519, May 1992.

[45] H. Chu and H. ElMaragfy, "Real-time multi-robot path planner based on a heuristic approach," in *Proceedings of IEEE Conference on Robotics and Automation* (Nice, France), pp. 475–480, May 1992.

[46] A. Rovetta and R. Sala, "Robot motion planning with parallel systems," in *Proceedings of IEEE Conference on Robotics and Automation* (Nice, France), pp. 2224–2229, May 1992.

[47] B. Etkin, *Dynamics of Atmospheric Flight*. London: John Wiley & Sons, 1972.

[48] R. Patel and N. Munro, *Multivariable System Theory and Design*. Oxford: Pergamon Press, 1982.

[49] R. Kalman, "On the general theory of control systems," in *Proceedings 1-st. IFAC Congress* (Moscow), London: Butterworths, pp. 695–703, 1960.

[50] S. Arimoto, F. Miyazaki, and S. Kawamura, "Cooperative motion control of multiple robot arms or fingers," in *Proceedings of IEEE Conference on Robotics and Automation* (Raleigh, USA), pp. 1407–1412, March–April 1987.

[51] C. A. Derventzis and E. J. Davison, "Robust motion force control of cooperative multi-arm systems," in *Proceedings of IEEE Conference on Robotics and Automation* (Nice, France), pp. 2230–2237, May 1992.

[52] J. Wen and K. Kreutz, "Motion and force control for multiple cooperative manipulators," in *Proceedings of IEEE Conference on Robotics and Automation* (Scottsdale, Arizona), pp. 1246–1251, May 1989.

[53] R. Murray and S. Sastry, "Control experiments in planar manipulation and grasping," in *Proceedings of IEEE Conference on Robotics and Automation* (Washington, USA), pp. 624–629, 1989.

[54] T. Yoshikawa and X. Zheng, "Coordinated dynamic hybrid position/force control for multiple robot manipulators handling one constrained object," in *Proceedings of IEEE Conference on Robotics and Automation* (Cincinnati, Ohio), pp. 1178–1183, May 1990.

[55] M. Uchiyama and P. Dauchez, "A symmetric hybrid position/force control scheme for the coordination of two robots," in *Proceedings of IEEE Conference on Robotics and Automation* (Philadelphia, USA), pp. 350–356, April 1988.

[56] J. J. Craig and M. H. Raibert, "A systematic method of hybrid position/force control of a manipulator," in *Proceedings of IEEE Computer Applications Conference* (Chicago, USA), 1979.

[57] M. Raibert and J. Craig, "Hybrid position/force control of manipulators," *Journal of Dynamic Systems, Measurement, and Control, Transactions of the ASME*, Vol. 102, pp. 126–133, June 1981.

INDEX

International Series on
MICROPROCESSOR-BASED AND
INTELLIGENT SYSTEMS ENGINEERING

Editor: Professor S. G. Tzafestas, *National Technical University, Athens, Greece*

International Series on
MICROPROCESSOR-BASED AND
INTELLIGENT SYSTEMS ENGINEERING

22. S.G. Tzafestas (ed.): *Computational Intelligence in Systems and Control Design and Applications*. 2000   ISBN 0-7923-5993-3
23. J. Harris: *An Introduction to Fuzzy Logic Applications*. 2000   ISBN 0-7923-6325-6
24. J.A. Fernández and J. González: *Multi-Hierarchical Representation of Large-Scale Space*. 2001   ISBN 1-4020-0105-3
25. D. Katic and M. Vukobratovic: *Intelligent Control of Robotic Systems*. 2003
                                                                      ISBN 1-4020-1630-1
26. M. Vukobratovic, V. Potkonjak and V. Matijevic: *Dynamics of Robots with Contact Tasks*. 2003   ISBN 1-4020-1809-6
27. M. Ceccarelli: *Fundamentals of Mechanics of Robotic Manipulation*. 2004
                                                                      ISBN 1-4020-1810-X
28. V.G. Ivancevic and T.T. Ivancevic: *Human-Like Biomechanics*. A Unified Mathematical Approach to Human Biomechanics and Humanoid Robotics. 2005
                                                                      ISBN 1-4020-4116-0
29. J. Harris: *Fuzzy Logic Applications in Engineering Science*. 2005
                                                                      ISBN 1-4020-4077-6
30. M.D. Zivanovic and M.K. Vukobratovic: *Multi-Arm Cooperating Robots*. Dynamics and Control. 2006   ISBN 1-4020-4268-X

www.springer.com